VOLUME ONE HUNDRED AND FIFTY THREE

INTERNATIONAL REVIEW OF NEUROBIOLOGY

Placebo Effects in Neurologic Disease

INTERNATIONAL REVIEW OF NEUROBIOLOGY

VOLUME 153

VOLUME ONE HUNDRED AND FIFTY THREE

INTERNATIONAL REVIEW OF NEUROBIOLOGY

Placebo Effects in Neurologic Disease

Edited by

NATALIE P. WITEK
Department of Neurological Sciences, Rush University Medical Center, Chicago, IL, United States

CHRISTOPHER G. GOETZ
Department of Neurological Sciences, Rush University Medical Center, Chicago, IL, United States

GLENN T. STEBBINS
Department of Neurological Sciences, Rush University Medical Center, Chicago, IL, United States

ELSEVIER

ACADEMIC PRESS
An imprint of Elsevier

Academic Press is an imprint of Elsevier
50 Hampshire Street, 5th Floor, Cambridge, MA 02139, United States
525 B Street, Suite 1650, San Diego, CA 92101, United States
The Boulevard, Langford Lane, Kidlington, Oxford OX5 1GB, United Kingdom
125 London Wall, London, EC2Y 5AS, United Kingdom

First edition 2020

ISBN: 978-0-12-821118-2
ISSN: 0074-7742

For information on all Academic Press publications
visit our website at https://www.elsevier.com/books-and-journals

Publisher: Zoe Kruze
Acquisitions Editor: Sam Mahfoudh
Editorial Project Manager: Hilal Johnson
Production Project Manager: Abdulla Sait
Cover Designer: Alan Studholme

Typeset by SPi Global, India

Contents

Contributors

Sharlet Anderson
Department of Neurological Sciences, Rush University Medical Center, Chicago, IL, United States

Marco Annoni
National Research Council of Italy (CNR), Institute of Biomedical Technologies (ITB), Fondazione Umberto Veronesi, Rome, Italy

Luana Colloca
Department of Pain Translational Symptoms Science, School of Nursing; Center to Advance Chronic Pain Research; Departments of Anesthesiology and Psychiatry, School of Medicine, University of Maryland, Baltimore, MD, United States

Jean-Christophe Corvol
Department of Neurology, Pitié-Salpêtrière Hospital, Sorbonne University, Assistance Publique Hôpitaux de Paris, Brain and Spine Institute, ICM, Inserm U 1127, CNRS UMR 7225, Paris, France

Joaquim J. Ferreira
Laboratory of Clinical Pharmacology and Therapeutics, Faculdade de Medicina, Universidade de Lisboa; Instituto de Medicina Molecular, Lisbon; CNS—Campus Neurológico Sénior, Torres Vedras, Portugal

Daniel M. Goldenholz
Beth Israel Deaconess Medical Center, Boston, MA, United States

Shira R. Goldenholz
Beth Israel Deaconess Medical Center, Boston, MA, United States

Kaori Ito
Applied Pharmacometrics & Research LLC, Londonderry, NH, United States

Jau-Shin Lou
Roger Gilbertson Endowed Chair in Neurology, Department of Neurology, University of North Dakota School of Medicine & Health Science, and Sanford Health, Fargo, ND, United States

Louise-Laure Mariani
Department of Neurology, Pitié-Salpêtrière Hospital, Sorbonne University, Assistance Publique Hôpitaux de Paris, Brain and Spine Institute, ICM, Inserm U 1127, CNRS UMR 7225, Paris, France

Tiago A. Mestre
University of Ottawa Brain and Mind Research Institute; The Ottawa Hospital Research Institute; Division of Neurology, Department of Medicine, The Ottawa Hospital, Ottawa, ON, Canada

Rema Raman
Alzheimer's Therapeutic Research Institute, University of Southern California, Los Angeles, CA, United States

Filipe B. Rodrigues
UCL Huntington's Disease Centre, UCL Queen Square Institute of Neurology, University College London, London, United Kingdom; Laboratory of Clinical Pharmacology and Therapeutics, Faculdade de Medicina, Universidade de Lisboa; Instituto de Medicina Molecular, Lisbon, Portugal

Klaus Romero
Critical Path Institute (C-Path), Tucson, AZ, United States

Glenn T. Stebbins
Department of Neurological Sciences, Rush University Medical Center, Chicago, IL, United States

A. Jon Stoessl
Pacific Parkinson's Research Centre and Djavad Mowafaghian Centre for Brain Health, University of British Columbia, Vancouver, BC, Canada

Preface

Throughout history, healers have used placebo to placate their patient's fears, concerns, and symptoms. The placebo has evolved overtime to be integral in revolutionizing modern medicine with its use in gold-standard randomized-controlled trials. Although the placebo effect is seen on one hand as a friend of the clinician, it is generally seen as a dangerous foe to the researcher. The strong placebo effect in neurologic disorders has been blamed for many negative clinical trials and has been subsequently studied carefully with the aim of minimizing its power.

Great efforts have been made in defining and understanding the observed improvement experienced by patients and study subjects who participate in the placebo arms of clinical trials.

In spite of the great depth of research on the impact of the placebo and the objective and subjective improvements experienced by the patient, the placebo effect in many circumstances remains unpredictable. However, the growing body of knowledge of the placebo effect has influenced current clinical trial design and can help shape future research. In an effort to summarize the existing body of knowledge regarding the placebo effect, this volume was compiled to share the knowledge of experts from several perspectives and is offered as a resource on the current status of the field. We hope it also prompts new ideas and testable hypotheses to move the field of placebo science forward.

This text is divided into two sections. The first is an overview of placebo in general practice and research. Specifically, the first section reviews the history of placebo, determinants of placebo effect, ways to minimize placebo in the research context, maximize it in clinical settings, statistical models to handle placebo, concepts of nocebo and lessebo, and finally the ethics of placebo's use. The second half reviews the impact that placebo effect has in specific neurological diseases, such as pain, Parkinson's disease, cognitive impairment, and epilepsy. Although it would be nearly impossible to perform an exhaustive review of placebo in every neurologic disease, we hope that you find this volume helpful in understanding impacts of placebo.

We would like to thank the many dedicated experts in the field who have contributed to this volume, and for each of their contributions they have made in moving our understanding of placebo forward through their prior works. We hope that you enjoy this volume and find it comprehensive and enjoyable.

Natalie P. Witek
Christopher G. Goetz
Glenn T. Stebbins

PART 1

Background and methods in placebo

CHAPTER ONE

Better than nothing: A historical account of placebos and placebo effects from modern to contemporary medicine

Marco Annoni*
National Research Council of Italy (CNR), Institute of Biomedical Technologies (ITB), Fondazione Umberto Veronesi, Rome, Italy
*Corresponding author: e-mail address: marco.annoni@cnr.it

Contents

Abstract

Throughout the history of medicine, multiple conceptions of "placebo" and "placebo effect" have often co-existed across different domains, and today the meaning of these concepts is still disputed. Against this background, this chapter provides a succinct account of the key events in the history of the concepts of "placebo," "placebo control," and "placebo effect." The first section reconstructs the etymology of the term "placebo" and its first introduction in medicine. The next sections provide an account of how placebos have been employed in both medical practice and scientific research in modern medicine. Later sections trace the emergence of the concepts of "placebo control" and "placebo effect" in the first half of the 20th century, from the first empirical studies investigating the effects of placebos up to the publication of Beecher's landmark article "The Powerful Placebo." Finally, the last two sections review the varieties of randomized, placebo-controlled trials (RCTs) in the second half of the 20th century,

International Review of Neurobiology, Volume 153
ISSN 0074-7742
https://doi.org/10.1016/bs.irn.2020.03.028

and the subsequent wave of empirical studies that, starting from the 1970s, have investigated the psychological, pharmacological and neurobiological mechanisms of placebo effects.

1. Introduction

With the possible exception of opium for pain and a few other remedies, prior to the 1930s, most therapies were helpful only insofar as they resulted in a placebo effect. Accordingly, Houston (1938, pp. 1416–8) observed that throughout history, the conscientious use of effective interventions has been like spotting "a whale in an otherwise unending green waste of water." Therefore, the great lesson "of medical history is that the placebo has always been the norm of medical practice, that it was only occasionally and at great intervals that anything really serviceable, such as the cure of scurvy with fresh fruits, was introduced into medical practice." Similarly, in their detailed reconstruction of the history of placebos from prehistory to the 20th century Shapiro and Shapiro (1997, p. 11) noted that "From Asclepius through Hippocrates to Galen, and until very recently, the history of medical treatment was largely the history of the placebo effect, because all medical treatments, with rare exceptions, were at best placebos, at worst unknowingly deadly."

Yet, while it is tempting to read the history of medicine solely as the history of placebos and the placebo effect, this view runs into two objections. First, the earliest use of the term "placebo" in medicine dates back to the seventeenth century, while the term "placebo effect" was likely coined only at the beginning of the 20th century. Thus, describing the history of medicine as the history of placebos and the placebo effect may constitute an instance of "historical presentism," i.e., the fallacious tendency to project backwards interpretative categories and concepts that did not exist in the past (Andersen, 2005, pp. 104–105). Second, and more importantly, adopting such a view may obscure the fact that also these concepts are the results of a fascinating and tumultuous process of historical evolution. In the past, multiple and conflictual conceptions of "placebo" and "placebo effect" have co-existed at the same time across different domains; and today the meaning of these concepts is still disputed in medicine as well as in philosophy (Blease & Annoni, 2019; Howick, 2017).

Against this background, this review provides a synthetic account of the key events in the history of the "placebo" and "placebo effect" concepts, from the first recorded use of the word "placebo" in medical practice, until the recent institution of a new and multidisciplinary field of academic research under the umbrella name of "placebo studies."

2. The etymology of "placebo" and its first recorded use in medicine (fourth century–1772)

The etymology of the term "placebo" leads back to the Hebrew Bible (Shapiro & Shapiro, 1997, p. 28). In the fourth century St. Jerome translated the word *ethalekh* with the word "placebo" meaning, "I shall please" (Jacobs, 2000). The word "placebo" acquired a more secular meaning only in 14th century France, when it was customary for mourning families to dispense a meal after the completion of the traditional prayer cycle, termed Office of the Dead. To access the banquet, distant relatives and uninvited persons simulated grief and desperation, joining the rest of the congregation in the singing the Psalm 116:9: "Placebo Domino in regione vivorum" which translates to, "I shall please the Lord in the land of the living." Accordingly, these people were called "placebo singers," or "placebo" (Shapiro, 1964). In the same period, Chaucer named "Placebo" a character in his *Canterbury Tales*. Like the French mourners, Placebo was essentially a sycophant—a false and parasitic person whose defining trait was that of trying "to please" his more powerful brother.

The first reported use of the word "placebo" in medical contexts dates to 1772, when Scottish professor of medicine William Cullen gave a series of clinical lectures at the University of Edinburg, then a leading center for medical education. In one of the manuscripts of these lectures, Cullen states that he had administered a remedy to a patient despite believing it to be inefficacious, adding that sometimes "it is necessary to give a medicine and [this is] what I call a placebo." Then, in another passage, Cullen further explains that he has given a similar "placebo" to another patient, noting "I considered him as absolutely incurable, and as hasting very fast to his fate, and I took him in hopes of making some observant upon his case, and even of learning something by his death, I prescribed therefore in pure placebo, but I make it a rule even in employing placebos to give what would have a tendency to be of use to the patient" (Cullen, 1772, p. 218f).

This latter quote is interesting for two reasons. First, it provides a snapshot of the early use of placebos in clinical practice. Interestingly, despite the use of the adjective "pure," it is likely that Cullen administered to his patients only what were later defined as "impure placebos," i.e., remedies or compounds believed to be effective, but only at higher dosages or for a different condition (Finniss, 2018, p. 3; Kerr, Milne, & Kaptchuk, 2008). Second, Cullen's quote neatly foreshadowed the double role that placebos would have played in medicine until mid of the 20th century. On the one hand, they served as epistemic tools to test and disprove claims of therapeutic effectiveness. On the other hand, they instead served as tools of the healing arts to please or comfort patients, and to ease doctors' work. To understand the subsequent evolution of the placebo construct, and the emergence of the concepts of "placebo control" and "placebo effect," it is thus crucial to follow this double path of clinical treatment and "control" for comparative experiments.

3. Placebos as treatments in clinical practice (1772–1954)

It is unclear whether Cullen had been the first to introduce the term "placebo" in medicine or whether his use of the word has been inspired by someone else. What is certain, however, is that, just a few years after his lectures, the term "placebo" had become sufficiently common as to deserve its first entries in the technical dictionaries of the time. In 1785 Motherby's *New Medical Dictionary* defined placebo as, "a common place method or medicine"; and then, in 1803, Fox's *New Medical Dictionary*, as "an epithet given to any medicine adopted more to please than to benefit the patient" (1803). As these dictionary entries demonstrate, by the end of the 18th century the word "placebo" had joined the medical jargon to indicate something that, despite being ineffective, was nonetheless administered to please and comfort patients.

Meanwhile, the administration of sham interventions became equally common on the other side of the Atlantic. In 1803, Thomas Jefferson famously noted that "one of the most successful physicians I have ever known has assured me that he used more bread pills, drops of colored water, and powder of hickory ashes, than of all other medication put together" (Jefferson, 1898). He defined this practice "the pious fraud." Then, a century after these remarks, Richard Cabot, a prominent physician at Harvard Medical School, recalled how he "was brought up, as I suppose every physician is, to use *placebos*, bread pills, water subcutaneously, and other devices" (Cabot, 1903, p. 348).

This trend continued at least until the mid-20th century. In 1945, one of the earliest scholarly papers devoted to the "the placebo" concluded that it was able to "smooth [the patient's] path"; that "it cannot harm and may comfort"; and that it was especially useful for the "ignorant ... disappointed and displeased ... hopeless, [and] incurable case" (Pepper, 1945). Remarkably, this passage reaffirmed a belief shared by many healers and physicians at least since the previous century: that the more a person was "unintelligent, neurotic, or inadequate," the more the placebo was likely to be effective (de Craen, Kaptchuk, Tijssen, & Kleijnen, 1999). In a similar vein, another article published in the *Lancet* shortly thereafter entitled "The humble humbug" provided another vivid characterization of how placebos were still pejoratively or at best with begrudging respect conceptualized and used in medicine in 1954. Here the placebo is portrayed as "a means of reinforcing a patient's confidence in his recovery, when the diagnosis is undoubted and no more effective treatment is possible; that for some unintelligent or inadequate patients life is made easier by a bottle of medicine to comfort their ego; that to refuse a placebo to a dying incurable patient may be simply cruel" (Anonymous, 1954, p. 321).

Thus, from the time of Cullen's first use of the term to the one of the "humble humbug," placebos in clinical settings have been mostly conceived as inert remedies unable to influence the underlying causes of patients' diseases. Accordingly, their primary role was that of providing mental relief, to preserve patients' hope, and to facilitate doctors' work in difficult or desperate cases.

4. Placebos as controls in scientific research (1772–1930s)

Prior to the systematic adoption of the randomized placebo-controlled trial (RCT) most of the evidence supporting the administration of medical interventions derived from non-comparative sources, such as the doctor's personal experience or the teachings of authoritative figures of the past, like Hippocrates or Galen. The use of sham interventions for comparative purposes, however, predates both the introduction of RCTs and the use of the word "placebo" in research. For instance, already in the 16th century, the Church used "fake trials" to distinguish between true and false cases of exorcism (Kaptchuk, Kerr, & Zanger, 2009). Individuals who claimed to be "possessed" were exposed to false holy objects and, in case they reacted as if these were "real" it was concluded that the possession was just in their imagination (Finniss, Kaptchuk, Miller, & Benedetti, 2010). However, soon after the use of similar techniques began to be adopted in experimental and scientific settings. Starting from the late 17th the use of "blind-assessment" became one of the available strategies to demonstrate whether the claim that a treatment was effective was "true" (and thus "scientific") or "false" (and thus a "fraud"). However, the use of blinding procedures was not yet articulated into a systematic methodology; rather, it "was regularly dusted off and used to either combat potential quackery or gain access to scientific respectability" (Kaptchuk, 1998b, pp. 393–394).

One of the earliest examples of a blindfolded experiment was conducted in France to test the claims of "mesmerism." Mesmerism was introduced by Franz Anton Mesmer and was one of the most influential healing systems of the late 18th century. It was based on the belief that it was possible to manipulate an invisible fluid—called "animal magnetism"—to promote healing and cure various diseases. Mesmer and other "mesmerists" conducted both personalized and collective healing rituals, which sometimes were quite intricate and involved previously "mesmerized" objects. Due to Mesmer's popularity and the pressure of public opinion, in 1784 Louis XVI appointed two Royal Commissions to investigate mesmerism. One of these commissions—which included Antoine-Laurent de Lavoisier and was headed by Benjamin Franklin—conducted several experiments with diverse blinding procedures. In one of them, a group of blindfolded women was asked to indicate the direction of magnetic fluid flow by a mesmerist. In another experiment, a patient was "mesmerized" after drinking plain water that had been deceptively presented as "mesmerized." In all cases, blinding procedures nullified the effects of mesmerism, and sham interventions were as effective as "mesmerized" objects in relieving or inducing effects. Accordingly, the Commission concluded that "this agent, the fluid, has no existence" and that its effects were due to the physical contacts between healer and patient, to imitation, and the effect of "imagination" (Franklin et al., 1785, p. 97).

After mesmerism, in the first half of the 19th century, blind controls were used to evaluate other contested remedies, such as Hahnemann's homeopathy and Perkins' tractors. Invented by Dr. Elisha Perkins, the "tractors" consisted of small metallic rods allegedly able to cure the sick by directing an "electroid fluid" away from their body. In 1799, John Haygarth, a British physician, investigated the effects of Perkins' tractors using blind controls. He constructed identical "tractors" but replaced the inner metallic rods with wooden ones. He then treated 15 patients, observing that they were all relieved from their "pain" or "paralysis." Therefore, Haygarth concluded that Perkins's tractors were effective only due to patients' "imagination." Hence, just like placebos, they were suited only for the "illiterate patients." He published his findings under a remarkable title, "Of the imagination, as a cause and as a cure of disorders of the body; Exemplified by fictitious tractors and epidemical convulsions" (Carlson & Simpson, 1970).

Other "techniques of intentional ignorance" (Kaptchuk, 1998a) were later employed to test less controversial remedies. For instance, in 1863, Austin Flint conducted one of the first trials investigating the natural history of a disease. Specifically, he wanted to study the "natural history" of articular rheumatism to judge whether the treatments then in use were effective. To this end, he observed the course of the disease in 13 untreated patients who received only "palliative measures" for pain. However, in order to give them the "moral impression" of receiving an active medicine, Flint also gave to all of them "a placebo which consisted, in nearly all the cases, of the tincture of quassia, very largely diluted. This was given regularly, and became well known in my wards as the *placeboic remedy* for rheumatism" (Flint, 1863, p. 21 italics in original). One of the results was that "the average intensity of the affection was not greater than a practical acquaintance with the disease would lead us to expect in the same number of cases treated with active remedies" (Flint, 1863, p. 34). In this respect, patients receiving the placebos reported improvements that were on a par with patients that Flint had treated with active medicines in a prior study and in clinical settings. This study represented one of the first instances in which a placebo was employed with the purpose of preventing patients from "dropping out" of a comparative trial upon realizing that they were not treated with a "real medication." However, as it was customary at the time, none of the patients was informed of being part of a clinical experimentation, as the theory and practice of informed consent emerged only in the second half of 20th century.

5. The "Conferences on Therapy" at Cornell University (1930s–1946)

In one of the earliest papers dedicated to the "placebo," the author notes that "apparently there has never been a paper published discussing the important subject of the placebo [...] Every one of us has often used the word and often prescribed the placebo, yet how many of us realize how little has been written and how little is known of the history of this word and this type of treatment?" (Pepper, 1945, p. 409). This quote underscores two interesting points. First, the use of placebos—and of the *word* "placebo"—had become common by the mid of the 20th century, at least in the United States. Second, the description of the placebo as a "type of treatment" hints at another crucial shift that has occurred at the beginning of the century: the invention of the "placebo effect."

While different factors may have conspired to this end, the emergence of the concept of "placebo effect" may likely be explained as an attempt to account for the improvements that were increasingly reported in the control groups of comparative experiments. For instance, in 1938, Diehl and colleagues tested the hypothesis that vaccines could have a positive effect on colds in a series of double-blind placebo-controlled experiments in which the students in the control groups "were treated in exactly the same manner as those in the experimental groups but received placebos instead of a vaccine" (Diehl, Baker, & Cowan, 1938). Contrary to the hypothesis, the study failed to show any superiority of the vaccines. Furthermore, the researchers found "a great reduction in the number of colds which the members of the control groups reported during the experimental period [...] these results were as good as many of those reported in uncontrolled studies which recommended the use of cold vaccines" (Diehl et al., 1938). The study failed not because vaccines did not reduce the number of colds; it failed because they reduced the colds just as much as the placebos did.

Indeed, the fact that sometimes people in control groups report significant improvements was not a novelty. However, these improvements were usually explained away as different kinds of effects attributed to patients' psychology. For instance, in a 1930 study, 40 hypertensive patients received a placebo, and the observed improvements were explained in terms of "the suggestion inherent in any drug" (Ayman, 1930, p. 249). Similarly, in 1936, two single-blind trials used saline injections as controls for ulcer disease. Again, the relief reported in the control groups was explained in terms of "psychic effects" (Sandweiss, 1936) and "suggestion" (Flood & Mullins, 1936).

A new phase began in the 1930s when a group of researchers started systematically analyzing the use of controls in clinical trials. The experiments conducted by Evans and Hoyle in 1933, and those conducted by Gold and colleagues in 1937 to test various treatments for *angina pectoris* were important precursors for three reasons. First, they made use of controls called "placebos"; second, they found no significant difference between placebos and tested drugs; third, placebos were employed with the articulated purpose to "equalize" the two arms of the trials. Interestingly, Gold and colleagues explained their use of placebos as a way of controlling not for "suggestion" but for other variables, including the "confidence aroused in the treatment," the "encouragement afforded by any new procedure," and "a change of the medical advisor" (Gold, Kwit, & Otto, 1937, p. 2178).

This new experimental and theoretical framework was further elaborated in 1946, when H. Gold organized with other colleagues—including H.G. Wolff, and E.E. DuBois—the "Conferences on therapy" at Cornell University. These events represented a watershed in the history of placebos for two reasons. First, the participants explicitly advocated the use of placebo controls in clinical trials. As explained by Gold (1954, p. 724) in a later article, a comparison was needed "between an allegedly potent agent and a blank of such physical properties as to render a distinction between the two impossible except through some pharmacologic potency which may exist [; the recommended] double-blind procedure which calls for an investigation in which neither the patient nor the doctor is aware of the identify of the two agents until the results are in and analyzed. This is imperative to avoid the influence of subconscious bias." To this end, they also stressed the importance of crafting placebo controls that were identical to tested interventions (Greiner et al., 1950, p. 153).

Second, the participants all agreed that placebos could have therapy-like effects, and thus they openly debated their possible uses in clinical practice. During a conference entitled "The use of placebos in therapy", Wolff noted that "the placebo is a potent agent and in its actions can resemble almost any drug." Furthermore, DuBois proposed an interesting division of "placebos" in three classes: (i) the pure placebos (e.g., lactose tablets); (ii) impure placebos (e.g. lower dosages of effective remedies); and the "placebo element," which consisted in all the rituals and aspects related to the prescription of a treatment, including the way in which it is presented and described.

The decisive terminological and conceptual step from the "placebo element" to the "placebo effect" was probably taken by Gold. In 1967 Mondel, one of his trainees, recalled that "He was the only one I ever heard refer to the placebo effect ... Others called it suggestion" (Shapiro & Shapiro, 1997, pp. 140–141). In any case, the shift from *imagination* via *suggestion* to the *placebo effect* represented a major milestone in the history of placebos, for it established the existence of an explicit correlation between the administration of a "placebo" and the observation of various kind of different and measurable effects, which seemed present both in clinical and research settings.

6. Beecher and the "powerful placebo" (1955)

The idea that of "placebo effect," and the necessity for adequate "controls" in clinical trials, inspired also Beecher's (1955) landmark article "The powerful placebo." In this article Beecher used mathematical methods to conduct a proto-systematic review of 15 placebo-controlled trials, concluding that the average number of people that had experienced relief in the placebo group was 35.2%. Retrospectively, Beecher's article proved crucial for how placebo and placebo effects came to be conceptualized in and outside biomedicine. It still represents the most frequently cited article in the placebo literature. However, Beecher's conclusions had severe methodological limitations and were based on unproven theoretical assumptions (Kienle & Kienle, 1997). For instance, 13 out of the 15 trials considered in his article lacked a no-treatment group, thus making it impossible to distinguish the improvements due to the natural history of the disease from the placebo effect. Also, Beecher deliberately excluded from his review studies with two controls (a placebo and a no-treatment group) in which the placebo arm showed no improvement. Nevertheless, these methodological flaws did not prevent his ideas from becoming widely popular.

Following the ideas of Gold and others at Cornell University in the 1940s, in this paper Beecher advanced a partially new conception of "placebo effect." Contrary to the idea that placebo effects were only "in the mind" of the patients, he claimed that they were able to "produce gross physical change" (Beecher, 1955). This meant that for many conditions, and not just for psychological ones, the placebo effect could represent a critical confounding factor in the assessment of the efficacy of interventions. A second feature assumed by Beecher was the idea that the placebo effect was constant and additive in the control and in the treatment group. As he wrote: "the placebo effect of active drug is masked by their active effects … [t]he total 'drug' effect is equal to its 'active' effect plus its placebo effect" (Beecher, 1955). On this "additive" view, the placebo effect measured in a RCT is something that is constant and evenly distributed between the experimental and the control arm. In the experimental arm, the total magnitude of the effect is thus given by the placebo effect *plus* the specific efficacy of the drug. Accordingly, to calculate the specific efficacy of an intervention one has to *subtract* from the total improvement measured in the experimental arm the improvement measured in the placebo or control arm. It follows that without a "placebo control" it is impossible to assess the true efficacy of an intervention. In Beecher's view placebo effects are thus powerful and ubiquitous and yet constant; therefore, they can and should be controlled through an appropriate, placebo-controlled, experimental design.

Like in the case of Gold and other researchers of the time, the conceptual redefinition operated by Beecher of the "placebo effect" can be interpreted as an attempt to create a persuasive rhetorical case for the systematic implementation of RCTs in clinical research (Kaptchuk, 1998b). To this end, the placebo effect was portrayed as a powerful threat requiring adequate controls.

7. The rise of RCTs and the first pioneering studies on placebo effects (1955–1970s)

Starting from the 1930s the diffusion of RCTs gained a decisive momentum by the mid of the 20th century. By 1980 the Food and Drug Administration required that evidence for the effectiveness of a drug be obtained from randomized double-blind placebo trials (Wampold, Imel, & Minami, 2007, p. 380). As a consequence of the recognition of the RCT as the new "golden standard" for clinical research, also the "new" concepts of "placebo controls" and "placebo effect" were increasingly adopted in biomedicine, first in clinical research and later in clinical practice.

The newfound popularity of these concepts, however, must not obscure the fact, at a closer look, the "popular view" that emerged at the beginning of second half of the 20th century of "placebo" and "placebo effect" was both precarious and deeply problematic—more akin to a conceptual patchwork than to a well-crafted and coherent construct. This was due to several reasons.

First, differently from clinical research, in clinical practice placebos were still prescribed primarily for the "mental relief" of the patient. With a few exceptions (Wolf, 1950; Wolff, DuBois, & Gold, 1946), only a few physicians held the view that the "placebo effect" could induce anything more than "psychological relief." Also, the studies that were then conducted to inquire into the positive effects of the doctor-patient relationship and communication were often not regarded as being studies on the "placebo effect" (Houston, 1938; Luparello, Leist, Lourie, & Sweet, 1970). Thus, the new conception of the "powerful placebo" as something capable of "great physical changes" had still to coexist with the older conception of a placebo as something "inert" whose effects were only, or mostly, of a psychological nature—either due to the "the power of suggestion" or to "imagination." Furthermore, the concomitant emergence of autonomy and informed consent as linchpins for medical ethics contributed to cast a negative light on the once accepted practice of deceiving patients "for their own good." Rather than being a "pious fraud," the deceptive administration of placebos came to be increasingly stigmatized as a relic of a bygone age of medical paternalism (Annoni & Miller, 2016).

Second, in clinical research the adoption of the new concept of "placebo effect" still incorporated a set of decisive but unscrutinized assumptions. One of such key-assumptions was that, since in clinical research the "placebo effect" was observed after the (deceptive) administration of a physical remedy, a placebo was needed "to cause" the placebo effect. In the following years this idea became so entrenched in biomedicine and the public imagery alike, that even today an authoritative dictionary like the OED still defines the "placebo effect" as "a beneficial effect produced by a placebo drug or treatment, which cannot be attributed to the properties of the placebo itself." Coupled with the assumption that a placebo is, by definition, something "inert," this way of conceptualizing placebo and placebo effects contributed to establish the paradoxical theory according to which a "placebo" is something "inert" and yet required to cause an effect, i.e., "the placebo effect" (Annoni & Blease, 2018; Howick, 2017; Moerman, 2002).

Third, even though the concept of the "placebo effect" eventually became pivotal for biomedicine, for many decades its *mechanisms* remained almost unexplored. This was an indirect consequence of the comparative logic beneath the RCT design for which it was introduced in the first place. By assuming the "additivity" assumption at the outset, the model of the RCTs proposed by Gold and then Beecher allowed to "black box" of all the different "non-specific" variables in the experimental and control group. As the placebo effect in the two arms of the trial was assumed to be constant and equal, the beautiful simplicity of the comparative logic of RCTs allowed for controlling it without knowing the real causes of those improvements. Due to this assumption, thus, it was possible to control for placebo effects without explaining their mechanisms, or disentangling them from other confounding factors, such as the natural history of diseases. As Kaptchuk has rightly observed, the new placebo effect "became both the *raison d'être* for, and the sacrificial victim of, the masked RCT" (Kaptchuk, 1998b, p. 1724).

In the following decades, this situation was challenged by a series of pioneering experiments. In 1950, Wolf studied the effects of suggestion and conditioning by administering various agents (urograstone, Benadryl, ipecac, prostigimine, atropine, lactose, and water) to observe their effects on the gastric motility of five subjects (Wolf, 1950). He concluded that "'placebo effects' which modify the pharmacological action of drugs or endow inert agents with potency are not imaginary, but may be associated with measurable changes at the end organs. These effects are at times more potent than the pharmacologic action customarily attributed to the agent" (Wolf, 1950, p. 88). These insightful remarks anticipated many of the ideas about the nature of placebo effects that were later empirically demonstrated in the following decades. Then, in 1954, Lasagna and colleagues investigated the possibility of identifying "placebo responders" or "reactors" by administering placebos and morphine in postoperative patients under different sequential order and dosage number. This study found that the propensity to respond to placebos depends on different variables—both physiological and non-physiological—thus raising complex issues about how such a variability ought to be harnessed in clinical research and practice (Lasagna, Mosteller, von Felsinger, & Beecher, 1954). Wolf, in another study conducted in 1957, similarly concluded that it was not possible to isolate "placebo reactors" and "non-reactors" as if they were abstract and distinct entities, i.e., independent from the context in which the therapy takes place (Wolf, Doering, Clark, & Hagans, 1957).

8. The empirical study of placebo effects and the golden age of placebo research (1970s–)

The next phase in the history of placebos initiated when a wave of laboratory experiments, clinical trials, and neurocognitive studies began to uncover the psychological and neurobiological mechanisms of placebo effects. These empirical studies proved crucial in turning the "placebo effect" from a curiosity at the fringe of biomedicine into a worthy subject of scientific and theoretical inquiry.

A first decisive step was taken in 1978, when Levine and colleagues published a pioneering study on the mechanisms of placebo analgesia (Levine, Gordon, & Fields, 1978). This study tested the hypothesis that endorphins mediate placebo analgesia using a double-blind, placebo-controlled design in dental postoperative pain 3 h or 4 h after surgery, patients received either a placebo or a pill containing naloxone (an opioid antagonist). Patients given naloxone reported significantly greater pain than those receiving the placebo. Furthermore, patients who reported a decrease of pain after taking the placebo were identified as placebo responders and received a second administration of naloxone. Again, naloxone was able to increase their pain, thus highlighting the role of endogenous opioids in placebo analgesia. Therefore, using this ingenious design, this study neatly demonstrated the existence of a key-placebo mechanism involving endogenous molecules.

In the following years, and despite some controversy (Howick et al., 2013; Hróbjartsson & Gøtzsche, 2001, 2004, 2010; Wampold, Imel, & Minami, 2007), a growing series of empirical studies have uncovered other mechanisms of placebo effects, confirming and extending these initial findings (Miller, Colloca, Crouch, & Kaptchuk, 2013). Initially, these studies focused on identifying and characterizing various psychological and pharmacological mechanisms of placebo effects. As for the psychological mechanisms, placebo effects has been typically explained in terms of expectancies and/or classical conditioning (Miller et al., 2013). Against this background, in the following decades, a series of important studies have suggested that expectancies and conditioning may both concur in modulating placebo responses; that other psychological mechanisms, such as social learning, may also play a role in placebo modulation; and that their interaction is not rigid and, in part, still in need of a shared conceptualization (Amanzio & Benedetti, 1999; Benedetti, Pollo, et al., 2003; Colloca & Benedetti, 2009; Montgomery & Kirsch, 1997; Voudouris, Peck, & Coleman, 1990). Furthermore, between 1970s and 1990s, the study of placebo effects and their mechanisms has been extended conditions other than pain, such as depression (Mayberg, Silva, Brannan, et al., 2002), irritable bowel-syndrome (IBS) (Kaptchuk et al., 2008), Parkinson's disease (de la Fuente-Fernandez et al., 2001), and drug addiction (Volkow, Wang, & Ma, 2003).

Even though a detailed reconstruction of these empirical discoveries and of their implications exceeds the purpose of the present historical review, it must nonetheless be emphasized that at least three important epistemological shifts have contributed in advancing and consolidating the empirical study of placebo effects in 50 years since the early pioneering studies.

A first important methodological insight has consisted in harnessing the comparative logic beneath RCTs not only to isolate the effects of tested interventions, but to disentangle and assess the magnitude of placebo effects. In its simplest form, this may entail adding a no-treatment arm to clinical trials to separate the natural history of disease and other from the placebo effect. On this issue, it is useful to introduce a distinction, which has gained currently only recently, between the "placebo response" and the "placebo effect" (Evers et al., 2018). According to this view, the *placebo and nocebo response* includes all health changes that result after administration of an inactive treatment—including natural history and regression to the mean. By contrast, the *placebo and nocebo effect*, instead, refers to the changes specifically attributable to placebo and nocebo mechanisms, including the neurobiological and psychological mechanisms of expectancies.

Another and more sophisticated strategy, however, consists in adopting ad hoc experimental designs. In this respect, an important epistemic innovation has been the adoption of "open-hidden" experiments. In this elegant trial design the same medication is delivered either in the full view of a clinician/nurse (expected open administration) or covertly, for example through an intravenous infusion machine (unexpected hidden administration) (Benedetti, Amanzio, & Maggi, 1995; Gracely, Dubner, Wolskee, & Deeter, 1983; Levine & Gordon, 1984; Levine, Gordon, Smith, & Fields, 1981). The scope of open-hidden experiments is to assess whether it makes a difference to administer a drug while removing some variables from the healing context (e.g., patient-doctor communication; bedside presence of caregivers, the awareness of receiving a treatment). In a later but classic experiment of this kind, Amanzio, Pollo, Maggi, and Benedetti (2001) administered four analgesics (buprenorphine, tramadol, ketorolac and metamizol) to patients in postoperative settings. In the open administration groups a doctor informed patients that the injection was a powerful painkiller and the pain will subside soon. In the hidden administration groups, patients received the exact same dose of analgesics through an automatic infusion machine and were thus unaware of receiving the painkiller. The difference in pain relief between the two groups provided a measure of the placebo effect. The study found that the dose of analgesic needed to reduce the pain by half was significantly greater in the hidden administration groups for all painkillers. Thus, the same dose of an analgesic had different effects depending on it being administered in an open or hidden way. Similar results have been replicated also in conditions other than pain, including anxiety (Benedetti, Giuliano, et al., 2003) and Parkinson's disease (Benedetti, Pollo, et al., 2003; Pollo et al., 2002).

In general, open-hidden experiments demonstrate that, for conditions amenable to placebo modulation, the effectiveness of a therapy depends both on *what* the therapy consists of and *how* is delivered. Between the 1980s and today, the importance of contextual factors in modulating placebo effects has been further demonstrated in other experiments isolating different variables of the healing context. For instance, in an important study, Kaptchuk et al. (2008) investigated the hypothesis that it was possible to isolate two additive placebo components related to (i) the ritual of the therapy and (ii) one to the interpersonal physician–patient relationship. The study was performed on 262 patients with IBS. Participants were randomized into three groups. The first arm was the one of "no treatment" or waiting list and was meant to operate as the control for factors such as regression to the mean and spontaneous remission. In the second arm, patients received placebo acupuncture, but their relationship with the physicians was curtailed to a minimum—limited interaction. Finally, in the third arm, patients received the same placebo acupuncture, but they found warm, confident, and emphatic physicians—augmented relationship. (Sham acupuncture in both groups was performed by the means of a 'Steinberger needle', a special device that looks and feels just like real acupuncture, except that it does not pierce the skin.) Trial outcomes were the same as the one required by the FDA to approve new drugs for IBS: global improvement scale, adequate relief of symptoms, symptom severity score, and quality of life. Results confirmed that for all outcomes people in the limited interaction group reported more relief than the ones in the waitlist group, but less than the ones in the augmented relationship group. Improvements were significant: after 3 weeks, the felt improvement was 27% in the control group, 43% for the limited groups, and 62% in the augmented group. This latter had a magnitude of effects comparable with those of *alosetron* and *tegaserod*, the two most commonly prescribed drugs for this condition which are known to have significant, even life-threatening, side effects. This study confirmed that non-specific effects can produce statistically and clinically significant outcomes and provided a first proof of principle that it is possible to isolate and combine diverse components of the healing context to modulate placebo effects.

More generally, this elegant study underscored two crucial points. First, "receiving a therapy" may induce significant placebo responses, even when the "therapy" is just a placebo. Since the late 1970s, this result has been corroborated by a series of empirical studies correlating the features of therapies (its color, shape, timing, etc.) to measurable differences attributed to placebo modulation. For instance, studies have found that branded placebos are more effective than generic ones (Branthwaite & Cooper, 1981; Waber, Shiv, Carmon, & Ariely, 2008); that sham surgery is more effective than sham injections, while sham injections are more effective than oral placebos (de Craen, Tijessen, de Gans, & Kleijnen, 2000); that blue and green placebos are better to induce sleep, while red ones are better as stimulants (de Craen, Roos, Leonard de Vries, & Kleijnen, 1996; Sallis & Buckalew, 1984); and that placebos administered frequently are better than those administered sparsely (de Craen et al., 2000).

Second, besides the content and the formal characteristics of a treatment, other contextual aspects of the clinical encounter can modulate placebo responses. In this respect, the doctor-patient relationship has emerged as a key-determinant of placebo responses. Like the healing context in general, in this historical phase several studies have analyzed the doctor-patient relationship as a set of different variables. Collectively, these studies vindicated the early intuitions by Huston and others that the doctor himself may indeed act as a powerful "therapeutic agent", and that the doctor-patient communication may drastically influence patients' response to several clinical outcomes (Egbert, Battit, Welch, & Bartlett, 1964; Gracely, Dubner, Deeter, & Wolskee, 1985; Houston, 1938).

A second methodological milestone for the study of placebo effects has been the introduction of brain imaging techniques, such as electroencephalography (EEG), the functional magnetic resonance imaging (fMRI), and the PET (positron emission tomography). The first studies with brain recording of the placebo effect were conducted with EEG in 1983 (Irwin & Fink, 1983; Levy & Jankovic, 1983), while the first one with PET was in 1996 (Schmidt et al., 1996) and with fMRI in 1999 (Enck, Horing, Broelz, & Weimer, 2018; Kleinschmidt et al., 1999). These techniques have allowed to explore the neurophysiological underpinnings of placebo effects, confirming and extending the results of prior pharmacological studies by describing the functional neuroanatomy of placebo responses in various conditions (Wager, Billing, Smith, et al., 2004). For instance, in the first brain imaging study of placebo and opioid analgesia, Petrovic and colleagues (Petrovic, Kalso, Petersson, & Ingvar, 2002) used PET to show that the same subsets of brain regions are affected either by a placebo or by the μ-opioid remifentanil, thus suggesting the existence of a shared mechanistic pathway between placebo-induced analgesia and an opioid-induced analgesia. Furthermore, brain imaging techniques have allowed neuroscientists to use the placebo effects as a model to "investigate the complex interaction between mental activities (such as expectancy) and various neuronal systems" in conditions other than pain (Miller et al., 2013, p. 77). The impact of brain imaging techniques for the empirical study of placebo effects has been so significant that Enck et al. (2018, p. 87), in their review of "knowledge gaps" in current placebo research, have identified the 1990s as the decade in which "placebo research" has been born as a research discipline, for those were the years in which the first wave of brain imaging studies was published.

On a related and important note, in these years the empirical study of placebo effects has shed light also on another aspect of the placebo phenomenon: the "nocebo effect." The word "nocebo" derives from the Latin for "I shall harm," and was first introduced in 1961 (Kennedy, 1961) to name "the other side" of the placebo phenomenon (Benedetti, Pollo, et al., 2003). In general, nocebo effects have been found to follow the same logic of their placebo counterparts. For instance, in a study by Kaptchuk et al. (2006), participants with persistent arm pain were randomized to receive either acupuncture or a pill. However, all treatments were just placebos: the pill did not contain any active molecule with analgesic powers, and the acupuncture was "sham" in that it was performed with retractable needles. Yet, many patients reported side effects, some even discontinued the participation in the trial. Interestingly, the nocebo side effects matched the descriptions that were given in informed consent documents. Similar effects were found by systematic reviews of anti-migraine trials, in which a high rate of side effects in the placebo control group matched those of real drugs (Amanzio, Corazzini, Vase, & Benedetti, 2003), and for sexual dysfunctions (Mondaini et al., 2007). However, to study nocebo effects in laboratory or clinical settings researchers must usually induce unpleasant or harmful stimuli in the participants. This creates an ethical conflict: on the one hand, the empirical research is needed to study the mechanisms of nocebo responses; on the other hand, these studies may entail harming trial participants. Research on nocebo effects have been thus comparatively more difficult to conduct, for they are more difficult to design and less likely to receive ethical approval.

Finally, a third crucial methodological turning point in the study of placebo effects has occurred with the first trial testing the efficacy of "open-label placebos." Since Cullen's first recorded use of the word "placebo," the administration of inert remedies has been associated with deception. Indeed, part of the negative connotation that has been attributed to placebos in clinical practice has been due to the idea that they must be given deceptively to be effective. This view has periodically resurfaced throughout the history of placebo effects, often being crystallized in phrases such as the "pious fraud" or the "humble humbug," and it has remained unchallenged until the second half of the 20th century, when Libarman (1962) and then Park and Covi (1965) finally put it to test. In their study, Park and Covi found that anxious and neurotic outpatients improved after taking placebos despite knowing that they were just inert pills, and thus that the "unawareness of the inert nature of the placebo is not an indispensable condition for improvement on placebo" (Park & Covi, 1965; cited in Miller et al., 2013, p. 73).

These conclusions have remained largely ignored until a landmark study by Katpchuk and collaborators (2010) in which placebo pills were administered in an open and non-deceptive way to IBS patients. The participants were all told that, despite being inert, placebos have been found to cause significant improvements by promoting mind-body self-healing processes (Annoni & Miller, 2016; Kaptchuk et al., 2010). Eligible patients were then randomized to two groups: a no-treatment control group and a group receiving placebo pills twice a day. Improvements were measured using the IBS Global Improvement Scale (IBS-GIS), the IBS Symptom Severity Scale (IBS-SSS), the IBS Adequate Relief (IBS-AR) questionnaire, and the IBS Quality of Life (IBS-QOL) questionnaire. Surprisingly, after both 11 and 21 days, people in the open placebo group reported significant improved IBS-GIS, IBS-SSS, and IBS-AR scores compared to no-treatment. Therefore, this study provided a compelling proof-of-principle that it is possible to elicit significant placebo effects by administering placebos without deception. Since this study, the same results have been replicated in other conditions known to be amenable to placebo modulation, such as depression, etc.

Moreover, this landmark study also demonstrated how ethical and regulatory issues (i.e., the need to avoid deception) may lead to a search for new experimental designs which, in turn, may illuminate new aspects of the placebo phenomenon. This approach hints at another recent turn in the history of placebos and placebo effects: i.e., the beginning of a new phase characterized by increasingly multi and inter-disciplinary research. This process, whereby different strands of previously independent researches have converged into a common area of inquiry has eventually reached an important tipping point in 2014, when the Society for the Interdisciplinary Placebo Studies (SIPS) was founded. Three years later, the SIPS has organized its first international conference, which hosted more than 200 participants from various disciplines (Enck et al., 2018; Evers et al., 2018).

9. Summary and conclusions

In the past centuries the concept of "placebo" has undergone several transformations. As argued in this review, at least five major turning points have marked its eventful history. The first key-event has been its introduction of the term "placebo" in medicine, which has occurred sometimes at the end of the 17th century. Then, a second crucial step has been the emergence of the concept of "placebo control" in clinical research. While the term was probably coined at the beginning of the 20th century, it represented the culmination of a rowdy process of historical evolution that began three centuries earlier with the first blindfolded comparative experiments. The emergence of the concept of placebo control has been instrumental also for the coming into being of the concept of "placebo effect". While the paternity of the term is uncertain, it was most likely created sometime in 1940s and it was more regularly in use by the time Beecher published his article on the "powerful placebo" in 1955. In the following decades, all these terms gained currency in and outside medicine as the RCT design became the "golden standard" of clinical research.

Then, a new chapter began with the first empirical studies investigating the psychological, pharmacological and neurobiological mechanisms of placebo effects. Preceded by a series of pioneering researches, this new phase was inaugurated by the landmark study in 1978 by Levine et al. in which the system of endogenous opioids was firstly identified as one of the key-mechanisms of placebo analgesia. In the following decades, an increasing and converging series of empirical studies employing a combination of diverse laboratory experiments, innovative experimental designs and brain imaging techniques has identified and characterized many and often overlapping mechanisms of placebo and nocebo effects. The single most important result achieved by these empirical studies has been the discovery there are *many placebo effects*, not just one; and that they are measurable objective psychobiological phenomena able to modulate symptoms across diverse settings and conditions.

In conclusion, the history of placebos and placebo effects reveals much about the past of medicine as it does about its possible future, as a better understanding of the placebo phenomenon may still yield significant insights not only for clinical practice and research, but also for other disciplines investigating the variety of human experiences with respect to health and disease.

Acknowledgment

This research has been founded by Fondazione Umberto Veronesi within the project DSB. AD008.458 "Care and integrity in research: perspectives and implications for bioethics and law."

References

Amanzio, M., & Benedetti, F. (1999). Neuropharmacological dissection of placebo analgesia: Expectation-activated opioid systems versus conditioning-activated specific subsystems. *Journal of Neuroscience, 19*, 484–494.

Amanzio, M., Corazzini, L. L., Vase, L., & Benedetti, F. (2003). A systematic review of adverse events in placebo groups of anti-migraine clinical trials. *Pain, 146*(3), 261–269.

Amanzio, M., Pollo, A., Maggi, G., & Benedetti, F. (2001). Response variability to analgesics: A role for non-specific activation of endogenous opioids. *Pain, 90*, 205–215.

Andersen, L. O. (2005). A note on the invention, invisibility and dissolution of the placebo effect. *Gesnerus, 62*, 102–110.

Annoni, M., & Blease, C. (2018). A critical (and cautiously optimistic) appraisal of Moerman's "meaning response" *Perspectives in Biology and Medicine, 61*(3), 379–387.

Annoni, M., & Miller, F. G. (2016). Placebo effects and the ethics of therapeutic communication. *Kennedy Institute of Ethics Journal, 26*(19), 79–103.

Anonymous. (1954). The humble humbug. *The Lancet, ii*, 321.

Ayman, D. (1930). An evaluation of therapeutic results in essential hypertension. *Journal of the American Medical Association, 95*(4), 246–249.

Beecher, H. K. (1955). The powerful placebo. *Journal of the American Medical Association, 159*, 1602–1606.

Benedetti, F., Amanzio, M., & Maggi, G. (1995). Potentiation of placebo analgesia by proglumide. *Lancet (London, England), 346*(8984), 1231.

Benedetti, F., Giuliano, M., Lopiano, L., Lanotte, M., Rainero, I., Vighetti, S., et al. (2003). Open versus hidden medical treatments: The patient's knowledge about a therapy affects the therapy outcome. *Prevention & Treatment, 6*, Article 1.

Benedetti, F., Pollo, A., Lopiano, L., Lanotte, M., Vighetti, S., & Raniero, I. (2003). Conscious expectation and unconscious conditioning in analgesic, motor and hormonal placebo/nocebo responses. *Journal of Neuroscience, 23*, 4315–4323.

Blease, C., & Annoni, M. (2019). Overcoming disagreement: A roadmap for placebo studies. *Biology and Philosophy, 34*(2), 18.

Branthwaite, A., & Cooper, P. (1981). Analgesic effects of branding in treatment of headaches. *British Medical Journal, 282*, 1576–1578.

Cabot, R. C. (1903). The use of truth and falsehood in medicine: An experimental study. *American Medicine, 5*, 3444–5349.

Carlson, E. T., & Simpson, M. M. (1970). Perkinism vs. mesmerism. *Journal of the History of the Behavioral Sciences, 6*, 16–24.

Colloca, L., & Benedetti, F. (2009). Placebo analgesia induced by social observational learning. *Pain, 2009*(144), 28–34.

Cullen, W. (1772). *Clinical lecture*. Edinburgh. February–April, 218f.

de Craen, A. J., Kaptchuk, T. J., Tijssen, J. G., & Kleijnen, J. (1999). Placebos and placebo effects in medicine: Historical overview. *Journal of the Royal Society of Medicine, 92*(10), 511–515.

de Craen, A. J., Roos, P. J., Leonard de Vries, A., & Kleijnen, J. (1996). Effect of colour of drugs: Systematic review of perceived effect of drugs and of their effectiveness. *British Medical Journal, 313*(7072), 1624–1626.

de Craen, A. J., Tijessen, J. G., de Gans, J., & Kleijnen, J. (2000). Placebo effect in the acute treatment of migraine: Subcutaneous placebos are better that Oral placebos. *Journal of Neurology*, *247*(3), 183–188.

de la Fuente-Fernandez, R., Ruth, T. J., Sossi, V., Schulzer, M., Calne, D. B., & Stoessl, A. J. (2001). Expectation and dopamine release: Mechanism of the placebo effect in Parkinson's disease. *Science*, *293*(5532), 1164–1166.

Diehl, H. S., Baker, A. B., & Cowan, D. W. (1938). Cold vaccines. An evaluation based on a controlled study. *Journal of the American Medical Association*, *11*, 1168–1173.

Egbert, L. D., Battit, G. E., Welch, C. E., & Bartlett, M. K. (1964). Reduction in postoperative pain by encouragement and instruction of patients—A study of doctor-patient rapport. *New England Journal of Medicine*, *270*(16), 825–827.

Enck, P., Horing, B., Broelz, E., & Weimer, K. (2018). Knowledge gaps in placebo research: With special reference to neurobiology. *International Review in Neurobiology*, *139*, 85–106.

Evers, A. W. M., Colloca, L., Blease, C., Annoni, M., Atlas, L. Y., et al. (2018). Implications of placebo and nocebo effects for clinical practice: Expert consensus. *Psychotherapy and Psychosomatics*, *87*(4), 204–210.

Finniss, D. G. (2018). Placebo effects: Historical and modern evaluation. *International Review in Neurobiology*, *139*, 1–26.

Finniss, D. G., Kaptchuk, T. J., Miller, F. G., & Benedetti, F. (2010). Biological, clinical, and ethical advances of placebo effects. *Lancet (London, England)*, *375*(9715), 686–695.

Flint, A. (1863). A contribution toward the natural history of articular rheumatism; consisting of a report of thirteen cases treated solely with palliative measures. *The American Journal of the Medical Sciences*, *46*, 17–23.

Flood, C. A., & Mullins, C. R. (1936). Treatment of peptic ulcer by means of injections. *American Journal of Digestive Diseases*, *3*, 303–305.

Franklin, B., Majault, L., Sallin, J. B., D'Arcet, de Bory, Guillotin, J.-I., et al. (1785). *Report of Dr. Benjamin Franklin, and other commissioners, charged by the king of France, with the examination of animal magnetism, as now practiced in Paris*. London: William Godwin J. Johnson.

Gold, H. (1954). How to evaluate a new drug. *American Journal of Medicine*, *17*, 722–727.

Gold, H., Kwit, N. T., & Otto, H. (1937). The xanthines (theobromine and aminophylline) in the treatment of cardiac pain. *Journal of the American Medical Association*, *108*, 2173–2179.

Gracely, R. H., Dubner, R., Deeter, W. D., & Wolskee, P. J. (1985). Clinician's expectations influence placebo analgesia. *Lancet (London, England)*, *325*(8419), 43.

Gracely, R. H., Dubner, R., Wolskee, P. J., & Deeter, W. R. (1983). Placebo and naloxone can alter post-surgical pain by separate mechanisms. *Nature*, *306*(5940), 264–265.

Greiner, T. H., Gold, H., Cattell, M., Travell, J., Bakst, H., Rinzler, S. J. H., et al. (1950). A method for the evaluation of effects of drugs on cardiac pain in patients with angina of effort: A study of Khellin (Visammin). *American Journal of Medicine*, *9*, 143–155.

Houston, W. R. (1938). The doctor himself as a therapeutic agent. *Annals of Internal Medicine*, *11*(8), 1416–1425.

Howick, J. (2017). The relativity of 'placebos': Defending a modified version of Grünbaum's definition. *Synthese*, *194*(4), 1363–1396.

Howick, J., Friedemann, C., Tsakok, M., Watson, R., Tsakok, T., Thomas, J., et al. (2013). Are treatments more effective than placebos? A systematic review and meta-analysis. *PLoS One*, *8*(5), e62599.

Hróbjartsson, A., & Gøtzsche, P. C. (2001). Is the placebo powerless? An analysis of clinical trials comparing placebo with no treatment. *New England Journal of Medicine*, *344*(21), 1594–1602.

Hróbjartsson, A., & Gøtzsche, P. C. (2004). Is the placebo effect powerless? Update of a systematic review with 52 new randomized trials comparing placebo with no treatment. *Journal of Internal Medicine*, *256*, 91–100. http://www.jameslindlibrary.org/haygarthj-1800/.

Hróbjartsson, A., & Gøtzsche, P. C. (2010). Placebo interventions for all clinical conditions. *Cochrane Database Systematic Review,* CD003974.
Irwin, P., & Fink, M. (1983). Familiarization session and placebo control in EEG studies of drug effects. *Neuropsychobiology, 10*(2–3), 173–177.
Jacobs, B. (2000). Biblical origins of placebo. *Journal of the Royal Society of Medicine, 93*(4), 213–214.
Jefferson, T. (1898). In P. L. Ford (Ed.), *v. 9. The writings of Thomas Jefferson.* New York: G. P. Putnam.
Kaptchuk, T. (1998a). Intentional ignorance: A history of blind assessment and placebo controls in medicine. *Bulletin of the History of Medicine, 72,* 389–433.
Kaptchuk, T. J. (1998b). Powerful placebo: The dark side of the randomised controlled trial. *Lancet (London, England), 351*(9117), 1722–1725.
Kaptchuk, T., Friedlander, E., Kelley, J. M., Sanchez, M. N., Kokkotou, E., Singer, J. P., et al. (2010). Placebos without deception: A randomized controlled trial in irritable bowel syndrome. *PLoS One, 5*(12), 15591.
Kaptchuk, T. J., Kelley, J. M., Conboy, L. A., Davis, R. B., Kerr, C. E., Jacobson, E. E., et al. (2008). Components of placebo effect: Randomised controlled trial in patients with irritable bowel syndrome. *British Medical Journal (Clinical Research Ed.), 336*(7651), 999–1003.
Kaptchuk, T. J., Kerr, C. E., & Zanger, A. (2009). Placebo controls, exorcisms and the devil. *Lancet, 374*(9697), 1234–1235.
Kaptchuk, T. J., Stason, W. B., Davis, R. B., Legedza, A. R. T., Schnyer, R. N., Kerr, C. E., et al. (2006). Sham device vs inert pill: Randomised controlled trial of two placebo treatments. *British Medical Journal, 332*(7538), 391–397.
Kennedy, W. P. (1961). The nocebo reaction. *Medicina Experimentalis International Journal of Experimental Medicine, 95,* 203–205.
Kerr, C. E., Milne, I., & Kaptchuk, T. J. (2008). William Cullen and a missing mind–body link in the early history of placebos. *Journal of the Royal Society of Medicine, 101*(2), 89–92.
Kienle, G. S., & Kienle, H. (1997). The powerful placebo effect: Fact of fiction. *Journal of Clinical Epidemiology, 50,* 1311–1318.
Kleinschmidt, A., Bruhn, H., Kruger, G., Merboldt, K. D., Stoppe, G., & Frahm, J. (1999). Effects of sedation, stimulation, and placebo on cerebral blood oxygenation: A magnetic resonance neuroimaging study of psychotropic drug action. *NMR in Biomedicine, 12*(5), 286–292.
Lasagna, L. C., Mosteller, F., von Felsinger, J. M., & Beecher, H. K. (1954). The study of the placebo response. *The American Journal of Medicine, 16*(6), 770–779.
Levine, J. D., & Gordon, N. C. (1984). Influence of the method of drug administration on analgesic response. *Nature, 312*(5996), 755–756.
Levine, J. D., Gordon, N. C., & Fields, H. L. (1978). The mechanism of placebo analgesia. *Lancet, 2*(8091), 654–657.
Levine, J. D., Gordon, N. C., Smith, R., & Fields, H. L. (1981). Analgesic responses to morphine and placebo in individuals with postoperative pain. *Pain, 10*(3), 379–389.
Levy, R. S., & Jankovic, J. (1983). Placebo-induced conversion reaction: A neurobehavioral and EEG study of hysterical aphasia, seizure, and coma. *Journal of Abnormal Psychology, 92*(2), 243–249.
Libarman, R. (1962). An analysis of the placebo phenomenon. *Journal of Chronic Diseases, 15,* 761–783.
Luparello, T. J., Leist, N., Lourie, C. H., & Sweet, P. (1970). The interaction of psychologic stimuli and pharmacologic agents on airway reactivity in asthmatic subjects. *Psychosomatic Medicine,* (32), 509–513.
Mayberg, H. S., Silva, J. A., Brannan, S. K., et al. (2002). The functional neuroanatomy of the placebo effect. *American Journal of Psychiatry, 159,* 728–737.
Miller, F. G., Colloca, L., Crouch, R. A., & Kaptchuk, T. J. (Eds.), (2013). *The placebo. A reader.* Baltimore: The John Hopkins University Press.

Moerman, D. E. (2002). *Meaning, medicine, and the "placebo effect" Vol. 28*. Cambridge: Cambridge University Press.
Mondaini, N., Gontero, P., Giubilei, G., Lombardi, G., Cai, T., Gavazzi, A., et al. (2007). Finasteride 5 mg and sexual side effects: How many of these are related to a nocebo phenomenon? *Journal of Sexual Medicine*, *6*, 1708–1712.
Montgomery, G. H., & Kirsch, I. (1997). Classical conditioning and the placebo effect. *Pain*, *72*, 107–113.
Park, L. C., & Covi, L. (1965). Nonblind placebo trial: An exploration of neurotic patients' response to placebo when its inert content is disclosed. *Archives of General Psychiatry*, *12*, 36–45.
Pepper, O. H. P. (1945). A note on the placebo. *American Journal of Pharmacology*, *117*, 409–412.
Petrovic, P., Kalso, E., Petersson, K. M., & Ingvar, M. (2002). Placebo and opioid analgesia—Imaging a shared neuronal network. *Science*, *295*(5560), 1737–1740.
Pollo, A., Torre, E., Lopiano, L., Rizzone, M., Lanotte, M., Cavanna, A., et al. (2002). Expectation modulates the response to subthalamic nucleus stimulation in parkinsonian patients. *Neuroreport*, *13*, 1383–1386.
Sallis, R. E., & Buckalew, L. W. (1984). Relation of capsule color and perceived potency. *Perceptual and Motor Skills*, *58*(3), 897–898.
Sandweiss, D. J. (1936). Treatment of gastroduodenal ulcer with histidine monohydrochloride [Larostindin]. *Journal of the American Medical Association*, *106*, 1452–1459.
Schmidt, M. E., Ernst, M., Matochik, J. A., Maisog, J. M., Pan, B. S., Zametkin, A. J., et al. (1996). Cerebral glucose metabolism during pharmacologic studies: Test-retest under placebo conditions. *Journal of Nuclear Medicine*, *37*(7), 1142–1149.
Shapiro, A. (1964). A historic and heuristic definition of the placebo. *Psychiatry*, *27*, 52–58.
Shapiro, A., & Shapiro, E. (1997). *The powerful placebo: From ancient priests to modern physicians*. Baltimore: John Hopkins University Press.
Volkow, N. D., Wang, G. J., & Ma, Y. (2003). Expectation enhances the regional brain metabolic and the reinforcing effects stimulants in cocaine abusers. *Journal of Neuroscience*, *23*, 11461–11468.
Voudouris, N. J., Peck, C., & Coleman, G. (1990). The role of conditioning and verbal expectancy in the placebo response. *Pain*, *43*, 121–128.
Waber, R. L., Shiv, B., Carmon, Z., & Ariely, D. (2008). Commercial features of placebo and therapeutic efficacy. *Journal of the American Medical Association*, *299*, 1016–1017.
Wager, T. D., Billing, J. K., Smith, E. E., et al. (2004). Placebo-induced changes in fMRI in the anticipation and experience of pain. *Science*, *303*, 1162–1166.
Wampold, B. E., Imel, Z. E., & Minami, T. (2007). The story of placebo effects in medicine: Evidence in context. *Journal of Clinical Psychology*, *63*(4), 379–390.
Wolf, S. (1950). Effects of suggestion and conditioning on the action of chemical agents in human subjects—The pharmacology of placebos. *Journal of Clinical Investigation*, *29*, 100–109.
Wolf, S., Doering, C. R., Clark, M. L., & Hagans, J. A. (1957). Chance distribution and the placebo "reactor" *Journal of Laboratory and Clinical Medicine*, *49*, 837–841.
Wolff, H. G., DuBois, E. F., & Gold, H. (1946). In Cornell conferences on therapy: "Use of placebos in therapy" *New York State Journal of Medicine*, *46*, 1718–1727.

CHAPTER TWO

Determinants of placebo effects

Sharlet Anderson*, Glenn T. Stebbins
Department of Neurological Sciences, Rush University Medical Center, Chicago, IL, United States
*Corresponding author: e-mail address: sharlet_a_anderson@rush.edu

Contents

Abstract

Determinants of placebo effects and placebo response can be considered from multiple intrinsic and extrinsic perspectives. Intrinsic factors may influence the patient and the clinician/researcher. Patient expectations and previous experiences are considered two of the major intrinsic determinants of placebo response. Other patient determinants include the neural systems under treatment/study, situational factors and reactions to the environment, and personality traits. Clinician/researcher determinants of placebo response include personality factors such as empathy, perceived expertise, the clinical relationship with the patient, and the clinician's belief in the efficacy of the treatment. Extrinsic determinants include the type of study design, influence of advertising or branding, and forces expressed by the cultural milieu. These determinants do not act in isolation, but rather form a complex interaction that ultimately impacts the promotion or deterrence of the placebo effect in clinical and research settings.

1. Introduction

The concept of placebo has evolved over time and with shifting contexts. The historical origin of the term *placebo* lies in the notion of pleasing

International Review of Neurobiology, Volume 153
ISSN 0074-7742
https://doi.org/10.1016/bs.irn.2020.03.029

others along with an element of deceit. In or around 400 AD, the Latin term "placebo" was the first word in a common prayer for the dead (Czerniak & Davidson, 2012). During that time, it was common practice for strangers off the street to join in on the recitation of the prayers, with hope that the family would reward them in some way, assuming they were friends or associates of the dearly departed. Similarly, family members of the departed often paid others to perform prayers for the deceased on their behalf. Thus, placebo came to be associated with deception early in its usage.

By the mid-1800s a placebo, as defined by Hooper's Medical Dictionary, was "any medicine adopted to please rather than benefit the patient" (Gross, 2017). With this definition in the context of medical care, comes the subtle assumption that any benefit to the patient may be incidental or separate from the intended purpose of pleasing them. Given that patients assume treatments are specifically intended to benefit them, the use of a treatment that has pleasing or perhaps placating the patient as the central intent is in some way deceitful. This underlying aspect of deceit carries over and is highlighted in more modern uses of the word placebo.

More recently, the American Society of Pain Management Nursing put forth a position statement strongly advising against the use of placebo in pain management, which is defined as "any sham medication or procedure designed to be void of any known therapeutic value" (Arnstein, Broglio, Wuhrman, & Kean, 2011, p. 226). This definition goes one step further in implying that the placebo was specifically designed NOT to have any therapeutic value. Furthermore, Arnstein and colleagues argue that using placebos in pain management is a failure in providing respectful care and a violation of both the principle of beneficence and of nonmaleficence. With this characterization of placebo and the use thereof, the element of deceit is central to the definition and is exhibited as something to avoid.

In an historical review of the use of placebo, Czerniak and Davidson (2012) define placebo as "a therapy used for its nonspecific psychophysiological or presumed effect but without any actual effect on the condition being treated." Another contemporary view defines the placebo effect as "the combined physical and emotional change occurring with inert treatment," which takes into account the perspective of the patient as an active agent in measuring outcome, given that emotions are internal and covert (Diederich & Goetz, 2008). This is unique in that it highlights what the recipient of the placebo brings to the treatment context, rather than the implied unidirectional determinations made by the clinician in previous definitions. Broadening even further, Oken (2008) referred to "a physiological state anticipating and contributing to the occurrence of some future health-related outcome through learning, conditioning, or other related process." In this view, not only is the patient an active agent in detecting a response, their previous experiences and expectations are also considered. Similarly, Kaptchuk and Miller (2015) define placebo effects as "improvements in patients" symptoms that are attributable to their participation in the therapeutic encounter, with its rituals, symbols, and interactions. Moerman (2000) described the placebo effect as "desirable psychological and physiological effects of meaning in the treatment of illness." As is illustrated, the focus has shifted away from the inert substance itself to that of the context surrounding the encounter with said inert substance, as well as the meaning the patient or participant assigns to those contextual factors.

Different terms have been used to describe the phenomenon of changes in patients' or participants' health status associated with receipt of a placebo: placebo effect and placebo response. On the surface it can seem that these terms are used interchangeably. However, with closer inspection of the particular variables studied, it is clear that they refer to different concepts (Benedetti & Amanzio, 2011). Czerniak and Davidson (2012) distinguish the placebo itself from the associated response and effect. They delineate the *placebo response* as any beneficial response to the placebo that is beyond what would be expected from the natural history of the disease/disorder. In contrast, they describe the *placebo effect* as any benefit attributable to the context of the treatment encounter. With this elaborated characterization, we see that contextual factors are taken into consideration as potentially beneficial or therapeutic. Benedetti and Amanzio (2011) define the placebo effect as that which is "observed in the placebo arm of a clinical trial" and includes natural history, regression to the mean, biases, and judgment errors in addition to the placebo response, which is the biological phenomenon alone. On the other hand, some definitions of the placebo response are less conservative as they include the "sum of the effects of optimized management plus the placebo" (Chaplin, 2006). Contextual factors often refer to all the factors apart from the drug itself that can influence patients' or participants' responses. More specifically, it can refer to treatment characteristics, patient characteristics, practitioner characteristics, healthcare setting, and patient-practitioner relationship (Blasi, Harkness, Ernst, Georgiou, & Kleijnen, 2001). Given that these are factors that are not specific to the drug itself, they are viewed as part of the placebo response. For the remainder of this chapter, the term placebo response will be used to refer to the phenomenon of the patient or participant experience with a placebo.

With the increasingly patient- and context-centered perspectives of placebo, placebo effect, and placebo response, comes a drive to identify the determinants thereof. As Czerniak and Davidson (2012) assert, the placebo can function as a "trusted ally to the practicing clinician and an unavoidable hindrance to the clinical investigator" (p. 771). Thus, whether the intent is to harness its potential healing power or to separate its effects from the treatment under investigation, it is useful to identify the determinants of the placebo effect.

2. Intrinsic factors impacting placebo response

2.1 Neural systems involved in the placebo response

There has been a substantial amount of attention focused on neural systems that underlie the placebo response in recent years. The psychological, intrinsic factors of expectation and reward are intimately intertwined with hypotheses that center upon dopaminergic systems. For example, Benedetti, Carlino, and Pollo (2011) describe that expectation of future events (placebo) can both modulate anxiety and trigger a reward response, even before the placebo itself has been administered (Benedetti et al., 2011). Parkinson's disease patients have demonstrated significant increases in dopamine release in PET imaging after receiving placebo injection, which was interpreted as response to expectation of symptom relief (de la Fuente-Fernandez, Ruth, Sossi, Calne, & Stoessl, 2001). This increase was seen in both patients who reported motor symptom improvement and those who did not. In another PET imaging study, patients with depression who were treated with placebo or fluoxetine, changes in the nucleus accumbens and orbitofrontal cortex were observed after only 1 week of treatment in both the placebo and active drug groups, which is before any therapeutic benefit would be expected (Mayberg et al., 2002). This suggests that the anticipation of symptom relief creates changes that are observable in neuroimaging studies.

In addition to the dopamine hypothesis of placebo response, much work has been conducted investigating the role of opioidergic systems in placebo analgesia and nocebo effects (Benedetti et al., 2011). Studies investigating levels of beta endorphin in cerebrospinal fluid, inhibition of colesystekinin, and activation of ronstral anterior cingulate and orbitofrontal cortex have implicated the contribution of opioidergic systems in placebo response (Benedetti & Amanzio, 2011).

2.2 Patient-based determinants

2.2.1 Personality factors

Identification of personality factors that impact placebo response would allow for the development of methods to steer participant selection so that placebo responses could be reduced, balanced across groups, and statistically managed in research efforts. Alternatively, with the same information clinical practice and patient care could be enhanced with the ability to harness the power of placebo effects (Ernst & Herxheimer, 1996). However, there is inconsistency in findings across studies of personality traits and placebo response. Price, Finniss, and Benedetti (2008) note that the mean effect size of a placebo group could be an inaccurate or incomplete metric against which to judge a medication's efficacy, given that the placebo group could contain a mix of large effects from responders to little or no effect from nonresponders. However, the relationship between personality traits and placebo response is likely more complex than separating responders from nonresponders (Price et al., 2008).

The Five Factor Model (Costa & McCrae, 1988) outlines the major personality traits through factor analysis and has been the basis of a wealth of research in human behavior. These factors are: openness, conscientiousness, extraversion, agreeableness, and neuroticism. With regard to the placebo response, agreeableness and extraversion have been investigated as potential mediators of the placebo response. As Jakšić and colleagues note, these personality traits are related to placebo response only in certain situations, likening them to moderators more than mediators (Jaksic, Aukst-Margetic, & Jakovljevic, 2013).

Many studies highlight the interaction between personality traits and situational factors that account for observed rates or magnitudes of placebo effect (Geers, Helfer, Kosbab, Weiland, & Landry, 2005; Geers, Wellman, Fowler, Helfer, & France, 2010; Oken, 2008). For example, Geers and colleagues (Geers et al., 2005) demonstrated that pessimists were more likely than optimists to experience negative symptoms from a placebo pill, but only when told they were receiving an active drug. There were no differences between pessimists and optimists in conditions when they were told that there was a 50–50 chance of receiving placebo or drug or 100% chance they would be receiving the placebo. A follow-up study showed that positive effects were more likely to be reported by participants rated high in optimism than those rated low, but only in the condition when they were informed that they would be receiving an active analgesic agent before a painful stimulus, not when they were informed that they had been given an inert substance (Geers et al., 2010).

Other personality traits besides those included in the Five Factor model have also been investigated. Colloca and Miller (2011) report that more robust placebo effects were observed in individuals who scored higher on measures of empathy, within the context of a study of social cues and analgesic effect of placebo (Colloca & Miller, 2011). In a review of 20 studies focused on the prediction of placebo response, some level of agreement was found for self-efficacy, locus of control, and contingency expectations (Horing, Weimer, Muth, & Enck, 2014). The authors note that these factors are largely cognition-based in contrast to the traditional personality traits often used to categorize individuals.

2.2.2 Genetic factors

Attempts to identify placebo responders have led to exploration of potential genetic factors. Benedetti and colleagues (2011) provide a review of the effect of placebo on the brain and highlight genetic variants in social anxiety disorder and depression that may contribute to placebo response. For example, reduced amygdala activity was found in homozygous carriers of the long allele of the 5-HTTLPR or G variant of the TPH2 G-703T polymorphism in participants with social anxiety disorder (Furmark et al., 2008). With regard to depression, monoamine oxidase A G/T polymorphisms coding for highest enzyme activity was associated with lower placebo response than with other genotypes (Leuchter, McCracken, Hunter, Cook, & Alpert, 2009). Peck and Coleman (1991) also reviewed mechanisms of placebo effect and their implications for treatment and research. They noted a polymorphism in the μ-opioid receptor gene (OPRM1) (Peciña, Love, Stohler, Goldman, & Zubieta, 2015) as well as a functional variant of amide hydrolase (Peciña et al., 2014) in placebo analgesia. The identification of potential genetic variations in dopaminergic, opioid, endocannabinoid, and serotoninergic pathways, has led to the proposal of a placebome that might be used to screen for placebo responders in clinical trials (Hall, Loscalzo, & Kaptchuk, 2015). However, the relationship between genetic variants and the behavioral manifestation of placebo response is likely to be complex and conclusions should be drawn cautiously given the potential for both Type I and Type II errors (Colagiuri, Schenk, Kessler, Dorsey, & Colloca, 2015).

2.2.3 Environmental factors

Expectancy is one of the major models for conceptualizing the placebo response and is perhaps the most thoroughly researched (Thompson, Ritenbaugh, & Nichter, 2009). Oken (2008) notes that although learned associations through conditioning may play a part in the placebo effect to a certain degree, what the patient brings to the treatment experience beyond an unconscious conditioned response is more likely to mediate the behavioral expression of the placebo response. When considering expectations that a patient or subject brings to the treatment situation, it is implied that that there is conscious awareness of the administration of some type of treatment. Benedetti and colleagues demonstrated that a more robust response was seen in patients who were openly administered a treatment than those who were covertly administered the same medication via hidden infusions (Benedetti et al., 2003). Thus, the patients' expectation, derived from their perception of having been treated, influenced their response. However, in this study, there were no placebos and thus, no placebo effect per se. Moerman and Jonas (2002) reviewed a set of studies that demonstrated that the meaning assigned to the color of a pill, the pill brand characteristics, and the specific instruction to expect a certain outcome of a treatment were the differentiating factors in the subjects' response (Moerman & Jonas, 2002). Thus, the meaning assigned to treatment itself or to the experience of receiving treatment, as well as expectations based on those meanings, can shape the mechanism of response (Moerman & Harrington, 2005).

Patients' beliefs regarding the treatment modality (surgery vs injection vs pill) and perceived costs (expensive vs cheap) may also play a role in the presence or magnitude of placebo response, based on their associated meanings and expectations. In some cultures, injections are perceived as powerful medicines, given their symbolic representation of the biomedical profession (Reeler, 2000). Beliefs and expectations regarding the efficacy of an injected substance may be an aspect of the placebo effect. In a review of several studies investigating enhanced placebo effect (Kaptchuk, Goldman, Stone, & Stason, 2000), several studies are described that demonstrate a stronger response to injected placebos than to orally administered placebos (Grenfell, Briggs, & Holland, 1961; Morison, Woodmansey, & Young, 1961; Traut & Passarelli, 1957). Though not specifically stated, it is likely that the participants had some prior exposure to and knowledge of injections and thus brought their own expectations to the experience. It is important to note that this perception of injections being more powerful than oral medications may not generalize across cultures (see Section 4 in this chapter). This difference in response to treatment modality has implications for the study design, given the resultant differences in effect size and thus the number of participants enrolled to achieve sufficient statistical power (Kaptchuk et al., 2000).

Meaning associations to the color of pills/capsules/tablets, and therefore expected effects, may also have an impact on whether the placebo was perceived as having a stimulant or anxiolytic effect (Buckalew & Coffield, 1982; de Craen et al., 1999). Expectations with regard to dosing and frequency may also influence the magnitude of the placebo response. For example, de Craen and colleagues (1999) conducted a systematic review of several placebo controlled trials of medications for duodenal ulcers that included either a twice a day or four times a day regimen. They found that rates of ulcer healing were greater in the four times a day regimen. They posit that participants in these arms of the trials may have believed they were receiving more treatment leading to a psychophysiological response. However, they note that it was possible that selection bias leading to inclusion of patients who are more compliant and thus more likely to stay with a demanding dosing schedule over time.

Experience shapes expectations, and expectations can create subjective experiences (Kirsch, 1985). Patient's assessment of the treatment is susceptible to expectations they bring with them, which in turn can be influenced by many factors: past experience with medication, interactions with physicians, beliefs about the potency of different forms of treatment, personality traits, and cultural factors in the experience of illness and healing. Thus, the relationship between patient expectation and the placebo response or lack thereof is likely quite complex. Blasi et al. (2001) conducted a meta-analysis examining studies that included provider behaviors specifically designed to influence participants' expectations of the treatment or illness. Ten of 19 studies demonstrated a significant relationship between providers' influential statements and healthcare outcomes. However, they noted that there were differences with regard to outcomes being rated as good, average, or poor, as well as potential differences according to the severity or chronicity of the condition and the prognosis. For example, patients with chronic illnesses may believe the treatment is effective for most people based on the clinician's assertion, but also believe that it will not work for them if they had been deemed treatment-resistant in the past. Indeed, a qualitative study investigating expectations in patients who had been in the placebo arm of a randomized controlled trial showed that having an expectation of no benefit was common (Stone, Kerr, Jacobson, Conboy, & Kaptchuk, 2005). In fact, several participants noted that their self-assessment over the course of the trial confirmed their initial expectation, as they did not experience any side effects nor did they detect symptom improvement. The authors also noted that past treatment failure was an inclusion criterion for some of these participants, which could have influenced their expectations.

Previous exposure to particular medications, or other medications perceived as being similar, can have a substantial effect on the placebo response via building an expectation of reward. Thus, if a patient or participant has positive associations to a particular medication or set of medications, it follows that they could expect this experience to be replicated in future administrations. For example, levels of dopamine release in response to placebo were comparable to that observed in response to levodopa in Parkinson's disease patients (de la Fuente-Fernandez et al., 2001). This finding was interpreted to be the expression of anticipation of reward (symptom relief), given prior experience with dopamine agonists and anticipation of similar effects. The placebo-reward hypothesis describes the activation of reward circuitry, in particular dopaminergic projections to the nucleus accumbens, as the mechanism for placebo response (de la Fuente-Fernández, 2009). This hypothesis can be applied to both research and clinical settings. In research settings, the response may be dependent upon perceived probability of receiving active drug. In clinical settings, the response triggered by the expectation of reward is related to the patient's perception of trustworthiness of the provider or providers in general, rather than probability of receiving treatment, which is 100% in the clinical setting (de la Fuente-Fernández, 2009). However, unless specifically queried, a positive previous response to dopamine agonist (or any drug perceived by the patient to be similar to the study drug) or to the healthcare relationship is assumed. In some cases, participants may have had past experiences that were neutral or negative with regard to certain drugs, clinicians, or medical settings that could color their expectations about any research or clinical treatment that they may try in the future.

Price et al. (2008) describe a model of placebo response determinants that includes an emotional aspect that interacts with cognitive factors. They illustrate how the combination of *desire* for analgesia combined with *expectation* of analgesia with conscious receipt of a placebo produces a robust response. In addition, they also describe how this synergy exists for appetitive or approach goals, as it does in desire for avoidance of a noxious experience.

Another hypothesis of determinants of the placebo response is related to classical conditioning. In general, the classical conditioning model of behavior learning involves the repeated pairing of two stimuli, one of which naturally produces a physiological response and the other that is neutral and does not produce a response. Over time and repeated pairings, the neutral stimulus will create the response, even in the absence of the original stimulus. In other words, if two phenomena co-occur frequently, the perceiver will respond to either one in the same way, even if the other is not present. In the context of placebo responses, tactile and gustatory properties of the treatments (placebo or otherwise) may be the stimuli that produces the response, based on prior experiences with similar feeling or tasting medications (Colloca & Miller, 2011). This is particularly relevant for patients who participate in multiple clinical trials. Thus, including questionnaires about participants' previous experience in clinical trials, the modes of treatment administration involved, and their perception of whether or not they received placebo or treatment in previous studies may be worth including.

2.3 Clinician/researcher based determinants

2.3.1 Personality factors

Characteristics of the healthcare provider can also have an influence on placebo response. For example, participants who received a placebo demonstrated more symptom reduction if their physician exuded warmth and empathy (Howe, Goyer, & Crum, 2017; Rakel et al., 2009) and higher competence (Howe et al., 2017). It is notable that differences in warmth, empathy, support, and patient engagement have been observed to be related to race and ethnicity of the patients as well (Friesen & Blease, 2018). For example, third party ratings of physician empathy were lower in interactions with Latino patients than with white patients (Hooper, Comstock, Goodwin, & Goodwin, 1982). However, in a different study, empathy ratings were similar in interactions with Latino vs white patients, though the number of positive expressions were lower (Sleath, Rubin, & Arrey-Wastavino, 2000). Thus, it is possible that physician characteristics, whether specifically manipulated in the study design or due to unconscious/implicit bias during the clinical encounter can have a significant impact on the participant's or patient's response to treatment (Cooper et al., 2012).

2.3.2 Environmental factors

Beliefs about the efficacy of an intervention may also influence clinician ratings of symptoms. In a study of Parkinson's disease patients' response to placebo, Goetz, Leurgans, Raman, and Stebbins (2000) show that placebo response could be objectively observed as improvement in motor symptoms in patients with Parkinson's disease (Goetz et al., 2000). Given that the participants were a subset of those who participated in a randomized controlled treatment trial in an era where disclosure of the possibility of receiving placebo in the informed consent process was mandatory, it is possible that the participants may have had some degree of expectation about change in symptoms. However, in a follow-up study, Goetz and colleagues (2002), showed that placebo response observed in ratings of motor symptoms were not associated with improved participant ratings of activities of daily living (Goetz et al., 2002). The interpretations include possible insensitivity of the ADL ratings, the magnitude of change attributed to placebo being within the limits of normal variability in motor symptoms over time, and potential placebo effect in clinician ratings. The latter interpretation is a unique hypothesis suggesting that at least some aspect of placebo effects could be driven by unconscious expectations on the part of the clinician, akin to the Rosenthal effect (Rosenthal & Jacobson, 1968), rather than the participant. In Rosenthal's original study, certain children were identified to teachers as having exceptional potential for accelerated academic growth based on their performance on an IQ test. In reality, the children were chosen at random. The children were re-tested at the end of the year and indeed, those who identified as accelerated had indeed lived up to the expectations of the teachers. This phenomenon was first demonstrated in lab rats by Rosenthal and Fode (1963). Experimenters were told that two groups of rats were specifically bred to be good maze runners or poor maze runners, when in reality they were all from the same litter. Experimenters' ratings of maze running ability of the two groups of rats confirmed these arbitrary assignments (Rosenthal & Fode, 1963). This phenomenon has been studied in numerous settings and is often referred to as experimenter expectancy effect or self-fulfilling prophesy.

The relationship between healthcare provider and patient can be a powerful source of healing and health change. It could be argued that before technological advances in medical treatments began, the relationship and healing rituals comprised the totality of "healthcare." In psychotherapy, the "therapeutic alliance" is a strong predictor of outcome (Ardito & Rabellino, 2011). Even with evidence-based advances in biomedical treatments and psychotherapy techniques, the provider-patient interaction or relationship is a contextual factor in responses and outcomes that warrants specific attention. Some argue that the therapeutic relationship, or the art of medicine, has been devalued or neglected in contemporary science (Miller & Kaptchuk, 2008). Blasi and colleagues (2001) conducted a meta-analysis of 19 randomized controlled trials specifically focused certain aspects of practitioner-patient interactions, one of which they name "cognitive care." This is defined as practices that are specifically intended to "influence patients' expectations about the illness or treatment." They found that of 19 studies that met their inclusion criteria, 10 showed that cognitive care had a significant impact on health outcome, thus demonstrating that practitioner-patient interactions can influence patient expectations, which in turn can have a direct impact on symptoms and outcome.

3. Extrinsic factors impacting placebo response

3.1 Study design

In this section, a brief introduction to placebo response related to aspects of study design is provided. A more extensive discussion of study design is covered in chapter "Statistical methods in handling placebo effect" by Raman of this volume. Given that the placebo effect can be viewed as all phenomena observed in the placebo arm of a clinical trial (Benedetti & Amanzio, 2011), the potential influence of the study design itself is a valid factor to examine. Without specific inquiry into participants' beliefs and expectations, some of which are likely deduced from the study design as described in the informed consent process, researchers are relegated to making assumptions about the potential drivers of the placebo response. Considerable variability in placebo response rates across studies in specific disease populations leads one to wonder how much of the variability is due to differences in study design vs other contextual or patient factors.

For example, Moerman (2000) conducted a systematic review of placebo response healing rate in ulcer patients and found that it ranged from 0% to 100%, which is considerably more variability than in active drug healing rates. He notes that differences in the length of the studies, among other factors, could have been one potential source of the variability. However, in this particular systematic review, the differences in study length were small (4 vs 6 weeks) and did not account for the variability observed. One consideration is that differences in study length may or may not be significant depending on the natural course of the disease. For example, a difference of 6 weeks in a study of ulcers may be significant, though in a more chronic condition such as Parkinson's disease, it may not be. Additionally placebo response can fluctuate from visit to visit in longitudinal studies, which can be further complicated by fluctuations in symptom severity as in Parkinson's disease (Goetz et al., 2000).

Some have suggested that in order to examine and separate the placebo response from other contextual factors, a three-arm study design is required.

This involves an arm for active drug, placebo, and for "no treatment." Others have suggested that a balanced placebo design is optimal (Oken, 2008). This includes two groups, one receiving placebo and one drug, and within each of those, there are two subgroups of those who were told they were receiving placebo and those who were told they were receiving drug. The end result is four total groups. However, there is also evidence that the number of arms in a study can influence the size of both placebo and drug response (Enck, Klosterhalfen, Weimer, Horing, & Zipfel, 2011), presumably because it creates expectations about the likelihood of receiving drug or placebo, even as early as in the informed consent process (Oken, 2008). Other study design factors such as the method or route of treatment administration can also have an impact on the magnitude of placebo response (see earlier section on patient expectations).

3.2 Advertising, branding, and exclusivity as enhancers of placebo effects

The type of information presented about a product or treatment can have a significant cultural impact on expectations regarding the quality or efficacy of that product. In a pair of studies conducted in China, the effects of brand awareness and price on brand attitudes and intention to purchase were examined (Hsiao, Hsu, Chu, & Fang, 2014). Several brands of ice cream were selected and determined to be low or high in brand awareness. The low brand awareness product was switched to the high brand awareness packaging and vice versa. Thus, the packaging was rendered arbitrary with regard to the product within. The authors emphasize that the expectation that well-known brands and expensive brands are often assumed to be superior in quality. Their results showed that individuals tended to judge the quality of the product based on brand awareness rather than their actual judgment from sampling the product, and that this effect was moderated by the price of the product. In another study of branding effects for medications, higher effectiveness in the treatment of headache was reported for both aspirin and a placebo labeled with a well-known brand name, as opposed to pills (both aspirin and placebo) that were presented as generic (Branthwaite & Cooper, 1981). These findings were later replicated in a study of university students with frequent headaches, who reported greater pain reduction and fewer side effects in pills (both placebo and ibuprofen) that were labeled with a brand name than those presented as generic (Faasse, Martin, Grey, Gamble, & Petrie, 2015). Additionally, patients with Parkinson's disease showed a greater response to placebo presented as an expensive treatment, than to a purportedly cheaper treatment (Espay et al., 2015). However, in a more fine-tuned analysis, negative beliefs about generic medications were found to be associated with lower income, lower education, nonwhite race, and having Medicare (Iosifescu, Halm, McGinn, Siu, & Federman, 2008). Negative beliefs about generic medications were also found in international studies. For example, a systematic review of studies conducted in several countries revealed that physicians and pharmacists held negative beliefs about generic medications, and that lack of trust in generic medications was associated with healthcare systems that were less mature (Toverud, Hartmann, & Hakonsen, 2015). The belief that generic drugs are inferior in quality or efficacy exists even in countries with single-payer healthcare systems (Duerden & Hughs, 2010). Thus, the consumer culture within which a study is conducted, or a treatment is rendered may also have an impact on the presence and magnitude of the placebo response depending on whether the treatment or placebo is perceived as a generic or brand-name medication. With the current state of advertising for medical products, both on television and online, the effect of this determinant on the placebo response needs study.

4. Culture

Many studies focusing on contextual factors in placebo effect and response overlook an important element that has potential to influence all the other factors described above: culture. The expression of illness and healing practices take place within a sociocultural context (Bhugra, Ventriglio, Till, & Malhi, 2015; Kirmayer, 2013) which shapes cognitive schema. Even the judgment of whether or not a medication is effective or tolerable is influenced by culture in that the meaning of drug effects can be symbolic. An example described by Lin (2010) contrasts drowsiness as an indicator that a medication is powerful and effective, against drowsiness as a sign of worsening symptoms or medication toxicity. For example, in a study of local interpretations of side effects of antimalarial drugs in the Peruvian Amazon, qualitative analyses of the reasons for treatment discontinuation were revealing (Muela Ribera, Hausmann-Muela, Gryseels, & Peeters Grietens, 2016). In the local culture objects, people, stages of life, illnesses, and many other things are assigned a quality of "hot" or "cold." Health is a state of balancing hot and cold, and sudden changes or "shocks" are generally viewed as unhealthy. Malaria symptoms often include sudden bouts of fever and chills. According to local beliefs, other things that "shock" are to be avoided while having malaria. This includes medication, alcohol, and certain foods. Thus, when locals experienced the side effect of dizziness from the anti-malarial drugs, it was interpreted as a "shock," which is something to be avoided when having malaria. This leads to a high rate of nonadherence. This concept of "hot" and "cold" qualities is found in some Asian cultures as well. In an older study of lithium therapy in Hong Kong Chinese psychiatric patients, participants reported that lithium was a "hot" substance due to the side effect of dry mouth. However, participants did not rate this side effect as bothersome, as it was easily remedied by drinking more water (a "cold" substance), which is viewed as a healthful behavior that should be done regardless (Lee, 1993).

Cross-cultural differences in placebo response have been observed, even within a specific disease population. For example, de Craen and colleagues (2000) showed that the higher reported relief from injections than pills in the United States did not occur in studies performed in Europe (de Craen et al., 2000), though they did not offer potential reasons for this difference. Similarly, a higher rate of placebo response was documented in patients presenting with acute mania in the United States than those in Russia and India (Vieta, Pappadopulos, Mandel, & Lombardo, 2011). The possibilty that both psychiatric treatment and clinical trials are more commonplace in the United States, and with increased awareness of the role of clinical trials in psychotropic medication comes a tendency to include participants with less severe symtpoms, which is associated with higher placebo response rate. Moerman (2000) notes that the results of several international studies of placebo response in ulcer disease suggest that Germany has the highest placebo response rate than most other countries, whereas Brazil had the lowest. However, in his analysis of several hypertension studies, Germany had the lowest placebo response. Moerman (2000) posits that while the high placebo response rates in ulcer disease in Germans is not clearly understood, the low placebo response rate in hypertension in Germany may be due to a culturally-based concern about the dangers of low blood pressure. Thus, differences in placebo response vary not only across cultures but also across disease populations within cultures.

A better understanding of culturally influenced meanings assigned to the symbolic aspect of treatments is necessary for accurate interpretation of placebo response. For example, in some cultures there may be a preference for treatment to be delivered in the form of a syrup, tincture, capsule, or injection (Bhugra & Ventriglio, 2015). Muela and colleagues describe superficial level knowledge of isolated cultural representations as being familiar with the "mushrooms" on the surface, whereas a "mycelium level" knowledge is required for communication and implementation of health interventions in nonmajority communities (Muela Ribera et al., 2016). This also applies for individuals who identify with more than one culture, are in transition from one cultural identity to another, or are ambivalent with regard to health related practices or beliefs from their culture (Kirmayer, 2013). Thus, for culturally competent patient care, it is imperative that clinicians are mindful of potential communication gaps that can occur when the importance of culturally influenced health beliefs is minimized (Lin, 2010). The communication gaps can lead to noncompliance with treatment recommendations. The intersection of culture and placebo effects is also important to recognize in research endeavors. Given that placebo response is partially mediated by contextual factors in the clinical encounter, healthcare disparities related to the quality of patient-physician interactions may affect the distribution of placebo effect unevenly across groups (Friesen & Blease, 2018).

Kirmayer (2013) and Lin (2010) both emphasize the role of increasing globalization and the need for cross-cultural approaches to patient care. Price et al. (2008) describe placebo effects as "reflections of mind-brain-body relationships," which may be difficult to explain to individuals from cultures that do not ascribe to dualistic conceptualizations. For these individuals, the lines between what is placebo vs what is treatment can be blurred and with regard to clinical trials, can affect the generalizability of findings. Indeed, Moerman (2000) posited that if the placebo effect results in a specific positive outcome, which may be at least partially mediated by the patient's meaning assigned to it, and the underlying mechanisms related to the expression of that effect are identified (at least partially), a paradox is introduced: At what point does this placebo cease to be a placebo?

5. Future directions

Given that both intrinsic and extrinsic factors can function as determinants of placebo response, exerting control over the multiple influences is a daunting challenge. Initial investigations aimed to identify personality factors that would make an individual a placebo responder, but this simplistic approach is insufficient to accommodate learning, changes over time, and the extrinsic factors that play into placebo responses in different situations. In parallel, early approaches examined study design, culture and the power of media exploitation as single determinants of placebo response, but again these approaches ignore the individual patient and clinician.

Whereas, investigators now accept that a multidimensional and prospectively followed longitudinal approach to placebo studies is essential for harnessing the benefits and minimizing the negative aspects of placebo reactions, the specific study design of such research is elusive. In spite of this challenge, both expectancy and classical conditioning were initially conceptualized as the major competing hypotheses to explain the placebo responses, and these well-studied disciplines remain worthy of further application. Neural systems involved in the placebo response have also been investigated, and appear to be intimately intertwined with psychological drivers, presenting opportunities especially in the realm of neuroimaging and connectivity analyses. Whereas studies have vividly described differences across cultures in terms of placebo responses, the greater globalization of modern culture could increase the complexity of these influences, and necessitate a more nuanced approach to cultural impact.

The evolution of conceptualizations of placebos and their effects reflects the increasing incorporation of biopsychosocial models of health and healthcare in Western approaches. Whether placebo effects are regarded as the clinician's ally or the researcher's hindrance (Czerniak & Davidson, 2012) hinges on accurate understanding of the determinants thereof. However, simple, straightforward views of placebo effects that fail to respect the interactional complexity of contextual, cultural, patient, and clinician driven factors can have a detrimental effect on both clinical care and research endeavors. In addition, exacerbation of health disparities is a potential downfall of marginalizing contextual factors to traditional, outdated, and oversimplified conceptualizations of the placebo response.

References

Ardito, R. B., & Rabellino, D. (2011). Therapeutic alliance and outcome of psychotherapy: Historical excursus, measurements, and prospects for research. *Frontiers in Psychology*, *2*(270), 270. https://doi.org/10.3389/fpsyg.2011.00270.

Arnstein, P., Broglio, K., Wuhrman, E., & Kean, M. B. (2011). Use of placebos in pain management. *Pain Management Nursing*, *12*(4), 225–229. https://doi.org/10.1016/j.pmn.2010.10.033.

Benedetti, F., & Amanzio, M. (2011). The placebo response: How words and rituals change the patient's brain. *Patient Education and Counseling*, *84*(3), 413–419. https://doi.org/10.1016/j.pec.2011.04.034.

Benedetti, F., Carlino, E., & Pollo, A. (2011). How placebos change the patient's brain. *Neuropsychopharmacology*, *36*(1), 339–354. https://doi.org/10.1038/npp.2010.81.

Benedetti, F., Maggi, G., Lopiano, L., Lanotte, M., Rainero, I., Vighetti, S., et al. (2003). Open versus hidden medical treatments: The patient's knowledge about a therapy affects the therapy outcome. *Prevention & Treatment*, *6*(1), Article 1a. https://doi.org/10.1037/1522-3736.6.1.61a.

Bhugra, D., & Ventriglio, A. (2015). Do cultures influence placebo response? *Acta Psychiatrica Scandinavica, 132*(4), 227–230. https://doi.org/10.1111/acps.12422.
Bhugra, D., Ventriglio, A., Till, A., & Malhi, G. (2015). Colour, culture and placebo response. *International Journal of Social Psychiatry, 61*(6), 615–617. https://doi.org/10.1177/0020764015591492.
Blasi, Z. D., Harkness, E., Ernst, E., Georgiou, A., & Kleijnen, J. (2001). Influence of context effects on health outcomes: A systematic review. *The Lancet, 357*(9258), 757–762. https://doi.org/10.1016/S0140-6736(00)04169-6.
Branthwaite, A., & Cooper, P. (1981). Analgesic effects of branding in treatment of headaches. *British Medical Journal, 282*(6276), 1576–1578. https://doi.org/10.1136/bmj.282.6276.1576.
Buckalew, L. W., & Coffield, K. E. (1982). An investigation of drug expectancy as a function of capsule color and size and preparation form. *Journal of Clinical Psychopharmacology, 2*(4), 245–248.
Chaplin, S. (2006). The placebo response: An important part of treatment. *Prescriber, 17*(5), 16–22. https://doi.org/10.1002/psb.344.
Colagiuri, B., Schenk, L. A., Kessler, M. D., Dorsey, S. G., & Colloca, L. (2015). The placebo effect: From concepts to genes. *Neuroscience, 307*, 171–190. https://doi.org/10.1016/j.neuroscience.2015.08.017.
Colloca, L., & Miller, F. G. (2011). How placebo responses are formed: A learning perspective. *Philosophical Transactions of the Royal Society of London. Series B, Biological Sciences, 366*(1572), 1859–1869. https://doi.org/10.1098/rstb.2010.0398.
Cooper, L. A., Roter, D. L., Carson, K. A., Beach, M. C., Sabin, J. A., Greenwald, A. G., et al. (2012). The associations of clinicians' implicit attitudes about race with medical visit communication and patient ratings of interpersonal care. *American Journal of Public Health, 102*(5), 979–987. https://doi.org/10.2105/ajph.2011.300558.
Costa, P. T., Jr., & McCrae, R. R. (1988). From catalog to classification: Murray's needs and the five-factor model. *Journal of Personality and Social Psychology, 55*(2), 258–265. https://doi.org/10.1037/0022-3514.55.2.258.
Czerniak, E., & Davidson, M. (2012). Placebo, a historical perspective. *European Neuropsychopharmacology, 22*(11), 770–774. https://doi.org/10.1016/j.euroneuro.2012.04.003.
de Craen, A. J., Moerman, D. E., Heisterkamp, S. H., Tytgat, G. N., Tijssen, J. G., & Kleijnen, J. (1999). Placebo effect in the treatment of duodenal ulcer. *British Journal of Clinical Pharmacology, 48*(6), 853–860. https://doi.org/10.1046/j.1365-2125.1999.00094.x.
de Craen, A. J., Tijssen, J. G., de Gans, J., & Kleijnen, J. (2000). Placebo effect in the acute treatment of migraine: Subcutaneous placebos are better than oral placebos. *Journal of Neurology, 247*(3), 183–188. https://doi.org/10.1007/s004150050560.
de la Fuente-Fernández, R. (2009). The placebo-reward hypothesis: Dopamine and the placebo effect. *Parkinsonism & Related Disorders, 15*, S72–S74. https://doi.org/10.1016/S1353-8020(09)70785-0.
de la Fuente-Fernandez, R., Ruth, T. J., Sossi, V., Schulzer, M., Calne, D. B., & Stoessl, A. J. (2001). Expectation and dopamine release: Mechanism of the placebo effect in Parkinson's disease. *Science, 293*(5532), 1164–1166. https://doi.org/10.1126/science.1060937.
Diederich, N. J., & Goetz, C. G. (2008). The placebo treatments in neurosciences. *New Insights From Clinical and Neuroimaging Studies, 71*(9), 677–684. https://doi.org/10.1212/01.wnl.0000324635.49971.3d.
Duerden, M. & Hughs, D. (2010). Generic and therapeutic substitutions in the UK: Are they a good thing? British Journal of Clinical Pharmacology 70(3), 335–341. doiI:https://doi.org/10.1111/j.1365-2125.2010.03718.x

Enck, P., Klosterhalfen, S., Weimer, K., Horing, B., & Zipfel, S. (2011). The placebo response in clinical trials: More questions than answers. *Philosophical Transactions of the Royal Society of London Series B, Biological Sciences*, *366*(1572), 1889–1895. https://doi.org/10.1098/rstb.2010.0384.

Ernst, E., & Herxheimer, A. (1996). The power of placebo. *BMJ (Clinical Research ed.)*, *313*(7072), 1569. https://doi.org/10.1136/bmj.313.7072.1569.

Espay, A. J., Norris, M. M., Eliassen, J. C., Dwivedi, A., Smith, M. S., Banks, C., et al. (2015). Placebo effect of medication cost in Parkinson disease: A randomized double-blind study. *Neurology*, *84*(8), 794–802. https://doi.org/10.1212/wnl.0000000000001282.

Faasse, K., Martin, L., Grey, A., Gamble, G., & Petrie, K. (2015). Impact of brand or generic labelling on medication effectiveness and side effects. *Health Psychology*, *35*(2), 187–190. https://doi.org/10.1037/hea0000282.

Friesen, P., & Blease, C. (2018). Placebo effects and racial and ethnic health disparities: An unjust and underexplored connection. *Journal of Medical Ethics*, *44*(11), 774–781. https://doi.org/10.1136/medethics-2018-104811.

Furmark, T., Appel, L., Henningsson, S., Ahs, F., Faria, V., Linnman, C., et al. (2008). A link between serotonin-related gene polymorphisms, amygdala activity, and placebo-induced relief from social anxiety. *Journal of Neuroscience*, *28*(49), 13066–13074. https://doi.org/10.1523/JNEUROSCI.2534-08.2008.

Geers, A. L., Helfer, S. G., Kosbab, K., Weiland, P. E., & Landry, S. J. (2005). Reconsidering the role of personality in placebo effects: Dispositional optimism, situational expectations, and the placebo response. *Journal of Psychosomatic Research*, *58*(2), 121–127. https://doi.org/10.1016/j.jpsychores.2004.08.011.

Geers, A. L., Wellman, J. A., Fowler, S. L., Helfer, S. G., & France, C. R. (2010). Dispositional optimism predicts placebo analgesia. *The Journal of Pain*, *11*(11), 1165–1171. https://doi.org/10.1016/j.jpain.2010.02.014.

Goetz, C. G., Leurgans, S., & Raman, R. (2002). Placebo-associated improvements in motor function: Comparison of subjective and objective sections of the UPDRS in early Parkinson's disease. *Movement Disorders*, *17*(2), 283–288. https://doi.org/10.1002/mds.10024.

Goetz, C. G., Leurgans, S., Raman, R., & Stebbins, G. T. (2000). Objective changes in motor function during placebo treatment in PD. *Neurology*, *54*(3), 710–714. https://doi.org/10.1212/wnl.54.3.710.

Grenfell, R. F., Briggs, A. H., & Holland, W. C. (1961). A double-blind study of the treatment of hypertension. *JAMA*, *176*(2), 124–128. https://doi.org/10.1001/jama.1961.03040150040010.

Gross, L. (2017). Putting placebos to the test. *PLoS Biology*, *15*(2). e2001998. https://doi.org/10.1371/journal.pbio.2001998.

Hall, K. T., Loscalzo, J., & Kaptchuk, T. J. (2015). Genetics and the placebo effect: The Placebome. *Trends in Molecular Medicine*, *21*(5), 285–294. https://doi.org/10.1016/j.molmed.2015.02.009.

Hooper, E. M., Comstock, L. M., Goodwin, J. M., & Goodwin, J. S. (1982). Patient characteristics that influence physician behavior. *Medical Care*, *20*(6), 630–638. https://doi.org/10.1097/00005650-198206000-00009.

Horing, B., Weimer, K., Muth, E. R., & Enck, P. (2014). Prediction of placebo responses: A systematic review of the literature. *Frontiers in Psychology*, *5*(1079). https://doi.org/10.3389/fpsyg.2014.01079.

Howe, L. C., Goyer, J. P., & Crum, A. J. (2017). Harnessing the placebo effect: Exploring the influence of physician characteristics on placebo response. *Health Psychology*, *36*(11), 1074–1082. https://doi.org/10.1037/hea0000499.

Hsiao, Y., Hsu, Y., Chu, S., & Fang, W. (2014). Discussion of whether brand awareness is a form of marketing. *International Journal of Business and Information*, *9*(1), 50–61.

Iosifescu, A., Halm, E., McGinn, T., Siu, A., & Federman, A. (2008). Beliefs about generic drugs among elderly adults in hospital based primary care practices. *Patient Education and Counseling*, *73*(2), 377–383. https://doi.org/10.1016/j.pec.2008.07.012.

Jaksic, N., Aukst-Margetic, B., & Jakovljevic, M. (2013). Does personality play a relevant role in the placebo effect? *Psychiatria Danubina*, *25*(1), 17–23.

Kaptchuk, T. J., Goldman, P., Stone, D. A., & Stason, W. B. (2000). Do medical devices have enhanced placebo effects? *Journal of Clinical Epidemiology*, *53*(8), 786–792. https://doi.org/10.1016/s0895-4356(00)00206-7.

Kaptchuk, T. J., & Miller, F. G. (2015). Placebo effects in medicine. *New England Journal of Medicine*, *373*(1), 8–9. https://doi.org/10.1056/NEJMp1504023.

Kirmayer, L. J. (2013). The cultural diversity of healing: meaning, metaphor,and mechanism. *Heart Views : The Official Journal of the Gulf Heart Association*, *14*(1), 39–40. https://doi.org/10.4103/1995-705x.107123.

Kirsch, I. (1985). Response expectancy as a determinant of experience and behavior. *American Psychologist*, *40*(11), 1189–1202. https://doi.org/10.1037/0003-066X.40.11.1189.

Lee, S. (1993). Side effects of chronic lithium therapy in Hong Kong Chinese: An ethnopsychiatric perspective. *Culture, Medicine and Psychiatry*, *17*, 301–320. https://doi.org/10.1007/BF01380006.

Leuchter, A., McCracken, J., Hunter, A., Cook, I., & Alpert, J. (2009). Monoamine oxidase a and catechol-o-methyltransferase functional polymorphisms and the placebo response in major depressive disorder. *Journal of Clinical Psychopharmacology*, *29*(4), 372–377. https://doi.org/10.1097/JCP.0b013e3181ac4aaf.

Lin, K. M. (2010). Cultural and ethnic issues in psychopharmacology. *Psychiatric Times*, *27*, 1–7.

Mayberg, H. S., Silva, J. A., Brannan, S. K., Tekell, J. L., Mahurin, R. K., McGinnis, S., et al. (2002). The functional neuroanatomy of the placebo effect. *The American Journal of Psychiatry*, *159*(5), 728–737. https://doi.org/10.1176/appi.ajp.159.5.728.

Miller, F. G., & Kaptchuk, T. J. (2008). The power of context: Reconceptualizing the placebo effect. *Journal of the Royal Society of Medicine*, *101*(5), 222–225. https://doi.org/10.1258/jrsm.2008.070466.

Moerman, D. E. (2000). Cultural variations in the placebo effect: Ulcers, anxiety, and blood pressure. *Medical Anthropology Quarterly*, *14*(1), 51–72. https://doi.org/10.1525/maq.2000.14.1.51.

Moerman, D. E., & Harrington, A. (2005). Making space for the placebo effect in pain medicine. *Seminars in Pain Medicine*, *3*(1), 2–6. https://doi.org/10.1016/j.spmd.2005.02.008.

Moerman, D. E., & Jonas, W. B. (2002). Deconstructing the placebo effect and finding the meaning response. *Annals of Internal Medicine*, *136*(6), 471–476. https://doi.org/10.7326/0003-4819-136-6-200203190-00011.

Morison, R. A. H., Woodmansey, A., & Young, A. J. (1961). Placebo responses in an arthritis trial. *Annals of the Rheumatic Diseases*, *20*(2), 179. https://doi.org/10.1136/ard.20.2.179.

Muela Ribera, J., Hausmann-Muela, S., Gryseels, C., & Peeters Grietens, K. (2016). Re-imagining adherence to treatment from the "other side": Local interpretations of adverse anti-malarial drug reactions in the Peruvian Amazon. *Malaria Journal*, *15*(1), 136. https://doi.org/10.1186/s12936-016-1193-x.

Oken, B. S. (2008). Placebo effects: Clinical aspects and neurobiology. *Brain*, *131*(11), 2812–2823. https://doi.org/10.1093/brain/awn116.

Peciña, M., Love, T., Stohler, C., Goldman, D., & Zubieta, J. (2015). Effects of the mu opioid receptor polymorphism (OPRM1 A118G) on pain regulation, placebo effects and associated personality trait measures. *Neuropsychopharmacology*, *40*(4), 957–965. https://doi.org/10.1038/npp.2014.272.

Peciña, M., Martínez-Jauand, M., Hodgkinson, C., Stohler, C., Goldman, D., & Zubieta, J. (2014). FAAH selectively influences placebo effects. *Molecular Psychiatry*, *19*(3), 385–391. https://doi.org/10.1038/mp.2013.124.

Peck, C., & Coleman, G. (1991). Implications of placebo theory for clinical research and practice in pain management. *Theoretical Medicine*, *12*(3), 247–270. https://doi.org/10.1007/BF00489609.

Price, D. D., Finniss, D. G., & Benedetti, F. (2008). A comprehensive review of the placebo effect: Recent advances and current thought. *Annual Review of Psychology*, *59*(1), 565–590. https://doi.org/10.1146/annurev.psych.59.113006.095941.

Rakel, D. P., Hoeft, T. J., Barrett, B. P., Chewning, B. A., Craig, B. M., & Niu, M. (2009). Practitioner empathy and the duration of the common cold. *Family Medicine*, *41*(7), 494–501.

Reeler, A. V. (2000). Anthropological perspectives on injections: A review. *Bulletin of the World Health Organization*, *78*(1), 135–143.

Rosenthal, R., & Fode, K. L. (1963). The effect of experimenter bias on the performance of the albino rat. *Behavioral Science*, *8*(3), 183–189. https://doi.org/10.1002/bs.3830080302.

Rosenthal, R., & Jacobson, L. (1968). Pygmalion in the classroom. *The Urban Review*, *3*(1), 16–20. https://doi.org/10.1007/BF02322211.

Sleath, B., Rubin, R. H., & Arrey-Wastavino, A. (2000). Physician expression of empathy and positiveness to Hispanic and non-Hispanic white patients during medical encounters. *Family Medicine*, *32*(2), 91–96.

Stone, D. A., Kerr, C. E., Jacobson, E., Conboy, L. A., & Kaptchuk, T. J. (2005). Patient expectations in placebo-controlled randomized clinical trials. *Journal of Evaluation in Clinical Practice*, *11*(1), 77–84. https://doi.org/10.1111/j.1365-2753.2004.00512.x.

Thompson, J. J., Ritenbaugh, C., & Nichter, M. (2009). Reconsidering the placebo response from a broad anthropological perspective. *Culture, Medicine and Psychiatry*, *33*(1), 112–152. https://doi.org/10.1007/s11013-008-9122-2.

Toverud, E., Hartmann, K., & Hakonsen, H. (2015). A systematic review of Physicians' and Pharmacists' perspectives on generic drug use: What are the global challenges? *Applied Health Econonomics and Health Policy*, *13*, S35–S45. https://doi.org/10.1007/s40258-014-0145-2.

Traut, E. F., & Passarelli, E. W. (1957). Placebos in the treatment of rheumatoid arthritis and other rheumatic conditions. *Annals of the Rheumatic Diseases*, *16*(1), 18–22. https://doi.org/10.1136/ard.16.1.18.

Vieta, E., Pappadopulos, E., Mandel, F. S., & Lombardo, I. (2011). Impact of geographical and cultural factors on clinical trials in acute mania: Lessons from a ziprasidone and haloperidol placebo-controlled study. *International Journal of Neuropsychopharmacology*, *14*(8), 1017–1027. https://doi.org/10.1017/s146114571100040x.

CHAPTER THREE

Strategies to minimize placebo effects in research investigations

Filipe B. Rodrigues[a,b,c], **Joaquim J. Ferreira**[b,c,d,*]
[a]UCL Huntington's Disease Centre, UCL Queen Square Institute of Neurology, University College London, London, United Kingdom
[b]Laboratory of Clinical Pharmacology and Therapeutics, Faculdade de Medicina, Universidade de Lisboa, Lisbon, Portugal
[c]Instituto de Medicina Molecular, Lisbon, Portugal
[d]CNS—Campus Neurológico Sénior, Torres Vedras, Portugal
*Corresponding author: e-mail address: jferreira@medicina.ulisboa.pt

Contents

Abstracts

Placebo-controlled trials are the research standard to evaluate new interventions for which there is no standard of care. While lessening performance and detection bias, such design provides a direct mode of comparison against the probed intervention. Still, using placebo arms may pose new challenges to the design, conduct and analysis of clinical trials. This is particularly relevant in circumstances of non-additivity between the therapeutic and the placebo effects, if the outcome of interest has floor or ceiling effects, or when the predicted effect size of the intervention is large and leads to small sample sizes. There are several possible strategies to mitigate the confounding effects of the placebo, each relevant to specific clinical trial designs. This chapter puts into context the new challenges created by the placebo effect, discusses possible ways around them, and explores the future of the field.

International Review of Neurobiology, Volume 153
ISSN 0074-7742
https://doi.org/10.1016/bs.irn.2020.04.002

1. Introduction

Drug development is a lengthy, expensive and broadly inefficient process (Hay, Thomas, Craighead, Economides, & Rosenthal, 2014). In clinical drug development success rate is frequently defined as the proportion of compounds that are ultimately approved. Across medicine, the probability of success for a given compound is low (10–13%) and meaningfully smaller in the central nervous system therapeutic area at 6–9% (Hay et al., 2014; Wong, Siah, & Lo, 2018). These figures seem to rise when stratifying participants with biomarkers and decline in orphan drugs for rare diseases (Wong et al., 2018).

Numerous reasons can be proposed to justify these facts: partly understood disease mechanisms; fragile associations between the therapeutic target, the disease pathogenesis and natural history; uninformative non-human models and inadequate evaluation of preclinical effects; suboptimal pharmacokinetic characteristics; limited clinical study designs; unenthusiastic commercial interests, among others.

With respect to study design, it is sometimes said that a trial failed due to an unexpectedly high placebo response (Demitrack, Faries, Herrera, DeBrota, & Potter, 1998). As an example, in a recently completed trial examining pridopidine for Huntington's disease, the placebo arm had the largest mean improvement from baseline in the primary outcome and the investigators suggested this contributed to the trial failure (Reilmann et al., 2019). A similar situation happened in a trial exploring pimavanserin for Parkinson's disease psychosis (Friedman et al., 2009).

In this chapter we will review some of the evidence around the placebo effect in clinical trials, with a special emphasis on why it may affect results and possible strategies to avoid it.

2. Concepts and definitions

Placebos are essential to determine the efficacy and safety of new interventions. Generally, randomized controlled trials compare one or more active arms against a control arm. In situations where there is no standard of care or comparator intervention, the control arm may take the form of a placebo or sham intervention. In particular cases, there may be a no-treatment arm, sometimes known as waiting room arm.

By definition, a placebo intervention is an "inert" intervention, deemed not to have any relevant direct pharmacological or physiological effect. That said, it may have indirect effects related to the act of receiving an intervention, and participating in a therapeutic encounter, with its rituals, symbols, and interactions (Kaptchuk & Miller, 2015). For the purpose of clinical trials, placebo interventions should be manufactured, stored and administered in a way that prevents study participants and research staff to differentiate them from active interventions. Conceptually they translate into the so-called placebo effect, which is measured as a magnitude of effect called the placebo response.

Additionally, in clinical trials encompassing a prospective no-treatment arm (as opposed to historically matched controls), study participants may have a response to the fact of being enrolled in a study and interacting with clinical trial staff. This is called the Hawthorn effect.

Epidemiology recognizes the placebo effect as a relevant effect modifier, changing the relationship between the causal factor being studied (i.e., a therapeutic intervention) and an effect (i.e., the studied outcome).

To accurately measure the placebo effect, trials need to be designed so that an active treatment arm, a placebo arm and a no-treatment arm are deployed—as in three-armed trials. The difference in responses between the placebo arm and the no-treatment arm is the placebo effect. Frequently this is not possible or feasible, and the combination between the Hawthorn and the placebo effect are used a surrogate for the placebo effect.

As a construct, the placebo effect is frequently assumed as a benefit attributable to partaking in a therapeutic encounter, but that is not always the case. Placebo interventions can associate with unfavorable outcomes—the nocebo effect—and this is well identifiable when analyzing the safety tables of placebo-controlled trials. Much alike the placebo effect, there is a caveat: most times the nocebo effect is quantified as a composite of the true nocebo and the Hawthorn effects.

Classically, in clinical trials the response to an active intervention follows an additive model (Fig. 1), whereby the main response is the arithmetic sum of the effects of several independent components, including drug-specific and non-specific effects (Enck, Bingel, Schedlowski, & Rief, 2013). The relative contribution of each effect is variable and depends on the characteristics of the population, the intervention, and condition being studied. In such a model each of the individual components are defined as below (Kirsch, 2000):

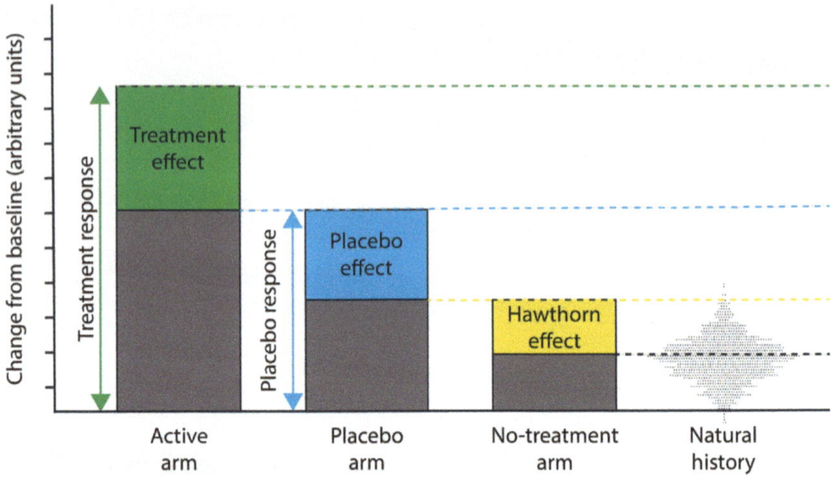

Fig. 1 Classical or additive model of the treatment, placebo and Hawthorn effects.

- the *treatment effect*, or the drug specific effect, is represented by the difference between what happens to participants in the active treatment arm (treatment response) and the what happens to participants in the placebo arm (placebo response) of a two- or three-armed trial;
- the *placebo effect* is the difference between the what happens to participants in the placebo arm (placebo response) and the no-intervention arm in a three-armed trial. As previously stated, this type of trial design is rarely used and most frequently the placebo response is a surrogate for the placebo effect;
- the *Hawthorn effect* is what happens to participants in the no-treatment arm in a three-armed trial and is very rarely measured. Sometimes data captured from prospective natural history studies can be used as a proxy for this effect;
- the *natural response* or the true natural history of the disease being studied. It is what happens to a patient (and not a participant) if they had not been recruited into a study. Currently our tools to observe this phenomenon are limited and biased. In theory retrospective natural history studies would provide an approximation of such occurrence;
- and random variation, and regression to the mean.

3. Why may the placebo effect be detrimental in clinical trials?

It is mentioned many times that the placebo effect influences all study arms in a clinical trial in the same way, as all study participants are subject to

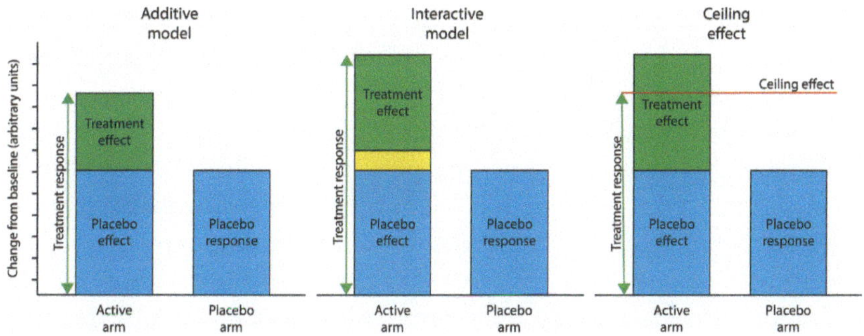

Fig. 2 When is the placebo effect problematic for clinical trials? Examples of the additive model, the interactive model, and the ceiling effect. Note that in the interactive model the overall treatment response can be equal, less or more that the arithmetic sum of the effects.

the same study procedures and are unaware of their treatment allocation provided that intervention blinding, and allocation concealment are adequately conducted. As so, from a methodological perspective, the placebo effect can only interfere with the efficiency of a clinical trial if a limited number of circumstances or assumptions are met. In an additive or independent model, the placebo effect is the same in the active and placebo arm (Fig. 2). Henceforth, whatever determinants modulate this effect, they will affect both the treatment and the placebo responses equivalently (Kirsch, 2000).

In a non-classical, non-additive model, where the components that constitute the treatment response interact with each other (Fig. 2), the main effect of their combination does not necessarily need to be numerically equal to their arithmetical sum (Enck et al., 2013; Kirsch, 2000). Three options are possible: the magnitude of the main effect can increase, decrease or remain the same.

Measurement is fundamental for clinical trials. A clinical trial is only useful if a certain phenomenon can be measured with an acceptable degree of accuracy. Choosing the measurement tools depends on external factors, like the trial design and population, and internal clinimetric characteristics. When a measurement tool is no longer able to capture change or variability in a dependent variable as a consequence of variation in an independent variable, it is said to have a ceiling (i.e., not able quantify high measurements above its limit of detection/measurement) or a floor effect (i.e., not able quantity low measurements below its limit of detection/measurement). When the tools used to measure what happens to participants in clinical trials suffer from either or both of these effects, it is also possible that the captured main effect may not correspond the arithmetical sum of the individual components of the treatment response. This is true in additive and interactive models.

Finally, modern clinical trials are based on the assumption that unbiased randomization and allocation of participants to treatment arms will result in an even distribution of measured and unmeasured participant characteristics. While being true if taken to a limit, this is often not the case when the predicted effect size of the therapeutic intervention is considerably large. Large effect sizes result in small sample sizes for an acceptable statistical power and type 1 error, and consequently a small number of participants per study arm. As a result, such a number may not be enough to assure an even distribution of characteristics across arms, and unbalance in placebo effect determinants (i.e., factors that modulate the magnitude and direction of the effect) may interfere with the magnitude of the placebo response in each of the study arms. This may have the potential to modify the magnitude of the main effect positively or negatively, can happen in additive and interactive models, and can be of further complexity in situations where a ceiling and/or floor effect are present.

While the first situation described below—non-additivity or interactivity—is problematic to empirically demonstrate, there is a strong theoretic rational to support it. Using balanced placebo designs, where deception is used to conceal allocation, would be the ultimate test, but this design poses several ethical question hardly acceptable nowadays (Kirsch, 1990). For the second (ceiling and floor effects) and the third situations (large effect sizes) there are vast examples in the literature. If any of these circumstances or a combination of them are present, the placebo effect can be problematic and act as a source of confounding, likely to contribute to biased estimates of the treatment effects. It can possibly act as either positive or negative confounder, and as partial or even complete confounder.

There are other situations related with the nature of the intervention where the placebo effect may impact the treatment outcome, but these are rarer and particular to niches of interest. An interesting exampling is the effect of parameter optimization sessions during the blinded phase of deep brain stimulation trials (Rodrigues, Duarte, Prescott, Ferreira, & Costa, 2019).

4. The impact

There is historical evidence that the treatment effects reported in clinical trials have declined over time (Ioannidis, 2008). At the same time, and across numerous spheres of medicine, the placebo responses have increased in magnitude (Dunlop et al., 2012; Kemp et al., 2010; Rheims, Perucca,

Cucherat, & Ryvlin, 2011; Tuttle et al., 2015; Undurraga & Baldessarini, 2012; Vieta & Cruz, 2008; Walsh, Seidman, Sysko, & Gould, 2002). In order to attain the desired statistical power—customarily set to 80% in clinical research and defined as the "probability of rejecting the null hypothesis when the alternative hypothesis is true," there seems to be a trend to devise clinical trials with larger sample sizes and more site numbers. The consequences of such approaches may be paradoxical: increasing the number of study sites and larger samples of participants may have a negative impact on quality control and increase variability. Also, some authors have found positive correlations between the number of study sites and the magnitude of the placebo effect (Yildiz, Vieta, Tohen, & Baldessarini, 2011).

Minimizing the placebo effect intends to optimize intervention-placebo differences, that is improving assay sensitivity (Enck et al., 2013). This will likely increase our ability to detect truly efficacious interventions and discriminate them from interventions without a clinically relevant impact. It has the potential to reduce the time, complexity and costs of clinical drug development, and curtail the number of study participants exposed to placebo-interventions. Overall we would expect the efficiency of clinical development programs to rise (Enck et al., 2013).

5. Placebo effect in neurology

The placebo effect is a broadly recognized phenomenon across medicine (Benedetti, 2014), neurological conditions not being an exception to this rule (Fig. 3) (Ferreira, Trenkwalder, & Mestre, 2018). Some placebo mechanisms are universal, such as the psychosocial context surrounding the patient, the professional and the environment (Finniss, Kaptchuk, Miller, & Benedetti, 2010), while others are disease specific. In the recent years, much as been discovered about the biological factors that mediate this effect: endogenous opioids and dopamine pathways are activated in placebo analgesia (Scott et al., 2008); basal ganglia neuronal activation and dopamine release happen after the administration of anti-parkinsonian placebos (de la Fuente-Fernandez et al., 2001); and brain activity changes in depression (Leuchter, Cook, Witte, Morgan, & Abrams, 2002). In neurology, the placebo phenomenon has been approached both from a mechanistic perspective, mostly in Parkinson's disease, but also from an epidemiologic angle. Dopamine has a direct implication in parkinsonism, and neuroimaging methods and deep brain stimulation recordings allowed us to establish robust mechanistic hypotheses. As for other conditions, where the specific

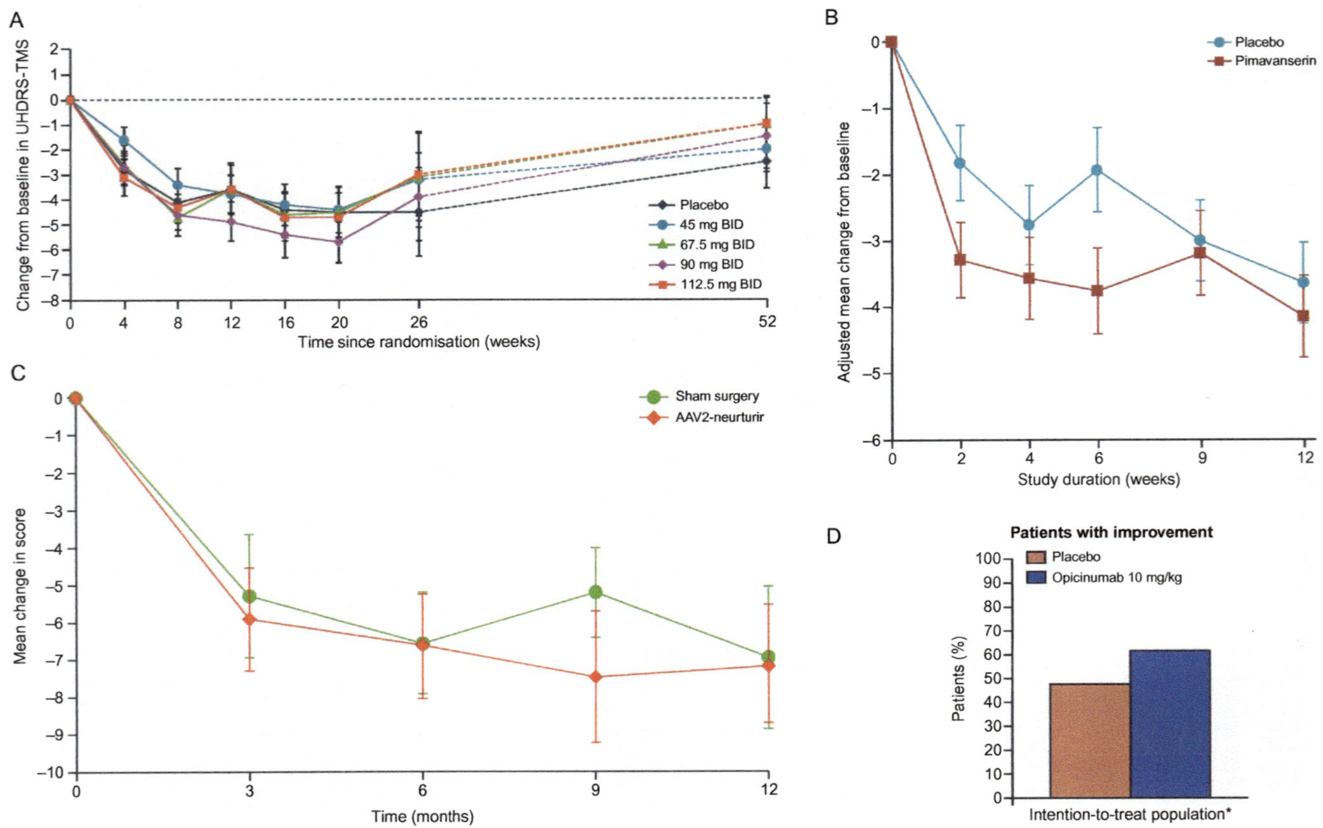

Fig. 3 Examples of the placebo effect in clinical trials of the central nervous systems therapeutic area. (A) Oral pridopidine for Huntington's disease (Reilmann et al., 2019). (B) Oral pimavanserine for psychosis in Alzheimer's disease (Ballard et al., 2018). (C) Putaminal injection of AAV2- neurturin for Parkinson's disease (Marks et al., 2010). (D) Intravenous infusion of opicinumab for relapsing multiple sclerosis (Cadavid et al., 2019).

mechanisms are less well known, secondary data generated from the placebo arms of clinical studies allowed us to look into the magnitude of the effect and to set the foundations for the exploration of its determinants (see chapter "Determinants of placebo effects" by Anderson, this volume).

The following list does not intend to be exhaustive but to provide the reader with a breath of conditions recently studied: Parkinson's disease (Goetz, Leurgans, Raman, & Stebbins, 2000; Rato, Duarte, Mestre, de Carvalho, & Ferreira, 2018); restless leg syndrome (Fulda & Wetter, 2008; Silva et al., 2017); multiple sclerosis (Papadopoulos & Mitsikostas, 2010); tic disorders (Cubo et al., 2013); neuropathic pain (Papadopoulos & Mitsikostas, 2012); dystonia (Duarte, Rodrigues, Ferreira, & Costa, 2018); Huntington's disease (Cubo et al., 2012); amyotrophic lateral sclerosis (Rato et al., 2018), progressive supranuclear palsy (Stamelou et al., 2016); migraine (de Craen, Tijssen, de Gans, & Kleijnen, 2000; Kam-Hansen et al., 2014); Alzheimer's disease (Hyde, May, Xue, & Zhang, 2017; Ito et al., 2013; Ozawa et al., 2017); epilepsy (Burneo, Montori, & Faught, 2002; Zaccara, Giovannelli, & Schmidt, 2015; Zhang et al., 2019); sleep disorders (McCall, D'Agostino, & Dunn, 2003; Walsh et al., 2000), among others.

6. Strategies to minimize the placebo effect in clinical trials

In order to optimize the efficiency of clinical trials, measures deemed to minimize the placebo effect should reduce the placebo response in a magnitude superior to their influence over the treatment response. While an ideal approach would reduce the placebo response without interfering with the treatment response, strategies that affect the placebo and the treatment responses equally are not fit for this purpose. Several measures have been proposed, focusing on different aspects of clinical trials, from study design to outcome selection and statistical analysis (Table 1). Below we list and expand on several of them.

However, it is of note that if the placebo effect is additive, the sample size is large and the measurement tools used do not depict ceiling or floor effects, then the presented measures are likely to be unfruitful, as they would be expected to reduce the placebo effect in equivalent magnitudes in both active and placebo arms.

6.1 Study design

There are multiple aspects of designing and planning clinical trials that have the potential to curtail the placebo effect. To better estimate the placebo effect,

Table 1 Strategies to minimize the impact of placebo effect in clinical trials.

Study design
3-armed trial designs
Pros: accurate measurement of the placebo effect
Cons: increased costs and recruitment burden
Placebo run-in/lead-in designs
Pros: early exclusion of placebo responders
Cons: larger sample sizes and ethical issues
Randomized withdrawal/discontinuation design
Pros: accurate measurement of the placebo effect
Cons: carry-over effect and ethical issues
Sequential parallel comparison design
Pros: accurate measurement of the placebo effect
Cons: complexity and novelty
Follow-up time/number of visits/schedule of assessments
Pros: ecological responses to interventions
Cons: less data collected and value for money
Statistical power
Minimal clinical important difference accounting for placebo
Pros: adequate statistical power
Cons: reduced predicted effect sizes leading to larger samples
Simulation-based power calculations
Pros: as above, inexpensive, more precise predictions, low risk
Cons: as above, computationally intense, require access to 'big data'
Patient selection
Population enrichment
Pros: smaller effect sizes, simple implementation
Cons: external validity issues, recruitment challenge
Allocation stratification
Stratified randomization
Pros: allows balancing of predefined characteristics across arms
Cons: requires understating of placebo determinants and well defined strata

Table 1 Strategies to minimize the impact of placebo effect in clinical trials.—cont'd

Adaptative randomization
Pros: as above, allows more stratification factors
Cons: as above, requires real-time expert input
Outcomes
Staff training
Pros: simple implementation
Cons: low efficacy, lack of systematization and homogeneity
Outcome selection
Pros: optimizes signal to noise ratio
Cons: outcome selection bias
Statistical analysis
Stratification
Pros: complements interpretation of full dataset
Cons: statistical power
Multivariate modeling
Pros: implementation after trial completion
Cons: requires understating of placebo determinants
Sub-group analysis
Pros: complements interpretation of main analysis
Cons: post hoc and statistical power

and hence the treatment effect, classical parallel studies can be added a waiting room arm. This group of study participants are allocated to a study arm that neither receive the active intervention nor a placebo intervention but undergoes all the other study procedures involved in the clinical trials. Alternatively, a historical (i.e., retrospective) or prospective cohort (modified Zelen design) matched to the clinical trials participants can also be used with the same intention, but with a higher degree of bias (Enck et al., 2013; Zelen, 1979).

Alternative study designs to the conventional parallel design have also been developed to address the issue of the placebo effect. Trials with a short (i.e., 1 or 2 week) placebo lead-in or run-in phases can select and exclude high placebo responders before the double-blinded phase of the trial. Alternatively, active intervention responders could also be identified and selected. Typically in these trials, the initial phase is single-blinded (i.e., investigator not blinded) posing selection and detection bias issues (Iovieno & Papakostas, 2012), although fully double-blinded versions have been used at the expense of being more resource intensive (Faries, Heiligenstein, Tollefson, & Potter, 2001). There are also ethical concerns with a design that recruits and consents participants to then exclude them from trial participation early in the course of the study. As an add on to the placebo lead-in/run-in design, or as an stand-alone strategy, a randomized withdrawal/discontinuation design period could also be useful to detect and minimize the impact of the placebo (Ivanova & Tamura, 2015).

Another variation is the sequential parallel comparison design, which is comprised of 2 phases with equal duration and unbalanced randomization (Fava, Evins, Dorer, & Schoenfeld, 2003). All participants complete the first phase and are unevenly randomized to the active intervention only, placebo followed by active intervention, and placebo only. Placebo responders are discontinued or offered an open-label extension at the end of the first phase, and non-responders continue through to the second phase maintaining their initial allocation. The final analysis pools data from both phases maximizing power while reducing the placebo effect.

Clinical trial duration, number of trials visits and intensity of the schedule of assessments are all hypothetically associated with the placebo effect. Hence, impoverishment of clinical trials (i.e., internet-based assessments and limited investigator-participant interaction) is an interesting proposal. As an addition, simplification may also increase external validity, as interactions and context in clinical research are very different from the standard clinical setting. Still, in many clinical trials in progressive neurological conditions it is clear that the response to placebo interventions abates with time, usually being more prominent during the first 2 to 8 weeks. Suggestions have been made to design trials with longer durations of follow-up (Enck et al., 2013). However, some condition present high placebo effects even in these circumstances (Potkin et al., 2011).

As pre-conditioning seems to be a relevant modulator of the magnitude of the placebo in certain conditions (de la Fuente-Fernandez, 2012; Frisaldi et al., 2017), in theory cross-over designs should be avoided (Suchman & Ader, 1992) and the fact that treatment-naïve participants seem to be less susceptible to the placebo effect should be considered while setting up a new trial.

6.2 Statistical power

Statistical methods allow us to draw inferences from a sample about a population, but they are not without limitations. Both false positive (i.e., rejecting a true null hypothesis) and false negative results (i.e., not rejecting the null when the alternative hypothesis is true) are possible. Statistical power and sample size calculations are necessary to define how many observations would be needed to answer the research question and should be performed during the planning phase the study. Several parameters need to be set to execute sample size calculations. First, they depend on the study design and outcome measure. The investigators need to consider what effect size is expected to be detected, and how much type 1 and type 2 error are they willing to accept. All these elements will allow the estimation of the adequate sample size to test a certain hypothesis, with a specific design, outcome, effect size, power and significance level (Ryan, 2013).

The estimation of the effect size is usually based on a known minimal clinically important difference derived from previous interventional or observational data, or an educated guest when no other evidence is available. The placebo effect is relevant as it is usually not accounted for in this process. Minimal clinically important differences should ideally be based on the true natural history of the condition and not on what would happen to the natural history given the Hawthorn or the placebo effects. The direct consequence being unintentionally underpowering a study design, as placebo-arm participants behave dissimilarly to observational study participants, and frequently show improvement, lessening the observed contrast between groups. One potential alternative would be to subtract the placebo effect from the minimal clinically important difference when estimating the expected effect size. The downside being greater sample sizes.

Using Monte Carlo simulation-based power calculations is a modern approach to sample size calculation (Landau & Stahl, 2013). It involves creating a synthetic dataset—based of previous knowledge of the disease behavior and what happens in the context of a clinical trial—and generate it multiple times adding a degree of natural variability. This approach allows modeling the placebo effect on the course of the synthetic trial (Muthen & Brown, 2009).

Both methods could lessen the underpowering of a study due to the placebo effect but are only unbiased and precise if an appropriate amount of data is available to generate them. They usually require individual-level data pooled from several clinical trials which may be difficult to generate and access (see chapter "Placebo in epilepsy" by Goldenholz, this volume, for an example employed in Alzheimer's disease).

6.3 Patient selection

Deep phenotyping of placebo responders and a better understanding of the disease natural history derived from large multicentric collaborative initiatives can be used to enrich clinical trials' populations. Selection of placebo non-responders and exclusion of responders can be attained by simply tailoring the inclusion and exclusion criteria. This has been attempted several times without success mainly in psychiatric indications, but this is a field where there are concerns about diagnostic heterogeneity and misclassification. The development of objective disease and staging criteria, and the use of imaging, biofluid, digital and genetic biomarkers, may facilitate and increase the success of this strategy (Kirsch et al., 2008; Kobak, Kane, Thase, & Nierenberg, 2007). Central to patient selection, is maintaining the external validity of a trial, and enrichment may be counter-productive in this context.

6.4 Allocation stratification

Randomization is at the core of modern clinical trials. This process facilitates the unpredictable allocation of participants to treatment groups, balancing participants' characteristics. There are several possible methods of randomization. Stratified and adaptative randomization methods can be used to ensure certain confounding characteristics are evenly distributed across arms, such as placebo effect determinants. Stratified randomization takes block randomization one step further, making sure groups are balanced both in terms of arm allocation and confounding variables. Adaptative randomization, also known as minimization, uses automated strategies to allocate participants to study arms leveraging certain predefined characteristics. Each time a participant is randomized, the chances of the following participant to be allocated to a certain group change so that differences across groups are minimized.

Furthermore, to address the "large effects, small sample" issue it is relevant that the investigators are aware of the population-, disease- and intervention-specific determinants of the placebo effect in their particular study. Small samples are vulnerable to confounding of treatment effect by uncontrolled causal variables (Krause & Howard, 2003). A method to study this phenomenon prior to study commencement would be to use simulations.

6.5 Outcomes and endpoints

The participant's expectations and beliefs about the trial outcome may have an impact on the magnitude of the placebo effect (Juergens, Seekatz,

Moosdorf, Petrie, & Rief, 2010; Rutherford et al., 2017). Similarly, researchers involved in outcome assessment may also be influence by their own preconceptions about the effects of an intervention. Indeed, the lower the test-retest reliability of the assessing tool, the more susceptible a trial is to this occurrence.

Supported by the biopsychosocial theory behind how placebo interventions work, in the setting of a clinical trial it is critical to inform and educate investigators and research staff about these phenomena. The placebo effect depends on contextual and environmental cues surrounding medical intervention and not only on dummy pills. Human interaction and ecological relationships are at the core of the genesis of this effect, and researchers should make efforts to neutralize these cues while also lessening their expectations of benefit and harm. As a methodological plus, training also reduces clinical site effect (i.e., inter-centre variability) by means of harmonizing and standardizing behaviors.

From our experience, clinical site staff and rater training is already frequently used specially in industry sponsored trials. It is often implemented during site initiation procedures, but rarely re-discussed throughout the trial duration, and usually lacks systematization or standardization. Also, from the health care practitioner's perspective it is challenging to dissociate the clinical interaction from the research interaction, as they often happen in the same temporal and physical space. With the advent of pragmatic clinical trials, an umbrella term that frequently encompasses setting up clinical trials interlaced within the usual clinical pathways, this will be of increased difficulty. For all these reasons such strategies have demonstrated little evidence of being able to trim the placebo effect (Schatzberg & Kraemer, 2000).

Outcome selection is also relevant. Patient reported outcomes seem to be sensitive to the placebo response and need to be supported by other more objective outcomes (Rief et al., 2009; Silva et al., 2017), ideally with biological correlation (Hrobjartsson & Gotzsche, 2010). An interesting proposal is to develop and use biomarkers of the placebo response to minimize the placebo in phases such as recruitments, allocation, and the statistical analysis. Opposite to this would be to use assessment tools that are less likely to respond to the placebo effect. Hard outcome measures such as dichotomous outcomes like hospitalizations, exacerbations, clinical diagnosis or death, and biomarker-based surrogate outcomes may be useful in this setting. As an example, neuroimaging-based lesion burden could be used in neuroinflammation, neurodegeneration could be tracked with biochemical markers of neuronal breakdown, and time to symptomatic onset is appealing for conditions with a premanifest phase.

6.6 Analysis

Randomization in theory avoids unmeasured confounding. However, it does not guarantee an even distribution of all characteristics across study arms. This is particularly true for studies with small sample sizes. Acknowledging the determinants of the placebo effect enables researchers to use statistical methods to account for such confounding of the treatment effect. The main methods used to adjust for covariate are stratification of the analyses (i.e., each individual analysis is repeated for each stratum) and using multivariate modeling where the confounded is accounted for. As an example, in a blood pressure lowering trial the FDA suggested investigators include in their analysis factors that modify the magnitude of the placebo response such as gender and ethnicity (NDA 21985).

A post-hoc strategy is to use subgroup analysis. Participants can be stratified based on responder analysis of placebo responsive outcomes, and then reanalyzed following the methods developed for the primary outcome. This approach is unlikely to be statistically powered but can be used as part of the armamentarium to understand if placebo determinants acted as confounding of the treatment effect (Cepeda, Berlin, Gao, Wiegand, & Wada, 2012).

7. Challenges and future direction

Drug development is expensive and takes a far-from-insignificant amount of time and effort by the academia, industry and patient populations. This toll needs to be lessened and increasing the efficiency of clinical trials is an integral component of an overarching strategy to accelerate drug development and bring innovation to those in need.

Although the placebo field has experienced substantial progress, much more is still needed to harness the placebo effect in clinical trials. So far, regardless of the methodology used, our ability to minimize the placebo effect has been limited. Published clinical trials have revealed the instability of this effect and the lack of tools available to predict it. Retrospective analyses of completed trials and their strategies haven't been particularly fruitful to help design "placebo-free" trials, and there is an unmet need for newer approaches.

Designing and erecting disease-specific clinical trial datasets, encompassing rigorous standardization and high-standard quality control processes, will allow the study of the magnitude and determinants of the placebo effect and associated phenomenon. This information is fundamental to plan trials aimed at efficaciously answer specific therapeutic questions, minimizing bias and confounding factors.

Comprehensiveness is of the highest importance. Such datasets should incorporate all available information including patient-, disease- and intervention-particular data. This will allow the study of individuals across the phenotypic spectrums, disease stages and demographic characteristics. Large samples sizes resembling the big data paradigm of *all data*, and multi-dimensionality allow for the application of complex modern statistical and data analysis methods. It will hasten our understanding of the disease natural history when affected by a therapeutic effect and allow us to better predict—using modeling and clinical trial simulation techniques—the likely outcomes of an objective clinical development plan.

Investigating the placebo has direct implications on the study design, statistical power, eligibility criteria, outcome measures, data analysis plan and interpretation. For a minority of neurological conditions there are already seminal work available that goes in line with this proposal, but for the majority this is lacking.

This won't be feasible without close collaboration and data-sharing between stakeholders, including the ones that traditionally have shown higher levels of resistance to publicly release data collected and generated by them, such as the ones acting on the biotechnology, pharmaceutical and regulatory arenas. Cooperative efforts like those promoted by third-party institutions will be vital to achieve a better understanding of the placebo effect and related effects.

Pushing newly developed knowledge and know-how into effect is always demanding. Some concepts used in modern clinical development programs have endured the test of time and are deeply rooted in the way we plan, execute and interpret clinical trials. However, scientific paradigms are in constant permutation and evolution. The same applies to measures to reduce the placebo effect. Further exploration and implementation of new avenues is not only necessary but also unavoidable.

Simple measures are already being commonly executed, while others need a risk-avid development environment to be deployed and have never been tested empirically. Clinical trial simulation is almost always possible and allows the in-silico implementation of high uncertainty methodologies in a low risk setting.

Finally, critical and independent assessment will be needed to underpin the advantage and disadvantages of such paradigms, to fine tune strategies that work better, and to restart an everlasting cycle of development-implementation-evaluation aimed at optimizing processes, and ultimately bring better health care to the populations.

References

Ballard, C., Banister, C., Khan, Z., Cummings, J., Demos, G., Coate, B., et al. (2018). Evaluation of the safety, tolerability, and efficacy of pimavanserin versus placebo in patients with Alzheimer's disease psychosis: A phase 2, randomised, placebo-controlled, double-blind study. *Lancet Neurology*, *17*(3), 213–222. https://doi.org/10.1016/s1474-4422(18)30039-5.

Benedetti, F. (2014). *Placebo effects*. USA: Oxford University Press.

Burneo, J. G., Montori, V. M., & Faught, E. (2002). Magnitude of the placebo effect in randomized trials of antiepileptic agents. *Epilepsy & Behavior*, *3*(6), 532–534. https://doi.org/10.1016/s1525-5050(02)00531-0.

Cadavid, D., Mellion, M., Hupperts, R., Edwards, K. R., Calabresi, P. A., Drulovic, J., et al. (2019). Safety and efficacy of opicinumab in patients with relapsing multiple sclerosis (SYNERGY): A randomised, placebo-controlled, phase 2 trial. *Lancet Neurology*, *18*(9), 845–856. https://doi.org/10.1016/s1474-4422(19)30137-1.

Cepeda, M. S., Berlin, J. A., Gao, C. Y., Wiegand, F., & Wada, D. R. (2012). Placebo response changes depending on the neuropathic pain syndrome: Results of a systematic review and meta-analysis. *Pain Medicine*, *13*(4), 575–595. https://doi.org/10.1111/j.1526-4637.2012.01340.x.

Cubo, E., Gonzalez, M., del Puerto, I., de Yebenes, J. G., Arconada, O. F., & Gabriel y Galan, J. M. (2012). Placebo effect characteristics observed in a single, international, longitudinal study in Huntington's disease. *Movement Disorders*, *27*(3), 439–442. https://doi.org/10.1002/mds.24062.

Cubo, E., Gonzalez, M., Singer, H., Mahone, E. M., Scahill, L., Muller-Vahl, K. R., et al. (2013). Impact of placebo assignment in clinical trials of tic disorders. *Movement Disorders*, *28*(9), 1288–1292. https://doi.org/10.1002/mds.25365.

de Craen, A. J., Tijssen, J. G., de Gans, J., & Kleijnen, J. (2000). Placebo effect in the acute treatment of migraine: Subcutaneous placebos are better than oral placebos. *Journal of Neurology*, *247*(3), 183–188. https://doi.org/10.1007/s004150050560.

de la Fuente-Fernandez, R. (2012). The powerful pre-treatment effect: Placebo responses in restless legs syndrome trials. *European Journal of Neurology*, *19*(10), 1305–1310. https://doi.org/10.1111/j.1468-1331.2012.03719.x.

de la Fuente-Fernandez, R., Ruth, T. J., Sossi, V., Schulzer, M., Calne, D. B., & Stoessl, A. J. (2001). Expectation and dopamine release: Mechanism of the placebo effect in Parkinson's disease. *Science*, *293*(5532), 1164–1166. https://doi.org/10.1126/science.1060937.

Demitrack, M. A., Faries, D., Herrera, J. M., DeBrota, D., & Potter, W. Z. (1998). The problem of measurement error in multisite clinical trials. *Psychopharmacology Bulletin*, *34*(1), 19–24.

Duarte, G. S., Rodrigues, F. B., Ferreira, J. J., & Costa, J. (2018). Adverse events with botulinum toxin treatment in cervical dystonia: How much should we blame placebo? *Parkinsonism & Related Disorders*, *56*, 16–19. https://doi.org/10.1016/j.parkreldis.2018.06.017.

Dunlop, B. W., Thase, M. E., Wun, C. C., Fayyad, R., Guico-Pabia, C. J., Musgnung, J., et al. (2012). A meta-analysis of factors impacting detection of antidepressant efficacy in clinical trials: The importance of academic sites. *Neuropsychopharmacology*, *37*(13), 2830–2836. https://doi.org/10.1038/npp.2012.153.

Enck, P., Bingel, U., Schedlowski, M., & Rief, W. (2013). The placebo response in medicine: Minimize, maximize or personalize? *Nature Reviews. Drug Discovery*, *12*(3), 191–204. https://doi.org/10.1038/nrd3923.

Faries, D. E., Heiligenstein, J. H., Tollefson, G. D., & Potter, W. Z. (2001). The double-blind variable placebo lead-in period: Results from two antidepressant clinical trials. *Journal of Clinical Psychopharmacology*, *21*(6), 561–568. https://doi.org/10.1097/00004714-200112000-00004.

Fava, M., Evins, A. E., Dorer, D. J., & Schoenfeld, D. A. (2003). The problem of the placebo response in clinical trials for psychiatric disorders: Culprits, possible remedies, and a novel study design approach. *Psychotherapy and Psychosomatics*, *72*(3), 115–127. https://doi.org/10.1159/000069738.

Ferreira, J. J., Trenkwalder, C., & Mestre, T. A. (2018). Placebo and nocebo responses in other movement disorders besides Parkinson's disease: How much do we know? *Movement Disorders*, *33*(8), 1228–1235. https://doi.org/10.1002/mds.113.

Finniss, D. G., Kaptchuk, T. J., Miller, F., & Benedetti, F. (2010). Biological, clinical, and ethical advances of placebo effects. *Lancet*, *375*(9715), 686–695. https://doi.org/10.1016/s0140-6736(09)61706-2.

Friedman, J., Ravina, B., Mills, R., Williams, H., Revell, S., Bahr, D., et al. (2009). A multi-centre, placebo-controlled, double-blind trial to examine the safety and efficacy of pimavanserin in the treatment of psychosis in Parkinson's disease. *European Journal of Neurology*, *16*(S3), 656.

Frisaldi, E., Carlino, E., Zibetti, M., Barbiani, D., Dematteis, F., Lanotte, M., et al. (2017). The placebo effect on bradykinesia in Parkinson's disease with and without prior drug conditioning. *Movement Disorders*, *32*(10), 1474–1478. https://doi.org/10.1002/mds.27142.

Fulda, S., & Wetter, T. C. (2008). Where dopamine meets opioids: A meta-analysis of the placebo effect in restless legs syndrome treatment studies. *Brain*, *131*(Pt 4), 902–917. https://doi.org/10.1093/brain/awm244.

Goetz, C. G., Leurgans, S., Raman, R., & Stebbins, G. T. (2000). Objective changes in motor function during placebo treatment in PD. *Neurology*, *54*(3), 710–714. https://doi.org/10.1212/wnl.54.3.710.

Hay, M., Thomas, D. W., Craighead, J. L., Economides, C., & Rosenthal, J. (2014). Clinical development success rates for investigational drugs. *Nature Biotechnology*, *32*(1), 40. Retrieved from https://www.nature.com/articles/nbt.2786.

Hrobjartsson, A., & Gotzsche, P. C. (2010). Placebo interventions for all clinical conditions. *Cochrane Database System Review*, *2010*(1), Cd003974. https://doi.org/10.1002/14651858.CD003974.pub3.

Hyde, A. J., May, B. H., Xue, C. C., & Zhang, A. L. (2017). Variation in placebo effect sizes in clinical trials of oral interventions for management of the behavioral and psychological symptoms of dementia (BPSD): A systematic review and meta-analysis. *The American Journal of Geriatric Psychiatry*, *25*(9), 994–1008. https://doi.org/10.1016/j.jagp.2017.02.022.

Ioannidis, J. P. (2008). Effectiveness of antidepressants: An evidence myth constructed from a thousand randomized trials? *Philosophy, Ethics, and Humanities in Medicine*, *3*, 14. https://doi.org/10.1186/1747-5341-3-14.

Iovieno, N., & Papakostas, G. I. (2012). Does the presence of an open-label antidepressant treatment period influence study outcome in clinical trials examining augmentation/combination strategies in treatment partial responders/nonresponders with major depressive disorder? *The Journal of Clinical Psychiatry*, *73*(5), 676–683. https://doi.org/10.4088/JCP.11r06978.

Ito, K., Corrigan, B., Romero, K., Anziano, R., Neville, J., Stephenson, D., et al. (2013). Understanding placebo responses in Alzheimer's disease clinical trials from the literature meta-data and CAMD database. *Journal of Alzheimer's Disease*, *37*(1), 173–183. https://doi.org/10.3233/jad-130575.

Ivanova, A., & Tamura, R. N. (2015). A two-way enriched clinical trial design: Combining advantages of placebo lead-in and randomized withdrawal. *Statistical Methods in Medical Research*, *24*(6), 871–890. https://doi.org/10.1177/0962280211431023.

Juergens, M. C., Seekatz, B., Moosdorf, R. G., Petrie, K. J., & Rief, W. (2010). Illness beliefs before cardiac surgery predict disability, quality of life, and depression 3 months later. *Journal of Psychosomatic Research*, *68*(6), 553–560. https://doi.org/10.1016/j.jpsychores.2009.10.004.

Kam-Hansen, S., Jakubowski, M., Kelley, J. M., Kirsch, I., Hoaglin, D. C., Kaptchuk, T. J., et al. (2014). Altered placebo and drug labeling changes the outcome of episodic migraine attacks. *Science Translational Medicine*, *6*(218), 218ra215. https://doi.org/10.1126/scitranslmed.3006175.

Kaptchuk, T. J., & Miller, F. G. (2015). Placebo effects in medicine. *New England Journal of Medicine*, *373*(1), 8–9. https://doi.org/10.1056/NEJMp1504023.

Kemp, A. S., Schooler, N. R., Kalali, A. H., Alphs, L., Anand, R., Awad, G., et al. (2010). What is causing the reduced drug-placebo difference in recent schizophrenia clinical trials and what can be done about it? *Schizophrenia Bulletin*, *36*(3), 504–509. https://doi.org/10.1093/schbul/sbn110.

Kirsch, I. (1990). *Changing expectations: A key to effective psychotherapy*. Thomson Brooks/Cole Publishing Co.

Kirsch, I. (2000). Are drug and placebo effects in depression additive? *Biological Psychiatry*, *47*(8), 733–735. https://doi.org/10.1016/s0006-3223(00)00832-5.

Kirsch, I., Deacon, B. J., Huedo-Medina, T. B., Scoboria, A., Moore, T. J., & Johnson, B. T. (2008). Initial severity and antidepressant benefits: A meta-analysis of data submitted to the Food and Drug Administration. *PLoS Medicine*, *5*(2). e45https://doi.org/10.1371/journal.pmed.0050045.

Kobak, K. A., Kane, J. M., Thase, M. E., & Nierenberg, A. A. (2007). Why do clinical trials fail? The problem of measurement error in clinical trials: Time to test new paradigms? *Journal of Clinical Psychopharmacology*, *27*(1), 1–5. https://doi.org/10.1097/JCP.0b013e31802eb4b7.

Krause, M. S., & Howard, K. I. (2003). What random assignment does and does not do. *Journal of Clinical Psychology*, *59*(7), 751–766. https://doi.org/10.1002/jclp.10170.

Landau, S., & Stahl, D. (2013). Sample size and power calculations for medical studies by simulation when closed form expressions are not available. *Statistical Methods in Medical Research*, *22*(3), 324–345. https://doi.org/10.1177/0962280212439578.

Leuchter, A. F., Cook, I. A., Witte, E. A., Morgan, M., & Abrams, M. (2002). Changes in brain function of depressed subjects during treatment with placebo. *American Journal of Psychiatry*, *159*(1), 122–129.

Marks, W. J., Jr., Bartus, R. T., Siffert, J., Davis, C. S., Lozano, A., Boulis, N., et al. (2010). Gene delivery of AAV2-neurturin for Parkinson's disease: A double-blind, randomised, controlled trial. *Lancet Neurology*, *9*(12), 1164–1172. https://doi.org/10.1016/s1474-4422(10)70254-4.

McCall, W. V., D'Agostino, R., Jr., & Dunn, A. (2003). A meta-analysis of sleep changes associated with placebo in hypnotic clinical trials. *Sleep Medicine*, *4*(1), 57–62. https://doi.org/10.1016/s1389-9457(02)00242-3.

Muthen, B., & Brown, H. C. (2009). Estimating drug effects in the presence of placebo response: Causal inference using growth mixture modeling. *Statistics in Medicine*, *28*(27), 3363–3385. https://doi.org/10.1002/sim.3721.

Ozawa, C., Roberts, R., Yoshida, K., Suzuki, T., Lebowitz, B., Reeves, S., et al. (2017). Placebo effects in the treatment of noncognitive symptoms of Alzheimer's disease: Analysis of the CATIE-AD data. *The Journal of Clinical Psychiatry*, *78*(9), e1204–e1210. https://doi.org/10.4088/JCP.17m11461.

Papadopoulos, D., & Mitsikostas, D. D. (2010). Nocebo effects in multiple sclerosis trials: A meta-analysis. *Multiple Sclerosis*, *16*(7), 816–828. https://doi.org/10.1177/1352458510370793.

Papadopoulos, D., & Mitsikostas, D. D. (2012). A meta-analytic approach to estimating nocebo effects in neuropathic pain trials. *Journal of Neurology*, *259*(3), 436–447. https://doi.org/10.1007/s00415-011-6197-4.

Potkin, S., Agid, O., Siu, C., Watsky, E., Vanderburg, D., & Remington, G. (2011). Placebo response trajectories in short-term and long-term antipsychotic trials in schizophrenia. *Schizophrenia Research*, *132*(2–3), 108–113. https://doi.org/10.1016/j.schres.2011.07.028.

Rato, M. L., Duarte, G. S., Mestre, T., de Carvalho, M., & Ferreira, J. J. (2018). Strong nocebo effect in amyotrophic lateral sclerosis trials might mask conclusions. *Lancet Neurology*, *17*(10), 842. https://doi.org/10.1016/s1474-4422(18)30310-7.

Reilmann, R., McGarry, A., Grachev, I. D., Savola, J. M., Borowsky, B., Eyal, E., et al. (2019). Safety and efficacy of pridopidine in patients with Huntington's disease (PRIDE-HD): A phase 2, randomised, placebo-controlled, multicentre, dose-ranging study. *Lancet Neurology*, *18*(2), 165–176. https://doi.org/10.1016/s1474-4422(18)30391-0.

Rheims, S., Perucca, E., Cucherat, M., & Ryvlin, P. (2011). Factors determining response to antiepileptic drugs in randomized controlled trials. A systematic review and meta-analysis. *Epilepsia*, *52*(2), 219–233. https://doi.org/10.1111/j.1528-1167.2010.02915.x.

Rief, W., Nestoriuc, Y., Weiss, S., Welzel, E., Barsky, A. J., & Hofmann, S. G. (2009). Meta-analysis of the placebo response in antidepressant trials. *Journal of Affective Disorders*, *118*(1–3), 1–8. https://doi.org/10.1016/j.jad.2009.01.029.

Rodrigues, F. B., Duarte, G. S., Prescott, D., Ferreira, J., & Costa, J. (2019). Deep brain stimulation for dystonia. *Cochrane Database System Review*, *1*, Cd012405. https://doi.org/10.1002/14651858.CD012405.pub2.

Rutherford, B. R., Wall, M. M., Brown, P. J., Choo, T. H., Wager, T. D., Peterson, B. S., et al. (2017). Patient expectancy as a mediator of placebo effects in antidepressant clinical trials. *The American Journal of Psychiatry*, *174*(2), 135–142. https://doi.org/10.1176/appi.ajp.2016.16020225.

Ryan, T. P. (2013). *Sample size determination and power*. John Wiley & Sons.

Schatzberg, A. F., & Kraemer, H. C. (2000). Use of placebo control groups in evaluating efficacy of treatment of unipolar major depression. *Biological Psychiatry*, *47*(8), 736–744. https://doi.org/10.1016/s0006-3223(00)00846-5.

Scott, D. J., Stohler, C. S., Egnatuk, C. M., Wang, H., Koeppe, R. A., & Zubieta, J.-K. (2008). Placebo and nocebo effects are defined by opposite opioid and dopaminergic responses. *Archives of General Psychiatry*, *65*(2), 220–231.

Silva, M. A., Duarte, G. S., Camara, R., Rodrigues, F. B., Fernandes, R. M., Abreu, D., et al. (2017). Placebo and nocebo responses in restless legs syndrome: A systematic review and meta-analysis. *Neurology*, *88*(23), 2216–2224. https://doi.org/10.1212/wnl.0000000000004004.

Stamelou, M., Schope, J., Wagenpfeil, S., Del Ser, T., Bang, J., Lobach, I. Y., et al. (2016). Power calculations and placebo effect for future clinical trials in progressive supranuclear palsy. *Movement Disorders*, *31*(5), 742–747. https://doi.org/10.1002/mds.26580.

Suchman, A. L., & Ader, R. (1992). Classic conditioning and placebo effects in crossover studies. *Clinical Pharmacology and Therapeutics*, *52*(4), 372–377. https://doi.org/10.1038/clpt.1992.157.

Tuttle, A. H., Tohyama, S., Ramsay, T., Kimmelman, J., Schweinhardt, P., Bennett, G. J., et al. (2015). Increasing placebo responses over time in U.S. clinical trials of neuropathic pain. *Pain*, *156*(12), 2616–2626. https://doi.org/10.1097/j.pain.0000000000000333.

Undurraga, J., & Baldessarini, R. J. (2012). Randomized, placebo-controlled trials of antidepressants for acute major depression: Thirty-year meta-analytic review. *Neuropsychopharmacology*, *37*(4), 851–864. https://doi.org/10.1038/npp.2011.306.

Vieta, E., & Cruz, N. (2008). Increasing rates of placebo response over time in mania studies. *The Journal of Clinical Psychiatry*, *69*(4), 681–682. author reply 682. https://doi.org/10.4088/jcp.v69n0423g.

Walsh, J. K., Roth, T., Randazzo, A., Erman, M., Jamieson, A., Scharf, M., et al. (2000). Eight weeks of non-nightly use of zolpidem for primary insomnia. *Sleep*, *23*(8), 1087–1096.

Walsh, B. T., Seidman, S. N., Sysko, R., & Gould, M. (2002). Placebo response in studies of major depression: Variable, substantial, and growing. *JAMA*, *287*(14), 1840–1847. https://doi.org/10.1001/jama.287.14.1840.

Wong, C. H., Siah, K. W., & Lo, A. W. (2018). Estimation of clinical trial success rates and related parameters. *Biostatistics*, *20*(2), 273–286. https://doi.org/10.1093/biostatistics/kxx069.

Yildiz, A., Vieta, E., Tohen, M., & Baldessarini, R. J. (2011). Factors modifying drug and placebo responses in randomized trials for bipolar mania. *The International Journal of Neuropsychopharmacology*, *14*(7), 863–875. https://doi.org/10.1017/s1461145710001641.

Zaccara, G., Giovannelli, F., & Schmidt, D. (2015). Placebo and nocebo responses in drug trials of epilepsy. *Epilepsy & Behavior*, *43*, 128–134. https://doi.org/10.1016/j.yebeh.2014.12.004.

Zelen, M. (1979). A new design for randomized clinical trials. *The New England Journal of Medicine*, *300*(22), 1242–1245. https://doi.org/10.1056/nejm197905313002203.

Zhang, N., Lv, Y., Li, H., Chen, J., Li, Y., Yin, F., et al. (2019). Quantifying placebo responses in clinical evaluation of neuropsychiatric symptoms in Alzheimer's disease. *European Journal of Clinical Pharmacology*, *75*(4), 497–509. https://doi.org/10.1007/s00228-018-02620-x.

CHAPTER FOUR

Maximizing placebo response in neurological clinical practice

Louise-Laure Mariani, Jean-Christophe Corvol*

Department of Neurology, Pitié-Salpêtrière Hospital, Sorbonne University, Assistance Publique Hôpitaux de Paris, Brain and Spine Institute, ICM, Inserm U 1127, CNRS UMR 7225, Paris, France

*Corresponding author: e-mail address: jean-christophe.corvol@aphp.fr

Contents

Abstract

The placebo effect is a widely recognized phenomenon in clinical research, with a negative perception that it could hide the "true" drug effect. In clinical care its positive potential to increase known drug effects has been neglected for too long. The placebo and nocebo responses have been described in many neurologic disorders such as Parkinson's, Huntington's and Alzheimer's diseases, restless leg syndrome, tics, essential tremor, dystonia, functional movement disorders, neuropathic pain, headaches, migraine, amyotrophic lateral sclerosis, myasthenia gravis, chronic inflammatory demyelinating polyneuropathy, multiple sclerosis and epilepsy. Knowledge regarding placebo mechanisms and their consequences on clinical outcome have greatly improved over the last two decades. This evolution has led to reconsiderations of the importance of placebo response in the clinic and has given several clues on how to improve it in daily practice.

In this chapter, we first illustrate "why," e.g. the reasons (relevance to clinical practice, help in differential diagnosis/treatment of psychogenic movements, clinical impact, proven neurobiological grounds, health economic potential), and "how," e.g. the means (increase patients' knowledge, increase learning, improve patient-doctor relationship, increase Hawthorne effect, increase positive/decrease negative expectations (the Rosenthal effect), personalize placebo response), the placebo should be maximized

International Review of Neurobiology, Volume 153
ISSN 0074-7742
https://doi.org/10.1016/bs.irn.2020.04.003

(and nocebo avoided) in neurological clinical practice. Future studies regarding more specific neurobiological mechanisms will allow a finer tuning of placebo response in clinical practice. The use of placebo in clinical practice raises ethical issues, and a recent expert consensus regarding placebo use in the clinic is a first step to future guidelines necessary to this field.

1. Introduction

Voltaire, the French philosopher, described the concept of placebo as early as the 18th century when he stated that "the art of medicine consists of amusing the patient while nature cures the disease." The word Placebo comes from the Latin, Placere, which means "I shall please." It was originally used in a catholic religious context but, by the end of the 18th century, its use was acknowledged in the medical context as a harmless treatment (Finniss, Kaptchuk, Miller, & Benedetti, 2010; Witek, Stebbins, & Goetz, 2018). Jean-Martin Charcot (1825–1893) used placebo effects in the 19th century to improve Parkinson's disease (PD) motor function with a vibration chair "fauteuil trépidant," and later, Gilles de la Tourette further used a vibration helmet in the same way in PD patients (Goetz, 2009; Kapur, Stebbins, & Goetz, 2012). The perception of Placebo treatments shifted from positive to neutral, or negative, as it became widely used in randomized controlled trials (RCT) since the 1950s, as an "inert" treatment in the placebo arm. Along with the Placebo arms observations in RCT, the Nocebo, "I shall harm," responses have also been characterized, often quantified as the rate of adverse events in those arms (Häuser, Hansen, & Enck, 2012; Howick, Webster, Kirby, & Hood, 2018).

The placebo effect is a widely recognized phenomenon in clinical research, with a negative perception that it always threatens to obscure the "true" effect of an investigative drug or intervention. It is recognized to a lesser extent, in clinical care, in which its positive potential to increase known drug effects, or to help in the event of no available drug with a pharmacodynamic effect, has been neglected for too long. This view tends to change as recent discoveries of neurobiological substrates underlying the placebo and nocebo responses are being reported and show that placebo is not functionally "inert" (Benedetti, 2014; Evers et al., 2018; Finniss et al., 2010; Oken, 2008; Quattrone, Barbagallo, Cerasa, & Stoessl, 2018; Schedlowski, Enck, Rief, & Bingel, 2015; Wager & Atlas, 2015). Its relevance to clinical practice has become more widely acknowledged and led to recent expert consensus guidelines (Evers et al., 2018). Placebo and Nocebo responses refer to the beneficial or adverse clinical response that occur, respectively, after administration of an inert treatment or as part of active treatments. They represent neurobiological modifications occurring to the patient's brain after the act of administration of a drug (inert or pharmacologically active) which comprises of both the pharmacological process and the rituals or psychosocial context accompanying the administration of any treatment. The psychosocial context can be modulated by multiple parameters including, but not only, the patient's expectancies, previous behavioral conditioning, social learning, reward, memory, quality of the patient–physician relationship and the Hawthorne effect, e.g. behavioral, clinical and physiological changes occurring in an individual in response to the awareness of being observed or under study. In the clinical research setting such as RCT, it can be hard to differentiate the genuine placebo or nocebo effects from other confounding factors, which include, for instance, the natural recovery or fluctuations of the disease, the regression to the mean, the Hawthorne effect. However, this is less relevant to clinical daily practice, which focuses on increasing the placebo response and avoiding the nocebo response resulting in a clinical improvement, which is particularly relevant in neurological diseases, in which pharmacological options are often limited. The neurobiological components underlying this results can be modulated in clinical practice to maximize the expected improvements (Benedetti, 2014; Benedetti, Carlino, & Piedimonte, 2016; Enck, Bingel, Schedlowski, & Rief, 2013; Evers et al., 2018; Finniss et al., 2010; Häuser et al., 2012; Schedlowski et al., 2015). In this chapter, we will first discuss "why," e.g. the reasons, and then "how," e.g. the means, whereby positive placebos effects can be maximized (and nocebo avoided) in neurological clinical practice (Fig. 1), and conclude on the future perspectives and challenges encountered to maximize placebo impact in patient care.

2. Why should placebo be maximized in the clinic?

Placebo and Nocebo responses are not the neuroscientists' latest obsession or *idée fixe* but correspond to an everyday real clinical practice. Indeed surveys have been conducted in various countries (Fässler, Meissner, Schneider, & Linde, 2010; Héron, Leroux, & Dubayle, 2019; Howick et al., 2013; Kermen, Hickner, Brody, & Hasham, 2010; Linde et al., 2018, 2014; Tilburt, Emanuel, Kaptchuk, Curlin, & Miller, 2008), among which United States and European countries show that placebo is

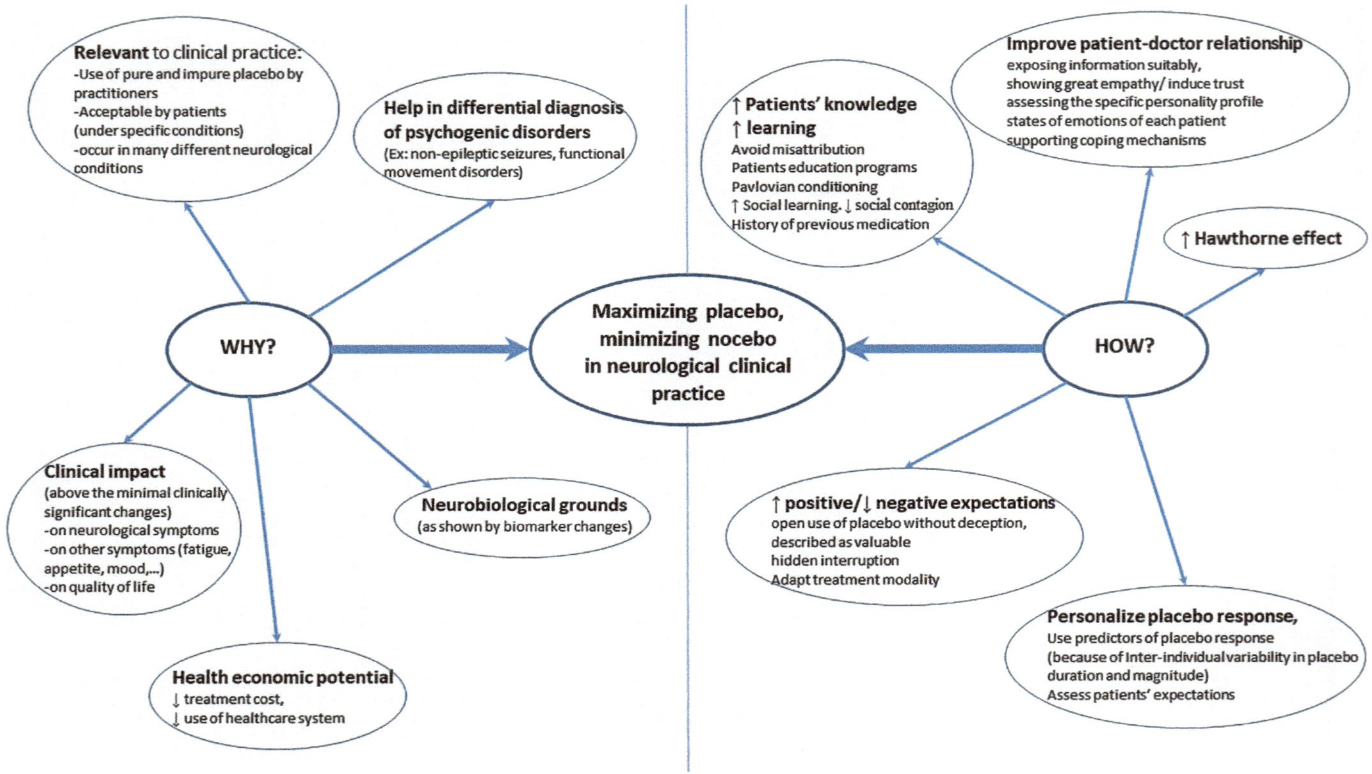

Fig. 1 Schematic drawing covering the Why and How of maximizing placebo benefits in clinical practice.

frequently used and professionals believe in the placebo effect. However, these surveys also demonstrate that these professionals often consider themselves to be insufficiently informed and trained in the competent use of placebo in current practice. Reviewed in a meta-analysis by Linde et al. (2018), more than half of the practitioners admitted using placebos, either "pure" as inert substances such as saline or more frequently nonspecific treatment such as vitamins, or "impure" as repurposing or misusing a drug which is usually prescribed for another indication (for instance, antibiotics) (Evers et al., 2018; Kermen et al., 2010; Linde et al., 2018). Placebo use also seems to be acceptable to patients as shown in a survey done in 2013 (Hull et al., 2013). Most respondents (50–84%) judged it acceptable for doctors to recommend placebo treatments under conditions that varied according to doctors' level of certainty about the benefits and safety of the treatment, the purpose of the treatment, and the transparency with which the treatment was described to patients. The conditions under which placebos are prescribed are essential regarding their acceptability by patients (Bishop, Aizlewood, & Adams, 2014). In RCT performed in PD patients, participation is viewed very positively even if the patients end up in the placebo arm, not only because of perceived clinical improvement under placebo exposure, but also because of other motivational aspects involved in an RCT such as helping other and helping science (Goetz, Janko, Blasucci, & Jaglin, 2003).

Another aspect of the relevance of placebo use in neurological disorders is that both placebo and nocebo responses have been shown to occur in many different neurological conditions (Table 1). In the field of movement disorders, the most investigated placebo and nocebo responses are in Parkinson's disease (PD) (Goetz, Laska, et al., 2008; Leal Rato et al., 2019; Mestre et al., 2016, 2018; Morberg et al., 2018; Stathis et al., 2013; Witek et al., 2018), but studies have also been performed in restless leg syndrome (RLS) (Fulda & Wetter, 2008; Silva et al., 2017), Huntington's disease (HD) (Cubo et al., 2012; Ferreira et al., 2018; HORIZON Investigators of the Huntington Study Group and European Huntington's Disease Network., 2013; Huntington Study Group, 2006), tics disorders (Cubo et al., 2013; Ferreira et al., 2018), essential tremor (ET) (Ferreira et al., 2018), dystonia (Ferreira et al., 2018) and functional movement disorders (Garcin, 2018). The area of neuropathic pain (Papadopoulos & Mitsikostas, 2012) has been extensively studied and placebo in headaches and Migraine (Amanzio et al., 2009; Kam-Hansen et al., 2014; Macedo et al., 2008, 2006; Meissner et al., 2013; Mitsikostas, 2016; Mitsikostas et al., 2011) has been also reported. Motor disorders such as Amyotophic

Table 1 Overview of neurologic disorders with specific reports of the placebo and nocebo responses.

Disorder	References
Parkinson's disease	Goetz, Laska, et al. (2008); Howick et al. (2018); Leal Rato et al. (2019); Mestre, Shamy, Benedetti, and Lang (2018), Mestre, Lang, and Okun (2016); Morberg et al. (2018); Stathis, Smpiliris, Konitsiotis, and Mitsikostas (2013); Witek et al. (2018)
Restless leg syndrome	Fulda and Wetter (2008); Howick et al. (2018); Silva et al. (2017)
Huntington's disease	Cubo et al. (2012); Ferreira, Trenkwalder, and Mestre (2018); HORIZON Investigators of the Huntington Study Group and European Huntington's Disease Network (2013); Huntington Study Group (2006)
Tics	Cubo et al. (2013); Ferreira et al. (2018)
Essential tremor	Ferreira et al. (2018)
Dystonia	Ferreira et al. (2018)
Functional movement disorders	Garcin (2018)
Neuropathic pain	Howick et al. (2018); Papadopoulos and Mitsikostas (2012)
Headaches and Migraine	Amanzio, Corazzini, Vase, and Benedetti (2009); Howick et al. (2018); Kam-Hansen et al. (2014); Macedo, Baños, and Farré (2008); Macedo, Farré, and Baños (2006); Meissner et al. (2013); Mitsikostas (2016); Mitsikostas, Mantonakis, and Chalarakis (2011)
Amyotrophic Lateral Sclerosis	Howick et al. (2018); Shafiq, Mitsikostas, and Zis (2017)
Myasthenia gravis	Varma and Zis (2019)
CIDP	Zis, Hadjivassiliou, Sarrigiannis, Jenkins, and Mitsikostas (2018)
Multiple sclerosis	Gklinos, Papadopoulos, and Mitsikostas, (2019); Howick et al., (2018); Papadopoulos and Mitsikostas, (2010)
Epilepsy	Howick et al. (2018); Rheims, Cucherat, Arzimanoglou, and Ryvlin (2008); Rheims and Ryvlin (2012); Zis, Shafiq, and Mitsikostas (2017)
Alzheimer's disease	Amanzio, Benedetti, and Vase (2012); Benedetti et al. (2006); Howick et al. (2018); Zis and Mitsikostas (2015)

CIDP, chronic inflammatory demyelinating polyneuropathy.

Lateral Sclerosis (Shafiq et al., 2017) and myasthenia gravis (Varma & Zis, 2019), inflammatory disorders such as chronic inflammatory demyelinating polyneuropathy (Zis et al., 2018) and multiple sclerosis (Gklinos et al., 2019; Papadopoulos & Mitsikostas, 2010), but also epilepsy (Rheims et al., 2008; Rheims & Ryvlin, 2012; Zis et al., 2017) and Alzheimer's disease (Amanzio et al., 2012; Benedetti et al., 2006; Zis & Mitsikostas, 2015), do not escape the impact of placebo and nocebo. Obviously, the frequency of placebo and nocebo responses in most trials, as they are performed now, include the true placebo/nocebo response, but also confounding factors and various contributing factors, such as natural course of the diseases (fluctuations, worsening, improvement), expectations, regression to the mean, response bias, the Hawthorne effect. Hence, the real rates and amplitude of placebo and nocebo responses are hard to evaluate, but there is no doubt about their existence in all of these neurological diseases. Since the placebo and pharmacological responses to an active drug are not simply additive, various new clinical trial designs have been proposed to better dissect the various components (pharmacological effect and placebo effect) of the responses to new treatments (Benedetti, 2014; Benedetti et al., 2016; Boehm, Berger, Weger, & Heusser, 2017; Enck et al., 2013; Weimer & Enck, 2014). These strategies range from either the modified Zelen's design or the balanced placebo design, to a multistage strategy for clinical trials. In the modified Zelen's design, an evaluation of natural history of the disease is performed by adding "no treatment" arms to trials. Other strategies such as putting patients on a waiting list have also been attempted, but remain controversial as they could induce either a placebo effect regarding positive expectations of being later included in the trial, or nocebo effect with worsening during the waiting period (Schedlowski et al., 2015). In the balanced placebo design, patients receiving the active medication or placebo are each divided into two groups with either deception or open prescription. Excluding placebo responders using trial design, such as with run-in placebo phase, is controversial because of the deception implied, the risk of mislabeling patients since nonresponders can turn to placebo-responders and vice-versa, and the inaccuracy regarding the generalizability of the population studied. Developing sequential trials in a multistep strategy allocates different aims at each step: initial efficacy studies in which placebo responses will try to be minimized, then effectiveness studies attempting to be closer to the real world everyday medical practice, which is the setting in which the medication will end up being used.

Another good reason for using and maximizing placebo in the clinical setting is its clinical impact and frequent benefit. Obviously, not all patients are responders, but among placebo responders, the effect can be clinically relevant. Indeed, studies have shown that the improvement observed in placebo arms of clinical trials can be above the minimal clinically significant changes (Ferreira et al., 2018). For instance, in HD, data from the TETRA-HD study showed some patients, albeit a minority, showed improvement in their motor symptoms both on UHDRS and CGI-Improvement scale (Ferreira et al., 2018; Huntington Study Group, 2006), and less cognitive worsening in the placebo arm as measured by the MMS scores in another study (HORIZON Investigators of the Huntington Study Group and European Huntington's Disease Network., 2013). In ET, clinically significant improvement has been reported after placebo intake. In a study assessing the efficacy of a 24-week treatment with topiramate in patients with moderate-to-severe ET, the 6-point change of placebo-treated patients on the Fahn-Tolosa-Marin Tremor Rating Scale score was within the 5 to 10 range for mild improvement, 15% of placebo-treated patients reported a good or very good outcome and 20% of placebo-treated patients were reported to present a good or very good improvement (Ferreira et al., 2018; Ondo et al., 2006). In PD, approximately 16% of placebo treated patients are good responders that may experience up to 50% reduction of the Unified Parkinson's Disease Rating Scale (UPDRS) (Benedetti et al., 2004, 2009; Frisaldi, Carlino, Lanotte, Lopiano, & Benedetti, 2014; Goetz, Wuu, et al., 2008). In tics, patients in the placebo arms of clinical trials can show a reduction in the Yale Global Tic Severity Scale Total Tic score close to the 25% improvement reported to be a minimal clinically important difference (MCID) (Cubo et al., 2013; Jeon et al., 2013). In RLS, different meta-analysis found a rate of 40.1% of responders with a placebo response effect size of −1.48 to −1.41 for the RLS-specific International Restless Legs Scale (IRLS) (Ferreira et al., 2018; Fulda & Wetter, 2008; Silva et al., 2017). The average mean reduction in the IRLS score was 6.58 points (95% CI: 4.86–8.29) in the placebo group, which is above the threshold for a MCID in the IRLS (Allen, 2013; Ondo, Grieger, Moran, Kohnen, & Roth, 2016). The placebo response was of smaller magnitude in polysomnographic measures in comparison with the IRLS. Placebo effects are more pronounced when assessed via patient-reported outcomes compared with biomarkers or disease markers (Hróbjartsson & Gøtzsche, 2010; Meissner, 2005; Schedlowski et al., 2015) which can impair proper measurement of the placebo effect in clinical trials. Biomarkers can reflect the biological grounds of the observed placebo clinical effects, such as the placebo-induced dopamine release shown on dopamine ligand PET studies in PD (de la Fuente-Fernández et al., 2001; Lidstone et al., 2010; Strafella, Ko, & Monchi, 2006). Nevertheless, in the clinical settings, the improvement felt by the patients and their quality of life is the outcome that physicians seek. For instance, in functional movement disorders, quality of life is reported to be as altered as in other movement disorders (Gendre et al., 2019) and can be significantly improved after Transcranial magnetic stimulation (TMS) (Garcin et al., 2013) or root magnetic stimulation (RMS) through a behavioral effect (Garcin et al., 2017). Placebo could also enhance cognition which would be particularly relevant to neurodegenerative diseases (Oken et al., 2008; Parker, Garry, Einstein, & McDaniel, 2011; Weger & Loughnan, 2013).

Whereas placebo can relieve patient symptoms, it will not cure neurodegenerative disorders. Even if placebos will not profoundly alter the pathophysiology of a disease, however, they can be considered as strong adjunctive symptomatic treatments. Given that positive expectations may reduce mood alterations and fatigue, particularly in the older population with neurological diseases, placebo treatment may help with symptoms that are not the core of the neurological disease but impair the quality of life, such as depression, anxiety or appetite modification (Kirsch, 2019). Recent studies have reported that placebo can activate reward mechanisms and induce changes in subjective and objective markers of appetite, such as plasma ghrelin levels and visual food cue reactivity (Hoffmann et al., 2018; Potthoff, Jurinec, & Schienle, 2019). Hence, even if these studies were performed in healthy volunteers, it could be hypothesized that placebo may help prevent undernutrition or binge eating in neurological patients.

Placebo and nocebo effects can also be helpful to the diagnosis of psychogenic disorders. Diagnosis of psychogenic nonepileptic seizures can be supported by induction techniques such as noninvasive verbal suggestion under prolonged EEG monitoring to identify pseudo-seizures (Gogia et al., 2019). In tremor, suggestion of worsening by hyperventilation and improvement by application of a vibrating tuning fork to the patient's forehead has been used to support diagnosis of psychogenic tremor (Galli, Béreau, Magnin, Moulin, & Aybek, 2019; Kenney et al., 2007).

Reducing various symptoms using placebo responses could also have health economic potential, because of reduced treatment cost and decreased use of the healthcare system. And this is actually what has been the case with the use of homeopathy and acupuncture in a pragmatic way. But this needs to be further investigated in cost-effectiveness studies dedicated to placebo use (Hamberger, Meissner, Hinterberger, Loew, & Weimer, 2019).

Importantly, as reviewed in many recent works (Benedetti, 2014; Benedetti et al., 2016; Evers et al., 2018; Finniss et al., 2010; Quattrone et al., 2018; Schedlowski et al., 2015; Wager & Atlas, 2015), there is no doubt today about the neurobiological components of the placebo effects underlying the placebo response, even though not all are yet known and many remain to discover and unravel. The changes observed from neuroimaging, single neuron recordings, deep brain stimulation (DBS) and repetitive TMS (rTMS) studies have shown an involvement of dopamine release in the dorsal and ventral striatum, modulated by prior exposure and learning strategies. These mechanisms have particularly been explored in PD and pain. But the dopaminergic hypothesis cannot explain all the observations. Levodopa-induced-dyskinesia can be reduced through placebo effect, which would be expected to increase if only mediated by increased dopamine transmission (Goetz, Laska, et al., 2008). In pain, the opioid system plays an important role and CB1 Cannabinoid receptors can also mediate placebo analgesia under specific circumstances (Amanzio & Benedetti, 1999; Benedetti, Amanzio, et al., 2011; Wrobel, Wiech, Forkmann, Ritter, & Bingel, 2014. Prefrontal connectivity has also been implicated in participating in the placebo response in Alzheimer's disease and pain (Benedetti et al., 2006; Krummenacher, Candia, Folkers, Schedlowski, & Schönbächler, 2010). Different polymorphisms have been found to be associated to low or high placebo responsiveness in various neurological and psychiatric conditions (Benedetti, 2014).

Hence, maximizing the placebo and reducing the nocebo responses in the clinic is clinically relevant and relies on serious biological grounds that can be considered as pharmacodynamics effects of symptomatic drugs and that can be helpful in the clinical setting of daily practice.

3. How should placebo be maximized in the clinic?

Since the use of placebo has been increasing and recognition of its clinical relevance growing, an Expert Consensus has been recently published with a list of "Do's" and "Don'ts" (Table 2), as a first step towards developing evidence-based and ethical recommendations about the implications of placebo and nocebo research for medical practice (Evers et al., 2018).

In daily practice, maximizing the placebo and avoiding the nocebo responses can focus on modulating expectations by increasing the positive

Table 2 "Do's" and "Don't's" in managing placebo and nocebo effects in the clinical practice.

	Do's	Don't's
1	Consider placebo effects as part of regular treatment	Do not take risks (e.g., prescribing invasive treatments) to maximize placebo effects
2	Inform patients about placebo and nocebo effects in such a way that treatment effects are maximized and side effects are minimized	Do not consider deception a necessary component of placebo effects
3	Ensure a patient-clinician relationship that is characterized by trust, warmth and empathy in order to maximize placebo effects and minimize nocebo effects	
4	Train health-care providers in patient-clinician communication to maximize placebo effects and minimize nocebo effects	
5	Prefer open-label rather than hidden placebo prescription in those cases where there is evidence for efficacy and where prescribing a placebo is legal	

Adapted from the expert consensus guidelines from Evers, A. W. M., Colloca, L., Blease, C., Annoni, M., Atlas, L. Y., Benedetti, F., et al. (2018). Implications of placebo and Nocebo effects for clinical practice: Expert consensus. *Psychotherapy and Psychosomatics, 87,* 204–210. doi:10.1159/000490354.

and decreasing the negative ones, and can act on the learning strategies, ranging from behavioral conditioning to social learning (Benedetti, 2014; Enck et al., 2013). The Hawthorne effect, which is not, per se, a placebo effect, could also be maximized at specific milestones of the disease, such as diagnosis time or symptomatic treatment introduction. This strategy can improve the patient's clinical state, but clearly extensive focusing on the patient can be time consuming and often less valued by the patient than a drug prescription, which offers the added potential benefit of inducing placebo effect (Fässler, 2011). Using placebo in the clinic could increase maximal drug efficacy while decreasing adverse effects and thereby reducing treatment doses and costs.

3.1 Personalizing placebo response and utilizing predictors of placebo response

Inter-individual variability is higher in placebo response than in drug response. The individual response to placebo can range from nonresponder to responder or good responder. For instance, data from Parkinsonian patients receiving either placebo or apomorphine showed that the motor response to apomorphine is virtually systematic after acute administration, whereas the response to placebo only affected half of the patients, and the magnitude of this response varied more in the placebo group (Benedetti, 2014). Other studies in pain and migraine reported variability in the placebo response (Macedo et al., 2006; Stein, Sprenger, Scholz, Wiech, & Bingel, 2012; Wager, Atlas, Leotti, & Rilling, 2011). The duration of placebo response is also variable. Whereas it is reported to be usually shorter than the response to pharmacological agents (Benedetti, 2014), some others have reported a possible long-lasting effect with placebo effects increasing over time (Tison et al., 2016). Hence, it is important to identify which specific patients are more prone to a placebo response.

A review by Horing and colleagues, focusing mainly on pain, identified expectations of how pain might change after treatment and expectations of how coping strategies impact pain as two modulators of placebo response (Horing, Weimer, Muth, & Enck, 2014). Hence, in the daily clinical practice, when assessing the patient's medical history, physician should always include a careful assessment of patients' expectations and pretreatment experience, because they could predict an enhanced placebo response, but also nocebo or suboptimal response in case of inadequate expectations (negative or overoptimistic). Pre-treatment training for more realistic expectations could be aimed at for these specific patients (see modulating expectations and patient-physician relationship subsections).

Predictors can vary across neurological diseases. In RLS, the placebo and nocebo responses seem greater in trials with longer duration, evaluating pharmacologic interventions, in idiopathic RLS, in industry-funded studies, and with unpublished results (Fulda & Wetter, 2008; Silva et al., 2017). In migraine RCT studies, a higher probability of receiving the drug, sham acupuncture and sham surgery increased placebo response (Diener, Dowson, Ferrari, Nappi, & Tfelt-Hansen, 1999; Meissner et al., 2013). But these predictors, although interesting in the context of RCT, are less relevant to daily clinical practice. Younger age seems to increase placebo response in epilepsy and tics (Cubo et al., 2013; Rheims et al., 2008), as well as higher baseline severity in PD (Goetz, Laska, et al., 2008). A meta-analysis of RCTs in PD that included 48 studies, consisting of 5618 participants on placebo, found that motor fluctuation and high baseline UPDRS part III score significantly increased the magnitude of the positive placebo response (Shin et al., 2016). Similar findings were observed in a study by Goetz et al., including 858 patients under placebo from 11 trials (Goetz, Wuu, et al., 2008). In the advanced fluctuating patients, low baseline daily levodopa dose and high baseline UPDRS motor score were significant predictors of placebo response in both studies (Goetz, Laska, et al., 2008; Goetz, Wuu, et al., 2008; Shin et al., 2016).

The symptoms that will predict better response to placebo can also vary. In PD, it seems that tremor and bradykinesia had better odds ratios for positive placebo response of 1.13 and 1.06 respectively, than rigidity and gait (Goetz, Wuu, et al., 2008; Witek et al., 2018). Another study in PD patients undergoing DBS surgery also found a significant better improvement when aware of the stimulation for bradykinesia than tremor and rigidity (Mercado et al., 2006).

Personality traits can also influence the potentiality to placebo response. An open personality on the NEO-Five Factor Inventory questionnaire can predispose to placebo response in PD (Lou et al., 2013). The same study also assessed perception and stress using the General Perceived Self-Efficacy Scale, and Perceived Stress Scale. Several studies in pain have shown that basal anxiety state and fear or personality with dispositional optimism will modulate placebo response (Baeken, Vanderhasselt, & De Raedt, 2011; Flaten, Aslaksen, Lyby, & Bjørkedal, 2011; Geers, Helfer, Kosbab, Weiland, & Landry, 2005; Geers, Wellman, Fowler, Helfer, & France, 2010; Lyby, Aslaksen, & Flaten, 2011; Morton, Watson, El-Deredy, & Jones, 2009). The Q-No questionnaire may also be used by the physician in order to assess the risk of nocebo effects that could limit adherence and treatment outcomes in headache for instance (Mitsikostas, 2016). In clinical practice, the use of such questionnaires could be helpful to assess patients' expectations and pretreatment experience.

The type of treatment introduced can also influence the rate of placebo or nocebo responses. For instance, in headaches and multiple sclerosis, preventive treatments are more associated with nocebo and treatment withdrawal (Mitsikostas et al., 2011; Papadopoulos & Mitsikostas, 2010). In multiple sclerosis, the pooled incidence of nocebo effect, assessed as the percentage of dropouts due to drug related adverse events, was 74.4% (95% CI: 69.92–88.30) in disease-modifying therapy trials and 25.3% (95% CI: 15.24–36.90) in symptomatic treatment trials (Papadopoulos & Mitsikostas, 2010). Oral disease-modifying treatments may be associated with a higher incidence and greater nocebo severity than newer injectables in multiple sclerosis (Gklinos et al., 2019). In migraine prophylaxis, sham acupuncture (proportion of responders, 0.38 [95% CI, 0.30–0.47]) and sham surgery (0.58 [0.37–0.77]) were associated with a more pronounced reduction of migraine frequency than oral pharmacological placebos (Meissner et al., 2013). Invasive treatment involving surgery are associated with a high placebo response as shown for instance in pain and PD studies (Goetz, Wuu, et al., 2008; Jonas, Crawford, et al., 2015; Mercado et al., 2006; Mestre et al., 2016; Pollo et al., 2002; Witek et al., 2018). But it would not justify invasive treatments as an option to potentiate placebo response in the daily clinical practice, and they are not recommended in a recent expert consensus (Evers et al., 2018).

The demographic factors influencing placebo response vary across neurological disease. Whereas children have been shown to have a higher placebo response than adults in epilepsy (Rheims et al., 2008) and tics (Cubo et al., 2013), this was not true for placebo response in regards to pain (Wrobel et al., 2015). Children did show a higher sensitivity to previous treatment experience (Wrobel et al., 2015) and younger age was associated to a higher placebo response in pain (Freeman, Emir, & Parsons, 2015), even if the elderly can still experience placebo analgesia (Wrobel, Fadai, Brassen, & Bingel, 2016). Older age predicts a higher placebo response in PD (Goetz, Wuu, et al., 2008). In tics, women are more prone to placebo response (Cubo et al., 2013).

Genetic polymorphism have also been implicated in the degree of placebo response (Benedetti, 2014; Colagiuri, Schenk, Kessler, Dorsey, & Colloca, 2015; Hall, Loscalzo, & Kaptchuk, 2015). These include polymorphisms from the dopamine (*COMT, MOA-A, DRD3, DBH* genes), opioid (*OPRM1* gene), serotonin (*TPH2, 5-HTTLPR, SLC6A4, HTR2A, NR3C1* genes), and endocannabinoid (*FAAH* gene) systems but these findings mainly concern pain, psychiatric disorders such as schizophrenia, depression and social anxiety. Gene association studies are lacking in neurological disorders and further studies will help to implement, in the future, clinical practice with these predictive factors.

Improve patient-doctor relationship to improve placebo effect in clinical practice.

Conditions under which placebo response can be maximized and nocebo avoided start with a patient-practitioner relationship, including exposing information suitably, showing great empathy, assessing the specific personality profile and states of emotions of each patient and supporting coping mechanisms.

In many health systems, patient-doctor interactions need further improvement regarding patient's understanding of their clinical situation.

In 2008 in the United States, 50% of patients did not have an adequate understanding of what their physician has told them following a visit (Bodenheimer, 2008; Enck et al., 2013). A study in the Finnish PD population, showed that the oldest subjects and those with a lower level of education have the greatest information needs to avoid therapeutic misconceptions (Reijula et al., 2017). Increasing patient understanding by using simpler wording can enhance a patient's adherence to treatments. Explaining to the patients what to expect in their disease and the concept of nocebo can also avoid misattribution of the symptoms experienced, which is the inappropriate attribution of improvement or worsening to a treatment when it was actually caused by the disorder's natural fluctuation of symptoms or other causes (Chavarria et al., 2017). Patient education programs developed in neurological disorders such as MS, HD, PD or epilepsy can improve adherence to treatments, but also improve coping self-efficacy and increase therapeutic effects of the prescribed drugs (A'Campo, Spliethoff-Kamminga, & Roos, 2012; A'Campo, Wekking, Spliethoff-Kamminga, Le Cessie, & Roos, 2010; Gutiérrez-Viedma et al., 2019; Shafer & Buchhalter, 2016; Wendebourg et al., 2016; Zimmer, Bläuer, Coslovsky, Kappos, & Derfuss, 2015).

Empathic behavior is crucial to the instatement of placebo effect and avoidance of nocebo effect. A study by Howe and colleagues, using a histamine skin prick test to induce an allergic response, showed the response to a placebo cream was greater when the provider acted both warmer and more competent and decreased when the provider acted colder and less competent (Howe, Goyer, & Crum, 2017). Trust in their health care professional leads the patients to more beneficial health behaviors, less symptoms and higher quality of life and to be more satisfied with their treatment (Birkhäuer et al., 2017). In the area of placebo response, this statement is even more biased towards a strong patient-to-doctor relationship. What could be seen as applying the Hawthorne effect in the daily clinical practice could sustain an improvement with the placebo effect (Morberg et al., 2018). But this approach, which seems obvious to many practitioners, can also be challenging because time-consuming, which can be hard to achieve in a realistic way on a daily basis. In a pragmatic way, increasing the patient follow-up at specific milestones, such as diagnosis time or introduction of a treatment, can be useful.

Open use of placebo without deception is also possible and efficient and more ethical than hidden uses (Kaptchuk & Miller, 2018). In irritable bowel syndrome, the induction of a placebo response remained as long as patients were informed that "the placebo, at least in some circumstances, works as a real drug through the release of endogenous-opioids" (Kaptchuk et al., 2010). In episodic migraine attacks, patients reported significant enhancement with 15% improvement in pain when taking the open label placebo and 15% worsening of pain with no treatment. In functional disorders (Kam-Hansen et al., 2014), TMS is used efficiently in the same way, with an opened explanation to the patients regarding the benefits expected in their condition in which there is no lesion in the neurological circuit or little changes in cortical excitability over time (Burke et al., 2018; Garcin et al., 2017, 2013; Pollak, Nicholson, Edwards, & David, 2014; Stone & Edwards, 2012). Indeed, informing the patients about the accuracy of their diagnosis and the potential reversibility of their symptoms is a crucial first-step to recovery in these patients (Stone & Hoeritzauer, 2019).

As the cost of treatment modulates the placebo effects and response, intervention should be described as valuable. Indeed, in a study in PD patients receiving a sham intravenous infusion of saline depicted as either a "cheap" or "expensive" new dopamine agonist, motor improvement was better with the expensive placebo compared to the cheap one (Espay et al., 2015). Furthermore, similar changes in fMRI activity and motor improvement on UPDRS were observed in both the levodopa and the "expensive" therapy.

3.2 Strategies that modulate expectations

Increasing positive and avoiding negative expectations are central to maximizing the positive clinical outcome. In PD, expectations will increase placebo response. In a study where patients were randomized into three groups, with 0%, 50%, and 100% expectancy of receiving levodopa, when in fact all received placebo, cortical excitability was linked to the degree of expectancy (Lou et al., 2013). In PD patients undergoing surgery for subthalamic nucleus DBS, the information about the condition of the stimulation enhanced the final clinical effect, with lower UPDRS motor score when the patients were aware of the stimulation (Mercado et al., 2006) and faster hand movement when the patients expected a good motor performance (Pollo et al., 2002). Examiner's belief about the effect of the intervention also influences the outcome. This phenomenon is called the Rosenthal or Pygmalion effect (Kohn, 1973; Learman, Avorn, Everitt, & Rosenthal, 1990). Thus positive examiner or caretakers expectations lead to positive effects in the subjects. This effect occurs without the overt awareness of the examiner (Learman et al., 1990).

The verbal induction of negative expectations can abolish the effect of potent opioid drugs like remifentanil (Bingel et al., 2011). When evaluating the profile of adverse events in placebo arms of clinical trials, the nocebo effect reported match the expected adverse events described in the information leaflets, reflecting the drug tested in the active arm (Colagiuri, McGuinness, Boakes, & Butow, 2012; Howick et al., 2018). For instance, in trials investigating migraine treatments, the placebo groups of nonsteroidal anti-inflammatory drug trials reported more gastrointestinal symptoms, whereas the placebo groups of anticonvulsant trials reported more neurological symptoms (Amanzio et al., 2009). In trials of anticonvulsants for migraine, patients receiving placebos report memory problems and anorexia, whereas in trials of triptans, patients receiving placebos report different adverse events (Kaptchuk & Miller, 2015). But the events seem to be higher in clinical trials than in the clinical daily practice, maybe because investigators in clinical trials may further look for adverse events possibly linked to the pharmacological properties of the medication under investigation. Nevertheless, changing the way the information is delivered is crucial to avoid negative expectations. Language and communication used by practitioners should use positive wordings, add meaning to any intervention and explain what is the nocebo phenomenon (Hansen & Zech, 2019). A recent study has shown that inducing a positive mood can block the formation of nocebo effects arising from knowledge of side effects (Geers, Close, Caplandies, & Vase, 2019). Since warning about adverse effects can increase their occurrence (Colagiuri et al., 2012), in the clinic, information needs to be delivered in a way that respects patient autonomy and right to knowledge yet does not introduce unnecessary harm. This balance can often be achieved by centering the information process so that frequent vs. rare adverse effects are made clear, but without forcing fear on patients. As mentioned in the above section, patients with inadequate expectations should also be trained, in order to rescale to adapted expectations, to avoid future disappointments that would lead to nocebo responses.

The key role of medical rituals and expectations can be also used via the open-hidden delivery paradigm (Benedetti, Carlino, & Pollo, 2011; Colloca, Lopiano, Lanotte, & Benedetti, 2004; Enck et al., 2013; Kaptchuk & Miller, 2015). Several studies with rizatriptan, morphine, fentanyl, and diazepam administered openly and covertly have shown differences in clinical outcomes (Colloca et al., 2004; Enck et al., 2013; Kam-Hansen et al., 2014; Kaptchuk & Miller, 2015). The use of open administration and hidden interruption of treatment can help in enhancing positive expectations at treatment initiation, decreasing negative expectations at time of treatment withdrawal, and reducing therapeutic drug doses. In pain, buprenorphine intake decreased by 33% in a group told that an intravenous saline was a powerful painkiller compared to a control group told they were receiving a rehydrating solution (Pollo et al., 2001). In PD, verbal communication about dose reduction is as effective as the 50% dose reduction itself, demonstrating that delivery of the information about the dose may have an important impact on the therapeutic outcome (Carlino et al., 2019).

Changing the aspect of the drug is not innocuous (Meissner & Linde, 2018). The switch from brand name to generic drugs with identical compounds is frequently associated with an increase in unwanted adverse effects, and therefore could lead to treatment discontinuation. However, more studies are needed to assess the exact nocebo size (Bingel and Placebo Competence Team, 2014; Spanou, Mavridis, & Mitsikostas, 2019). In migraine therapy, the size of Rnon-specific effects can vary depending of treatment modality with a higher placebo response with interventional therapies, such as acupuncture and surgery than with oral placebos (Meissner et al., 2013). Similar phenomena can be observed regarding nocebo response such as in multiple sclerosis (Gklinos et al., 2019). Hence, patients expectations regarding treatment modality should be controlled for before switching a treatment, since patients' and practitioners' preferences can shape a clinical response (Rheims & Ryvlin, 2012).

Another way to augment positive expectations could be the use of active placebos, which mimic the adverse effects of the active drug. For example, using caffeine tablets instead of sugar pills, which may be more convincing to the patients that they are under an active treatment (Bjørkedal & Flaten, 2011). Studies in pain have also shown that adverse effects can enhance treatment response through expectations (Berna et al., 2017).

3.3 Impact of learning

Learning is an important parameter in the placebo response that can easily translate into clinical practice. There are various learning processes ranging from Pavlovian conditioning to social learning. Learning strategies could be useful to turn nonresponders into responders, or to increase the amplitude of the placebo response and turn responders into good responders.

Conditioning can cause significant clinical improvement. For instance, placebo effects induced by auditory cues decrease parkinsonian rigidity in patients with subthalamic stimulation (Rätsep & Asser, 2016). Preconditioning PD patients with apomorphine resulted in significant improvement of bradykinesia which did not occur when verbal suggestions were solely performed (Frisaldi et al., 2017). A patient's prior exposure to medications influences future expectations. For instance, the molecular pathways involved in pain change depending on the previous treatments applied. Prior exposure to nonopioid treatment, such as ketorolac, switches to CB1 cannabinoid receptors' implication instead of only opioid pathway involvement. Indeed, the specific CB1 cannabinoid receptor antagonist, rimonabant or SR141716, blocks nonopioid placebo analgesic responses, but has no effect on opioid placebo responses (Benedetti, Amanzio, et al., 2011). In an analysis of two RLS trials, dopamine agonist pre-treatment tends to increase the apparent effect of new dopaminergic drugs by decreasing the placebo effect in the placebo arm without substantially modifying the placebo effect in the active treatment arm (de la Fuente-Fernández, 2012). A study in placebo analgesia combining classic conditioning and expectation showed that the influence of previous experience on subsequent treatment outcome was stronger in children than in adults (Wrobel et al., 2015). Negative carry-over effects can result from previous treatment failure, as reported in prior studies analyzing pain, for instance (Kessner et al., 2014).

The placebo-controlled drug reduction (PCDR) paradigm can help to decrease the doses necessary to maintain the same efficacy (Doering & Rief, 2012). It consists in the use of full-dose medication for a set period of time (acquisition period), followed by a maintenance or evocation period with interspersed placebo treatment (Carlino et al., 2019; Enck et al., 2013). This has already been tried successfully in various domains, such as in pain (Amanzio & Benedetti, 1999; Benedetti, Amanzio, et al., 2011) and Parkinson's disease (Benedetti et al., 2004; de la Fuente-Fernández et al., 2001).

In line with the "patient-oriented" medicine, learning about the various treatment options and repositioning the choice of treatment in the patient's hands can lead to future treatment success. Choice of treatment alternatives improves treatment effectiveness by enhancing personal control (Geers et al., 2013). In pain, the possibility to choose the treatment enhanced placebo analgesia as compared to conditions in which no choice was given (Rose, Geers, Rasinski, & Fowler, 2012). The typical application of this principle in neurological disease would be in patients with multiple sclerosis, in which multiple treatments are available. In PD, patient and caregiver expectations are recognized to be important for the choice of second line, nonoral, therapies (Odin et al., 2015). Patients need previous knowledge of the pros and cons of each of these treatments to allow for an informed choice.

In the modern era, particularly following the generalization of social media and online search engine usage, social learning has greatly increased and its clinical impact grown to extent that are not fully appreciated yet. It is particularly important to control for these factors in clinical practice since learning about other patients' experiences may influence their own (Benedetti, 2014; Craig & Prkachin, 1978). Nocebo response can be increased by observation and the social contagion of negative emotions (Benedetti, 2013; Webster, Weinman, & Rubin, 2016). Avoiding misattribution is crucial, as patients could wrongfully attribute the symptoms caused to the therapy, which in turn may lead to treatment interruption. But social learning and web-based learning can also be looked at as an opportunity to improve the clinical outcome by promoting social contact between successfully treated patients (Bishop et al., 2019).

4. Future perspectives

Knowledge regarding placebo mechanisms and their consequences on clinical outcome have greatly improved over the last two decades. This has led to several clues on how to improve placebo effect in the daily practice. Most of the data were generated from clinical trials and/or in laboratory settings. There is a need for more observational studies to better assess the extent of placebo in the clinic and its potential as a possible disease-modifying therapy. Future results regarding more specific neurobiological mechanisms will fine-tune placebo response in the clinical daily practice. For instance, in view of a potential personalized medicine, the genetic variants implicated in placebo response have been studied in other diseases but data in neurological disorders are still scarce and remain to be further explored. Considering the use of placebo in clinical practice raises many ethical issues, and the recent expert consensus regarding placebo use in the clinic is a first step to future guidelines necessary to this field.

Funding

No specific funding.

Conflict of interest

No conflict of interest relevant to this manuscript.

References

A'Campo, L. E. I., Spliethoff-Kamminga, N. G. A., & Roos, R. A. C. (2012). The patient education program for Huntington's disease (PEP-HD). *Journal of Huntington's Disease*, *1*, 47–56. https://doi.org/10.3233/JHD-2012-120002.

A'Campo, L. E. I., Wekking, E. M., Spliethoff-Kamminga, N. G. A., Le Cessie, S., & Roos, R. A. C. (2010). The benefits of a standardized patient education program for patients with Parkinson's disease and their caregivers. *Parkinsonism & Related Disorders*, *16*, 89–95. https://doi.org/10.1016/j.parkreldis.2009.07.009.

Allen, R. P. (2013). Minimal clinically significant change for the international restless legs syndrome study group rating scale in clinical trials is a score of 3. *Sleep Medicine*, *14*, 1229. https://doi.org/10.1016/j.sleep.2013.08.001.

Amanzio, M., & Benedetti, F. (1999). Neuropharmacological dissection of placebo analgesia: Expectation-activated opioid systems versus conditioning-activated specific subsystems. *The Journal of Neuroscience*, *19*, 484–494.

Amanzio, M., Benedetti, F., & Vase, L. (2012). A systematic review of adverse events in the placebo arm of donepezil trials: The role of cognitive impairment. *International Psychogeriatrics*, *24*, 698–707. https://doi.org/10.1017/S1041610211002420.

Amanzio, M., Corazzini, L. L., Vase, L., & Benedetti, F. (2009). A systematic review of adverse events in placebo groups of anti-migraine clinical trials. *Pain*, *146*, 261–269. https://doi.org/10.1016/j.pain.2009.07.010.

Baeken, C., Vanderhasselt, M. A., & De Raedt, R. (2011). Baseline "state anxiety" influences HPA-axis sensitivity to one sham-controlled HF-rTMS session applied to the right dorsolateral prefrontal cortex. *Psychoneuroendocrinology*, *36*, 60–67. https://doi.org/10.1016/j.psyneuen.2010.06.006.

Benedetti, F. (2013). Responding to nocebos through observation: Social contagion of negative emotions. *Pain*, *154*, 1165. https://doi.org/10.1016/j.pain.2013.05.012.

Benedetti, F. (2014). Placebo effects: From the neurobiological paradigm to translational implications. *Neuron*, *84*, 623–637. https://doi.org/10.1016/j.neuron.2014.10.023.

Benedetti, F., Amanzio, M., Rosato, R., & Blanchard, C. (2011). Nonopioid placebo analgesia is mediated by CB1 cannabinoid receptors. *Nature Medicine*, *17*, 1228–1230. https://doi.org/10.1038/nm.2435.

Benedetti, F., Arduino, C., Costa, S., Vighetti, S., Tarenzi, L., Rainero, I., et al. (2006). Loss of expectation-related mechanisms in Alzheimer's disease makes analgesic therapies less effective. *Pain*, *121*, 133–144. https://doi.org/10.1016/j.pain.2005.12.016.

Benedetti, F., Carlino, E., & Piedimonte, A. (2016). Increasing uncertainty in CNS clinical trials: The role of placebo, nocebo, and Hawthorne effects. *The Lancet Neurology*, *15*, 736–747. https://doi.org/10.1016/S1474-4422(16)00066-1.

Benedetti, F., Carlino, E., & Pollo, A. (2011). Hidden administration of drugs. *Clinical Pharmacology and Therapeutics*, *90*, 651–661. https://doi.org/10.1038/clpt.2011.206.

Benedetti, F., Colloca, L., Torre, E., Lanotte, M., Melcarne, A., Pesare, M., et al. (2004). Placebo-responsive Parkinson patients show decreased activity in single neurons of subthalamic nucleus. *Nature Neuroscience*, *7*, 587–588. https://doi.org/10.1038/nn1250.

Benedetti, F., Lanotte, M., Colloca, L., Ducati, A., Zibetti, M., & Lopiano, L. (2009). Electrophysiological properties of thalamic, subthalamic and nigral neurons during the anti-parkinsonian placebo response. *Journal of Physiology (London)*, *587*, 3869–3883. https://doi.org/10.1113/jphysiol.2009.169425.

Berna, C., Kirsch, I., Zion, S. R., Lee, Y. C., Jensen, K. B., Sadler, P., et al. (2017). Side effects can enhance treatment response through expectancy effects: An experimental analgesic randomized controlled trial. *Pain*, *158*, 1014–1020. https://doi.org/10.1097/j.pain.0000000000000870.

Bingel, U., & Placebo Competence Team (2014). Avoiding nocebo effects to optimize treatment outcome. *JAMA*, *312*, 693–694. https://doi.org/10.1001/jama.2014.8342.

Bingel, U., Wanigasekera, V., Wiech, K., Ni Mhuircheartaigh, R., Lee, M. C., Ploner, M., et al. (2011). The effect of treatment expectation on drug efficacy: Imaging the analgesic benefit of the opioid remifentanil. *Science Translational Medicine, 3*, 70ra14. https://doi.org/10.1126/scitranslmed.3001244.

Birkhäuer, J., Gaab, J., Kossowsky, J., Hasler, S., Krummenacher, P., Werner, C., et al. (2017). Trust in the health care professional and health outcome: A meta-analysis. *PLoS One, 12*, e0170988. https://doi.org/10.1371/journal.pone.0170988.

Bishop, F. L., Aizlewood, L., & Adams, A. E. M. (2014). When and why placebo-prescribing is acceptable and unacceptable: A focus group study of patients' views. *PLoS One, 9*, e101822. https://doi.org/10.1371/journal.pone.0101822.

Bishop, F. L., Greville-Harris, M., Bostock, J., Din, A., Graham, C. A., Lewith, G., et al. (2019). Informing adults with Back pain about placebo effects: Randomized controlled evaluation of a new website with potential to improve informed consent in clinical research. *Journal of Medical Internet Research, 21*, e9955. https://doi.org/10.2196/jmir.9955.

Bjørkedal, E., & Flaten, M. A. (2011). Interaction between expectancies and drug effects: An experimental investigation of placebo analgesia with caffeine as an active placebo. *Psychopharmacology, 215*, 537–548. https://doi.org/10.1007/s00213-011-2233-4.

Bodenheimer, T. (2008). The future of primary care: Transforming practice. *The New England Journal of Medicine, 359*, 2086–2089. https://doi.org/10.1056/NEJMp0805631.

Boehm, K., Berger, B., Weger, U., & Heusser, P. (2017). Does the model of additive effect in placebo research still hold true? A narrative review. *JRSM Open, 8*, https://doi.org/10.1177/2054270416681434. 2054270416681434.

Burke, M. J., Isayama, R., Jegatheeswaran, G., Gunraj, C., Feinstein, A., Lang, A. E., et al. (2018). Neurostimulation for functional neurological disorder: Evaluating longitudinal neurophysiology. *Movement Disorders Clinical Practice, 5*, 561–563. https://doi.org/10.1002/mdc3.12651.

Carlino, E., Piedimonte, A., Romagnolo, A., Guerra, G., Frisaldi, E., Vighetti, S., et al. (2019). Verbal communication about drug dosage balances drug reduction in Parkinson's disease: Behavioral and electrophysiological evidences. *Parkinsonism & Related Disorders, 65*, 184–189. https://doi.org/10.1016/j.parkreldis.2019.06.015.

Chavarria, V., Vian, J., Pereira, C., Data-Franco, J., Fernandes, B. S., Berk, M., et al. (2017). The placebo and Nocebo phenomena: Their clinical management and impact on treatment outcomes. *Clinical Therapeutics, 39*, 477–486. https://doi.org/10.1016/j.clinthera.2017.01.031.

Colagiuri, B., McGuinness, K., Boakes, R. A., & Butow, P. N. (2012). Warning about side effects can increase their occurrence: An experimental model using placebo treatment for sleep difficulty. *Journal of Psychopharmacology (Oxford, England), 26*, 1540–1547. https://doi.org/10.1177/0269881112458730.

Colagiuri, B., Schenk, L. A., Kessler, M. D., Dorsey, S. G., & Colloca, L. (2015). The placebo effect: From concepts to genes. *Neuroscience, 307*, 171–190. https://doi.org/10.1016/j.neuroscience.2015.08.017.

Colloca, L., Lopiano, L., Lanotte, M., & Benedetti, F. (2004). Overt versus covert treatment for pain, anxiety, and Parkinson's disease. *Lancet Neurology, 3*, 679–684. https://doi.org/10.1016/S1474-4422(04)00908-1.

Craig, K. D., & Prkachin, K. M. (1978). Social modeling influences on sensory decision theory and psychophysiological indexes of pain. *Journal of Personality and Social Psychology, 36*, 805–815. https://doi.org/10.1037//0022-3514.36.8.805.

Cubo, E., González, M., del Puerto, I., de Yébenes, J. G., Arconada, O. F., Gabriel y Galán, J. M. T., et al. (2012). Placebo effect characteristics observed in a single, international, longitudinal study in Huntington's disease. *Movement Disorders, 27*, 439–442. https://doi.org/10.1002/mds.24062.

Cubo, E., González, M., Singer, H., Mahone, E. M., Scahill, L., Müller-Vahl, K. R., et al. (2013). Impact of placebo assignment in clinical trials of tic disorders: Placebo in Tic disorders. *Movement Disorders, 28*, 1288–1292. https://doi.org/10.1002/mds.25365.

de la Fuente-Fernández, R. (2012). The powerful pre-treatment effect: Placebo responses in restless legs syndrome trials. *European Journal of Neurology, 19*, 1305–1310. https://doi.org/10.1111/j.1468-1331.2012.03719.x.

de la Fuente-Fernández, R., Ruth, T. J., Sossi, V., Schulzer, M., Calne, D. B., & Stoessl, A. J. (2001). Expectation and dopamine release: Mechanism of the placebo effect in Parkinson's disease. *Science, 293*, 1164–1166. https://doi.org/10.1126/science.1060937.

Diener, H. C., Dowson, A. J., Ferrari, M., Nappi, G., & Tfelt-Hansen, P. (1999). Unbalanced randomization influences placebo response: Scientific versus ethical issues around the use of placebo in migraine trials. *Cephalalgia, 19*, 699–700. https://doi.org/10.1046/j.1468-2982.1999.019008699.x.

Doering, B. K., & Rief, W. (2012). Utilizing placebo mechanisms for dose reduction in pharmacotherapy. *Trends in Pharmacological Sciences, 33*, 165–172. https://doi.org/10.1016/j.tips.2011.12.001.

Enck, P., Bingel, U., Schedlowski, M., & Rief, W. (2013). The placebo response in medicine: Minimize, maximize or personalize? *Nature Reviews Drug Discovery, 12*, 191–204. https://doi.org/10.1038/nrd3923.

Espay, A. J., Norris, M. M., Eliassen, J. C., Dwivedi, A., Smith, M. S., Banks, C., et al. (2015). Placebo effect of medication cost in Parkinson disease: A randomized double-blind study. *Neurology, 84*, 794–802. https://doi.org/10.1212/WNL.0000000000001282.

Evers, A. W. M., Colloca, L., Blease, C., Annoni, M., Atlas, L. Y., Benedetti, F., et al. (2018). Implications of placebo and Nocebo effects for clinical practice: Expert consensus. *Psychotherapy and Psychosomatics, 87*, 204–210. https://doi.org/10.1159/000490354.

Fässler, M. (2011). Clinical strategies for maximizing the placebo effect. *Family Medicine, 43*, 126.

Fässler, M., Meissner, K., Schneider, A., & Linde, K. (2010). Frequency and circumstances of placebo use in clinical practice—A systematic review of empirical studies. *BMC Medicine, 8*, 15. https://doi.org/10.1186/1741-7015-8-15.

Ferreira, J. J., Trenkwalder, C., & Mestre, T. A. (2018). Placebo and nocebo responses in other movement disorders besides Parkinson's disease: How much do we know? *Movement Disorders, 33*, 1228–1235. https://doi.org/10.1002/mds.113.

Finniss, D. G., Kaptchuk, T. J., Miller, F., & Benedetti, F. (2010). Biological, clinical, and ethical advances of placebo effects. *The Lancet, 375*, 686–695. https://doi.org/10.1016/S0140-6736(09)61706-2.

Flaten, M. A., Aslaksen, P. M., Lyby, P. S., & Bjørkedal, E. (2011). The relation of emotions to placebo responses. *Philosophical Transactions of the Royal Society of London Series B, Biological Sciences, 366*, 1818–1827. https://doi.org/10.1098/rstb.2010.0407.

Freeman, R., Emir, B., & Parsons, B. (2015). Predictors of placebo response in peripheral neuropathic pain: Insights from pregabalin clinical trials. *Journal of Pain Research, 8*, 257–268. https://doi.org/10.2147/JPR.S78303.

Frisaldi, E., Carlino, E., Lanotte, M., Lopiano, L., & Benedetti, F. (2014). Characterization of the thalamic-subthalamic circuit involved in the placebo response through single-neuron recording in Parkinson patients. *Cortex, 60*, 3–9. https://doi.org/10.1016/j.cortex.2013.12.003.

Frisaldi, E., Carlino, E., Zibetti, M., Barbiani, D., Dematteis, F., Lanotte, M., et al. (2017). The placebo effect on bradykinesia in Parkinson's disease with and without prior drug conditioning. *Movement Disorders, 32*, 1474–1478. https://doi.org/10.1002/mds.27142.

Fulda, S., & Wetter, T. C. (2008). Where dopamine meets opioids: A meta-analysis of the placebo effect in restless legs syndrome treatment studies. *Brain, 131*, 902–917. https://doi.org/10.1093/brain/awm244.

Galli, S., Béreau, M., Magnin, E., Moulin, T., & Aybek, S. (2019). Functional movement disorders. *Revue Neurologique, 176*, 244–251, S0035378719305417. https://doi.org/10.1016/j.neurol.2019.08.007.

Garcin, B. (2018). Motor functional neurological disorders: An update. *Revue Neurologique (Paris), 174*, 203–211. https://doi.org/10.1016/j.neurol.2017.11.003.

Garcin, B., Mesrati, F., Hubsch, C., Mauras, T., Iliescu, I., Naccache, L., et al. (2017). Impact of Transcranial magnetic stimulation on functional movement disorders: Cortical modulation or a behavioral effect? *Frontiers in Neurology, 8*, 338. https://doi.org/10.3389/fneur.2017.00338.

Garcin, B., Roze, E., Mesrati, F., Cognat, E., Fournier, E., Vidailhet, M., et al. (2013). Transcranial magnetic stimulation as an efficient treatment for psychogenic movement disorders. *Journal of Neurology, Neurosurgery, and Psychiatry, 84*, 1043–1046. https://doi.org/10.1136/jnnp-2012-304062.

Geers, A. L., Close, S., Caplandies, F. C., & Vase, L. (2019). A positive mood induction for reducing the formation of nocebo effects from side effect information. *Annals of Behavioral Medicine, 53*, 999–1008. https://doi.org/10.1093/abm/kaz005.

Geers, A. L., Helfer, S. G., Kosbab, K., Weiland, P. E., & Landry, S. J. (2005). Reconsidering the role of personality in placebo effects: Dispositional optimism, situational expectations, and the placebo response. *Journal of Psychosomatic Research, 58*, 121–127. https://doi.org/10.1016/j.jpsychores.2004.08.011.

Geers, A. L., Rose, J. P., Fowler, S. L., Rasinski, H. M., Brown, J. A., & Helfer, S. G. (2013). Why does choice enhance treatment effectiveness? Using placebo treatments to demonstrate the role of personal control. *Journal of Personality and Social Psychology, 105*, 549–566. https://doi.org/10.1037/a0034005.

Geers, A. L., Wellman, J. A., Fowler, S. L., Helfer, S. G., & France, C. R. (2010). Dispositional optimism predicts placebo analgesia. *The Journal of Pain, 11*, 1165–1171. https://doi.org/10.1016/j.jpain.2010.02.014.

Gendre, T., Carle, G., Mesrati, F., Hubsch, C., Mauras, T., Roze, E., et al. (2019). Quality of life in functional movement disorders is as altered as in organic movement disorders. *Journal of Psychosomatic Research, 116*, 10–16. https://doi.org/10.1016/j.jpsychores.2018.11.006.

Gklinos, P., Papadopoulos, D., & Mitsikostas, D. D. (2019). Nocebo in multiple sclerosis trials: A meta-analysis on oral and newer injectable disease-modifying treatments. *Multiple Sclerosis and Related Disorders, 36*, 101389. https://doi.org/10.1016/j.msard.2019.101389.

Goetz, C. G. (2009). Jean-Martin Charcot and his vibratory chair for Parkinson disease. *Neurology, 73*, 475–478. https://doi.org/10.1212/WNL.0b013e3181b1640b.

Goetz, C. G., Janko, K., Blasucci, L., & Jaglin, J. A. (2003). Impact of placebo assignment in clinical trials of Parkinson's disease. *Movement Disorders, 18*, 1146–1149. https://doi.org/10.1002/mds.10504.

Goetz, C. G., Laska, E., Hicking, C., Damier, P., Müller, T., Nutt, J., et al. (2008). Placebo influences on dyskinesia in Parkinson's disease. *Movement Disorders, 23*, 700–707. https://doi.org/10.1002/mds.21897.

Goetz, C. G., Wuu, J., McDermott, M. P., Adler, C. H., Fahn, S., Freed, C. R., et al. (2008). Placebo response in Parkinson's disease: Comparisons among 11 trials covering medical and surgical interventions. *Movement Disorders, 23*, 690–699. https://doi.org/10.1002/mds.21894.

Gogia, B., Rai, P. K., Matthys, S. A., Mong, E. R., Rodriguez, R., Yassin, A., et al. (2019). Use of noninvasive induction techniques in the diagnosis of PNES. *Epilepsy & Behavior, 99*, 106491. https://doi.org/10.1016/j.yebeh.2019.106491.

Gutiérrez-Viedma, Á., Sanz-Graciani, I., Romeral-Jiménez, M., Parejo-Carbonell, B., Serrano-García, I., Cuadrado, M.-L., et al. (2019). Patients' knowledge on epilepsy and SUDEP improves after a semi-structured health interview. *Epilepsy & Behavior, 99*, 106467. https://doi.org/10.1016/j.yebeh.2019.106467.

Hall, K. T., Loscalzo, J., & Kaptchuk, T. J. (2015). Genetics and the placebo effect: The placebome. *Trends in Molecular Medicine, 21*, 285–294. https://doi.org/10.1016/j.molmed.2015.02.009.

Hamberger, J., Meissner, K., Hinterberger, T., Loew, T., & Weimer, K. (2019). Placebo economics: A systematic review about the economic potential of utilizing the placebo effect. *Frontiers in Psychiatry, 10*, 653. https://doi.org/10.3389/fpsyt.2019.00653.

Hansen, E., & Zech, N. (2019). Nocebo effects and negative suggestions in daily clinical practice–forms, impact and approaches to avoid them. *Frontiers in Pharmacology, 10*, 77. https://doi.org/10.3389/fphar.2019.00077.

Häuser, W., Hansen, E., & Enck, P. (2012). Nocebo phenomena in medicine: Their relevance in everyday clinical practice. *Deutsches Ärzteblatt International, 109*, 459–465. https://doi.org/10.3238/arztebl.2012.0459.

Héron, A., Leroux, C., & Dubayle, D. (2019). Le placebo à l'hôpital: Regard sur son utilisation dans les services de médecine polyvalente. *Medical Science (Paris), 35*, 674–681. https://doi.org/10.1051/medsci/2019127.

Hoffmann, V., Lanz, M., Mackert, J., Müller, T., Tschöp, M., & Meissner, K. (2018). Effects of placebo interventions on subjective and objective markers of appetite–a randomized controlled trial. *Frontiers in Psychiatry, 9*, 706. https://doi.org/10.3389/fpsyt.2018.00706.

Horing, B., Weimer, K., Muth, E. R., & Enck, P. (2014). Prediction of placebo responses: A systematic review of the literature. *Frontiers in Psychology, 5*, 1079. https://doi.org/10.3389/fpsyg.2014.01079.

HORIZON Investigators of the Huntington Study Group and European Huntington's Disease Network. (2013). A randomized, double-blind, placebo-controlled study of Latrepirdine in patients with mild to moderate Huntington disease. *JAMA Neurology, 70*, 25. https://doi.org/10.1001/2013.jamaneurol.382.

Howe, L. C., Goyer, J. P., & Crum, A. J. (2017). Harnessing the placebo effect: Exploring the influence of physician characteristics on placebo response. *Health Psychology, 36*, 1074–1082. https://doi.org/10.1037/hea0000499.

Howick, J., Bishop, F. L., Heneghan, C., Wolstenholme, J., Stevens, S., Hobbs, F. D. R., et al. (2013). Placebo use in the United Kingdom: Results from a National Survey of primary care practitioners. *PLoS One, 8*, e58247. https://doi.org/10.1371/journal.pone.0058247.

Howick, J., Webster, R., Kirby, N., & Hood, K. (2018). Rapid overview of systematic reviews of nocebo effects reported by patients taking placebos in clinical trials. *Trials, 19*, 674. https://doi.org/10.1186/s13063-018-3042-4.

Hróbjartsson, A., & Gøtzsche, P. C. (2010). Placebo interventions for all clinical conditions. *Cochrane Database System Review* (1), Art. No. CD003974. https://doi.org/10.1002/14651858.CD003974.pub3.

Hull, S. C., Colloca, L., Avins, A., Gordon, N. P., Somkin, C. P., Kaptchuk, T. J., et al. (2013). Patients' attitudes about the use of placebo treatments: Telephone survey. *BMJ, 347*, f3757. https://doi.org/10.1136/bmj.f3757.

Huntington Study Group. (2006). Tetrabenazine as antichorea therapy in Huntington disease: A randomized controlled trial. *Neurology, 66*, 366–372. https://doi.org/10.1212/01.wnl.0000198586.85250.13.

Jeon, S., Walkup, J. T., Woods, D. W., Peterson, A., Piacentini, J., Wilhelm, S., et al. (2013). Detecting a clinically meaningful change in tic severity in Tourette syndrome: A comparison of three methods. *Contemporary Clinical Trials, 36*, 414–420. https://doi.org/10.1016/j.cct.2013.08.012.

Jonas, W. B., Crawford, C., Colloca, L., Kaptchuk, T. J., Moseley, B., Miller, F. G., et al. (2015). To what extent are surgery and invasive procedures effective beyond a placebo response? A systematic review with meta-analysis of randomised, sham controlled trials. *BMJ Open, 5*, e009655. https://doi.org/10.1136/bmjopen-2015-009655.

Kam-Hansen, S., Jakubowski, M., Kelley, J. M., Kirsch, I., Hoaglin, D. C., Kaptchuk, T. J., et al. (2014). Altered placebo and drug labeling changes the outcome of episodic migraine attacks. *Science Translational Medicine, 6*, https://doi.org/10.1126/scitranslmed.3006175. 218ra5-218ra5.

Kaptchuk, T. J., Friedlander, E., Kelley, J. M., Sanchez, M. N., Kokkotou, E., Singer, J. P., et al. (2010). Placebos without deception: A randomized controlled trial in irritable bowel syndrome. *PLoS One, 5*, e15591. https://doi.org/10.1371/journal.pone.0015591.

Kaptchuk, T. J., & Miller, F. G. (2015). Placebo effects in medicine. *The New England Journal of Medicine, 373*, 8–9. https://doi.org/10.1056/NEJMp1504023.

Kaptchuk, T. J., & Miller, F. G. (2018). Open label placebo: Can honestly prescribed placebos evoke meaningful therapeutic benefits? *BMJ, 363*, k3889. https://doi.org/10.1136/bmj.k3889.

Kapur, S. S., Stebbins, G. T., & Goetz, C. G. (2012). Vibration therapy for Parkinson's disease: Charcot's studies revisited. *Journal of Parkinson's Disease, 2*, 23–27. https://doi.org/10.3233/JPD-2012-12079.

Kenney, C., Diamond, A., Mejia, N., Davidson, A., Hunter, C., & Jankovic, J. (2007). Distinguishing psychogenic and essential tremor. *Journal of the Neurological Sciences, 263*, 94–99. https://doi.org/10.1016/j.jns.2007.06.008.

Kermen, R., Hickner, J., Brody, H., & Hasham, I. (2010). Family physicians believe the placebo effect is therapeutic but often use real drugs as placebos. *Family Medicine, 42*, 636–642.

Kessner, S., Forkmann, K., Ritter, C., Wiech, K., Ploner, M., & Bingel, U. (2014). The effect of treatment history on therapeutic outcome: Psychological and neurobiological underpinnings. *PLoS One, 9*, e109014. https://doi.org/10.1371/journal.pone.0109014.

Kirsch, I. (2019). Placebo effect in the treatment of depression and anxiety. *Frontiers in Psychiatry, 10*, 407. https://doi.org/10.3389/fpsyt.2019.00407.

Kohn, P. M. (1973). Relationships between expectations of teachers and performance of students (Pygmalion in the classroom). *The Journal of School Health, 43*, 498–503. https://doi.org/10.1111/j.1746-1561.1973.tb03741.x.

Krummenacher, P., Candia, V., Folkers, G., Schedlowski, M., & Schönbächler, G. (2010). Prefrontal cortex modulates placebo analgesia. *Pain, 148*, 368–374. https://doi.org/10.1016/j.pain.2009.09.033.

Leal Rato, M., Duarte, G. S., Ferreira, A. N., Alves, M., Mainoli, B., Teodoro, T., et al. (2019). Nocebo response in Parkinson's disease: A systematic review and meta-analysis. *Parkinsonism & Related Disorders*. https://doi.org/10.1016/j.parkreldis.2019.04.015.

Learman, L. A., Avorn, J., Everitt, D. E., & Rosenthal, R. (1990). Pygmalion in the nursing home. The effects of caregiver expectations on patient outcomes. *Journal of the American Geriatrics Society, 38*, 797–803. https://doi.org/10.1111/j.1532-5415.1990.tb01472.x.

Lidstone, S. C., Schulzer, M., Dinelle, K., Mak, E., Sossi, V., Ruth, T. J., et al. (2010). Effects of expectation on placebo-induced dopamine release in Parkinson disease. *Archives of General Psychiatry, 67*, 857–865. https://doi.org/10.1001/archgenpsychiatry.2010.88.

Linde, K., Atmann, O., Meissner, K., Schneider, A., Meister, R., Kriston, L., et al. (2018). How often do general practitioners use placebos and non-specific interventions? Systematic review and meta-analysis of surveys. *PLoS One, 13*, e0202211. https://doi.org/10.1371/journal.pone.0202211.

Linde, K., Friedrichs, C., Alscher, A., Wagenpfeil, S., Meissner, K., & Schneider, A. (2014). The use of placebo and non-specific therapies and their relation to basic professional attitudes and the use of complementary therapies among German physicians—A cross-sectional survey. *PLoS One, 9*, e92938. https://doi.org/10.1371/journal.pone.0092938.

Lou, J.-S., Dimitrova, D. M., Hammerschlag, R., Nutt, J., Hunt, E. A., Eaton, R. W., et al. (2013). Effect of expectancy and personality on cortical excitability in Parkinson's disease. *Movement Disorders, 28*, 1257–1262. https://doi.org/10.1002/mds.25471.

Lyby, P. S., Aslaksen, P. M., & Flaten, M. A. (2011). Variability in placebo analgesia and the role of fear of pain—An ERP study. *Pain, 152*, 2405–2412. https://doi.org/10.1016/j.pain.2011.07.010.

Macedo, A., Baños, J.-E., & Farré, M. (2008). Placebo response in the prophylaxis of migraine: A meta-analysis. *European Journal of Pain, 12*, 68–75. https://doi.org/10.1016/j.ejpain.2007.03.002.

Macedo, A., Farré, M., & Baños, J.-E. (2006). A meta-analysis of the placebo response in acute migraine and how this response may be influenced by some of the characteristics of clinical trials. *European Journal of Clinical Pharmacology, 62*, 161–172. https://doi.org/10.1007/s00228-005-0088-5.

Meissner, K. (2005). Preferring patient-reported to observer-reported outcomes substantially influences the results of the updated systematic review on placebos by Hróbjartsson and Gøtzsche. *Journal of Internal Medicine, 257*, 394. author reply 395-396. https://doi.org/10.1111/j.1365-2796.2005.01470.x.

Meissner, K., Fässler, M., Rücker, G., Kleijnen, J., Hróbjartsson, A., Schneider, A., et al. (2013). Differential effectiveness of placebo treatments: A systematic review of migraine prophylaxis. *JAMA Internal Medicine, 173*, 1941. https://doi.org/10.1001/jamainternmed.2013.10391.

Meissner, K., & Linde, K. (2018). Are blue pills better than green? How treatment features modulate placebo effects. In *International Review of Neurobiology* (pp. 357–378). Elsevier. https://doi.org/10.1016/bs.irn.2018.07.014.

Mercado, R., Constantoyannis, C., Mandat, T., Kumar, A., Schulzer, M., Stoessl, A. J., et al. (2006). Expectation and the placebo effect in Parkinson's disease patients with subthalamic nucleus deep brain stimulation. *Movement Disorders, 21*, 1457–1461. https://doi.org/10.1002/mds.20935.

Mestre, T. A., Lang, A. E., & Okun, M. S. (2016). Factors influencing the outcome of deep brain stimulation: Placebo, nocebo, lessebo, and lesion effects: Factors influencing deep brain stimulation outcome. *Movement Disorders, 31*, 290–298. https://doi.org/10.1002/mds.26500.

Mestre, T. A., Shamy, M., Benedetti, F., & Lang, A. E. (2018). Harnessing the power of placebos in movement disorders: Insights from Parkinson's disease in clinical research and practice: Harnessing the power of placebos. *Movement Disorders, 33*, 1195–1203. https://doi.org/10.1002/mds.27409.

Mitsikostas, D. D. (2016). Nocebo in headache. *Current Opinion in Neurology, 29*, 331–336. https://doi.org/10.1097/WCO.0000000000000313.

Mitsikostas, D. D., Mantonakis, L. I., & Chalarakis, N. G. (2011). Nocebo is the enemy, not placebo. A meta-analysis of reported side effects after placebo treatment in headaches. *Cephalalgia, 31*, 550–561. https://doi.org/10.1177/0333102410391485.

Morberg, B. M., Malling, A. S., Jensen, B. R., Gredal, O., Wermuth, L., & Bech, P. (2018). The Hawthorne effect as a pre-placebo expectation in Parkinsons disease patients participating in a randomized placebo-controlled clinical study. *Nordic Journal of Psychiatry*, *72*, 442–446. https://doi.org/10.1080/08039488.2018.1468480.
Morton, D. L., Watson, A., El-Deredy, W., & Jones, A. K. P. (2009). Reproducibility of placebo analgesia: Effect of dispositional optimism. *Pain*, *146*, 194–198. https://doi.org/10.1016/j.pain.2009.07.026.
Odin, P., Ray Chaudhuri, K., Slevin, J. T., Volkmann, J., Dietrichs, E., Martinez-Martin, P., et al. (2015). Collective physician perspectives on non-oral medication approaches for the management of clinically relevant unresolved issues in Parkinson's disease: Consensus from an international survey and discussion program. *Parkinsonism & Related Disorders*, *21*, 1133–1144. https://doi.org/10.1016/j.parkreldis.2015.07.020.
Oken, B. S. (2008). Placebo effects: Clinical aspects and neurobiology. *Brain*, *131*, 2812–2823. https://doi.org/10.1093/brain/awn116.
Oken, B. S., Flegal, K., Zajdel, D., Kishiyama, S., Haas, M., & Peters, D. (2008). Expectancy effect: Impact of pill administration on cognitive performance in healthy seniors. *Journal of Clinical and Experimental Neuropsychology*, *30*, 7–17. https://doi.org/10.1080/13803390701775428.
Ondo, W. G., Grieger, F., Moran, K., Kohnen, R., & Roth, T. (2016). Post hoc analysis of data from two clinical trials evaluating the minimal clinically important change in international restless legs syndrome sum score in patients with restless legs syndrome (Willis-Ekbom disease). *Journal of Clinical Sleep Medicine*, *12*, 63–70. https://doi.org/10.5664/jcsm.5396.
Ondo, W. G., Jankovic, J., Connor, G. S., Pahwa, R., Elble, R., Stacy, M. A., et al. (2006). Topiramate in essential tremor: A double-blind, placebo-controlled trial. *Neurology*, *66*, 672–677. https://doi.org/10.1212/01.wnl.0000200779.03748.0f.
Papadopoulos, D., & Mitsikostas, D. D. (2010). Nocebo effects in multiple sclerosis trials: A meta-analysis. *Multiple Sclerosis*, *16*, 816–828. https://doi.org/10.1177/1352458510370793.
Papadopoulos, D., & Mitsikostas, D. D. (2012). A meta-analytic approach to estimating nocebo effects in neuropathic pain trials. *Journal of Neurology*, *259*, 436–447. https://doi.org/10.1007/s00415-011-6197-4.
Parker, S., Garry, M., Einstein, G. O., & McDaniel, M. A. (2011). A sham drug improves a demanding prospective memory task. *Memory*, *19*, 606–612. https://doi.org/10.1080/09658211.2011.592500.
Pollak, T. A., Nicholson, T. R., Edwards, M. J., & David, A. S. (2014). A systematic review of transcranial magnetic stimulation in the treatment of functional (conversion) neurological symptoms. *Journal of Neurology, Neurosurgery, and Psychiatry*, *85*, 191–197. https://doi.org/10.1136/jnnp-2012-304181.
Pollo, A., Amanzio, M., Arslanian, A., Casadio, C., Maggi, G., & Benedetti, F. (2001). Response expectancies in placebo analgesia and their clinical relevance. *Pain*, *93*, 77–84. https://doi.org/10.1016/s0304-3959(01)00296-2.
Pollo, A., Torre, E., Lopiano, L., Rizzone, M., Lanotte, M., Cavanna, A., et al. (2002). Expectation modulates the response to subthalamic nucleus stimulation in Parkinsonian patients. *Neuroreport*, *13*, 1383–1386. https://doi.org/10.1097/00001756-200208070-00006.
Potthoff, J., Jurinec, N., & Schienle, A. (2019). Placebo effects on visual food Cue reactivity: An eye-tracking investigation. *Frontiers in Psychiatry*, *10*, 525. https://doi.org/10.3389/fpsyt.2019.00525.
Quattrone, A., Barbagallo, G., Cerasa, A., & Stoessl, A. J. (2018). Neurobiology of placebo effect in Parkinson's disease: What we have learned and where we are going: Placebo effect in PD. *Movement Disorders*, *33*, 1213–1227. https://doi.org/10.1002/mds.27438.

Rätsep, T., & Asser, T. (2016). Placebo effects induced by auditory cues decrease parkinsonian rigidity in patients with subthalamic stimulation. *Behavioural Brain Research, 301*, 27–32. https://doi.org/10.1016/j.bbr.2015.12.021.

Reijula, E., Pietilä, A.-M., Halkoaho, A., Selander, T., Martikainen, K., Kälviäinen, R., et al. (2017). Clinical features of Parkinson's disease patients are associated with therapeutic misconception and willingness to participate in clinical trials. *Trials, 18*, 444. https://doi.org/10.1186/s13063-017-2174-2.

Rheims, S., Cucherat, M., Arzimanoglou, A., & Ryvlin, P. (2008). Greater response to placebo in children than in adults: A systematic review and meta-analysis in drug-resistant partial epilepsy. *PLoS Medicine, 5*, e166. https://doi.org/10.1371/journal.pmed.0050166.

Rheims, S., & Ryvlin, P. (2012). Patient and physician preferences: Impact on treatment effectiveness. *Epileptic Disorders, 14*, 242–247. https://doi.org/10.1684/epd.2012.0533.

Rose, J. P., Geers, A. L., Rasinski, H. M., & Fowler, S. L. (2012). Choice and placebo expectation effects in the context of pain analgesia. *Journal of Behavioral Medicine, 35*, 462–470. https://doi.org/10.1007/s10865-011-9374-0.

Schedlowski, M., Enck, P., Rief, W., & Bingel, U. (2015). Neuro-bio-behavioral mechanisms of placebo and nocebo responses: Implications for clinical trials and clinical practice. *Pharmacological Reviews, 67*, 697–730. https://doi.org/10.1124/pr.114.009423.

Shafer, P. O., & Buchhalter, J. (2016). Patient education: Identifying risks and self-management approaches for adherence and sudden unexpected death in epilepsy. *Neurologic Clinics, 34*, 443–456. ix. https://doi.org/10.1016/j.ncl.2016.01.001.

Shafiq, F., Mitsikostas, D.-D., & Zis, P. (2017). Nocebo in motor neuron disease: Systematic review and meta-analysis of placebo-controlled clinical trials. *Amyotrophic Lateral Sclerosis & Frontotemporal Degeneration, 18*, 576–582. https://doi.org/10.1080/21678421.2017.1335325.

Shin, C. W., Hahn, S., Park, B.-J., Kim, J.-M., Park, E. O., & Jeon, B. (2016). Predictors of the placebo response in clinical trials on Parkinson's disease: A meta-analysis. *Parkinsonism & Related Disorders, 29*, 83–89. https://doi.org/10.1016/j.parkreldis.2016.05.019.

Silva, M. A., Duarte, G. S., Camara, R., Rodrigues, F. B., Fernandes, R. M., Abreu, D., et al. (2017). Placebo and nocebo responses in restless legs syndrome: A systematic review and meta-analysis. *Neurology, 88*, 2216–2224. https://doi.org/10.1212/WNL.0000000000004004.

Spanou, I., Mavridis, T., & Mitsikostas, D. D. (2019). Nocebo in biosimilars and generics in neurology: A systematic review. *Frontiers in Pharmacology, 10*, 809. https://doi.org/10.3389/fphar.2019.00809.

Stathis, P., Smpiliris, M., Konitsiotis, S., & Mitsikostas, D. D. (2013). Nocebo as a potential confounding factor in clinical trials for Parkinson's disease treatment: A meta-analysis. *European Journal of Neurology, 20*, 527–533. https://doi.org/10.1111/ene.12014.

Stein, N., Sprenger, C., Scholz, J., Wiech, K., & Bingel, U. (2012). White matter integrity of the descending pain modulatory system is associated with interindividual differences in placebo analgesia. *Pain, 153*, 2210–2217. https://doi.org/10.1016/j.pain.2012.07.010.

Stone, J., & Edwards, M. (2012). Trick or treat? Showing patients with functional (psychogenic) motor symptoms their physical signs. *Neurology, 79*, 282–284. https://doi.org/10.1212/WNL.0b013e31825fdf63.

Stone, J., & Hoeritzauer, I. (2019). How do I explain the diagnosis of functional movement disorder to a patient? *Movement Disorders Clinical Practice, 6*, 419. https://doi.org/10.1002/mdc3.12785.

Strafella, A. P., Ko, J. H., & Monchi, O. (2006). Therapeutic application of transcranial magnetic stimulation in Parkinson's disease: The contribution of expectation. *NeuroImage, 31*, 1666–1672. https://doi.org/10.1016/j.neuroimage.2006.02.005.

Tilburt, J. C., Emanuel, E. J., Kaptchuk, T. J., Curlin, F. A., & Miller, F. G. (2008). Prescribing "placebo treatments": Results of national survey of US internists and rheumatologists. *BMJ*, *337*, a1938. https://doi.org/10.1136/bmj.a1938.

Tison, F., Keywood, C., Wakefield, M., Durif, F., Corvol, J.-C., Eggert, K., et al. (2016). A phase 2A trial of the novel mGluR5-negative allosteric modulator dipraglurant for levodopa-induced dyskinesia in Parkinson's disease: Dipraglurant in levodopa-induced dyskinesia. *Movement Disorders*, *31*, 1373–1380. https://doi.org/10.1002/mds.26659.

Varma, A., & Zis, P. (2019). Nocebo effect in myasthenia gravis: Systematic review and meta-analysis of placebo-controlled clinical trials. *Acta Neurologica Belgica*, *119*, 257–264. https://doi.org/10.1007/s13760-019-01143-1.

Wager, T. D., & Atlas, L. Y. (2015). The neuroscience of placebo effects: Connecting context, learning and health. *Nature Reviews Neuroscience*, *16*, 403–418. https://doi.org/10.1038/nrn3976.

Wager, T. D., Atlas, L. Y., Leotti, L. A., & Rilling, J. K. (2011). Predicting individual differences in placebo analgesia: Contributions of brain activity during anticipation and pain experience. *The Journal of Neuroscience*, *31*, 439–452. https://doi.org/10.1523/JNEUROSCI.3420-10.2011.

Webster, R. K., Weinman, J., & Rubin, G. J. (2016). A systematic review of factors that contribute to nocebo effects. *Health Psychology*, *35*, 1334–1355. https://doi.org/10.1037/hea0000416.

Weger, U. W., & Loughnan, S. (2013). Mobilizing unused resources: Using the placebo concept to enhance cognitive performance. *The Quarterly Journal of Experimental Psychology (Hove)*, *66*, 23–28. https://doi.org/10.1080/17470218.2012.751117.

Weimer, K., & Enck, P. (2014). Traditional and innovative experimental and clinical trial designs and their advantages and pitfalls. *Handbook of Experimental Pharmacology*, *225*, 237–272. https://doi.org/10.1007/978-3-662-44519-8_14.

Wendebourg, M. J., Feddersen, L. K., Lau, S., Köpke, S., Moss-Morris, R., Heesen, C., et al. (2016). Development and feasibility of an evidence-based patient education program for managing fatigue in multiple sclerosis: The "fatigue management in MS" program (FatiMa). *International Journal of MS Care*, *18*, 129–137. https://doi.org/10.7224/1537-2073.2014-105.

Witek, N., Stebbins, G. T., & Goetz, C. G. (2018). What influences placebo and nocebo responses in Parkinson's disease?: Placebo and nocebo response in PD. *Movement Disorders*, *33*, 1204–1212. https://doi.org/10.1002/mds.27416.

Wrobel, N., Fadai, T., Brassen, S., & Bingel, U. (2016). Preserved capacity for placebo analgesia in the elderly. *The Journal of Pain*, *17*, 1318–1324. https://doi.org/10.1016/j.jpain.2016.08.012.

Wrobel, N., Fadai, T., Sprenger, C., Hebebrand, J., Wiech, K., & Bingel, U. (2015). Are children the better placebo analgesia responders? An experimental approach. *The Journal of Pain*, *16*, 1005–1011. https://doi.org/10.1016/j.jpain.2015.06.013.

Wrobel, N., Wiech, K., Forkmann, K., Ritter, C., & Bingel, U. (2014). Haloperidol blocks dorsal striatum activity but not analgesia in a placebo paradigm. *Cortex*, *57*, 60–73. https://doi.org/10.1016/j.cortex.2014.02.023.

Zimmer, A., Bläuer, C., Coslovsky, M., Kappos, L., & Derfuss, T. (2015). Optimizing treatment initiation: Effects of a patient education program about fingolimod treatment on knowledge, self-efficacy and patient satisfaction. *Multiple Sclerosis and Related Disorders*, *4*, 444–450. https://doi.org/10.1016/j.msard.2015.06.010.

Zis, P., Hadjivassiliou, M., Sarrigiannis, P. G., Jenkins, T. M., & Mitsikostas, D.-D. (2018). Nocebo in chronic inflammatory demyelinating polyneuropathy; a systematic review and meta-analysis of placebo-controlled clinical trials. *Journal of the Neurological Sciences*, *388*, 79–83. https://doi.org/10.1016/j.jns.2018.03.009.

Zis, P., & Mitsikostas, D.-D. (2015). Nocebo in Alzheimer's disease; meta-analysis of placebo-controlled clinical trials. *Journal of the Neurological Sciences*, *355*, 94–100. https://doi.org/10.1016/j.jns.2015.05.029.

Zis, P., Shafiq, F., & Mitsikostas, D.-D. (2017). Nocebo effect in refractory partial epilepsy during pre-surgical monitoring: Systematic review and meta-analysis of placebo-controlled clinical trials. *Seizure*, *45*, 95–99. https://doi.org/10.1016/j.seizure.2016.12.003.

CHAPTER FIVE

Statistical methods in handling placebo effect

Rema Raman*
Alzheimer's Therapeutic Research Institute, University of Southern California, Los Angeles, CA, United States
*Corresponding author: e-mail address: remar@usc.edu

Contents

Abstract

A critical issue facing the therapeutic area of neurological diseases is the large number of failed randomized clinical trials, especially when moving from promising Phase 2 trials to failed Phase 3 trials. A common cited reason for these failures is a high placebo response rate that thereby reduces the observed treatment effect. Explanations for this higher than anticipated placebo response include small sample sizes, inadequate study designs and/or analytic methods, baseline characteristics of the trial sample, possible investigator bias and a participant's own expectations and conditional learning. Several innovative study designs and new methodological approaches to statistical analyses have been proposed to handle placebo effects anticipated or observed in double blind, randomized clinical trials (RCT's). This chapter examines current study designs being used to reduce the observed placebo response and statistical analysis methods being employed for addressing this problem in neuroscience clinical trials.

International Review of Neurobiology, Volume 153
ISSN 0074-7742
https://doi.org/10.1016/bs.irn.2020.04.004

1. Introduction

The double blind, placebo-controlled, randomized clinical trial (RCT), the gold standard in trial design, continues to be the trial design of choice to study the safety and efficacy of a new intervention in the area of neurosciences therapeutic research. This therapeutic area, however, has faced a large number of failed and negative clinical trials resulting in a dearth of available treatments. Several reasons have been attributed for these failures, including untested mechanism of action and use of biomarkers that do not reliably predict clinical outcomes. Another common reason cited is the excessively high placebo response observed in these trials across multiple therapeutic areas in neurosciences, especially pain, Parkinson's disease and depression (Fournier et al., 2010; Freeman et al., 1999; Khan, Schwartz, Kolts, Ridgway, & Lineberry, 2007; Quessy & Rowbotham, 2008; Stolk, Ten Berg, Hemels, & Einarson, 2003; Tarpey & Petkova, 2010; Turner, Deyo, Loeser, Von Korff, & Fordyce, 1994; Tuttle et al., 2015; Walsh, Seidman, Sysko, & Gould, 2002), with the observed placebo response being heterogeneous in some instances (Enck, Klosterhalfen, Weimer, Horing, & Zipfel, 2011). Explanations for this include increased trial length, the use of patient reported outcomes (PRO), changing definitions of disease, subjective and noisy endpoints, site differences and publication bias.

This chapter will focus on the novel approaches in study design and statistical analysis that have been proposed to improve the overall efficiency and effectiveness of RCT's. The first part of the chapter will describe the study designs commonly proposed or utilized within this therapeutic area with the primary goal of minimizing or understanding the placebo response through increasing power and lowering sample size. The second part of this chapter will provide an overview of the various statistical methods being applied to clinical trial data to permit an accurate quantification of placebo and active treatment effects, and ultimately obtain an assessment of the "true" active treatment effect. The chapter will end with practical considerations on addressing this issue when designing, implementing and analyzing neurosciences clinical trials.

2. Trial designs

One area of substantial research has been in studying the specific study design elements that impact the placebo rates and modifying and developing trial designs that maximize the ability to detect a treatment effect with the smallest sample size. These fall into two main categories: (a) study designs that eliminate or reduce placebo responses, and (b) study designs that permit estimation of placebo effects.

2.1 Designs to eliminate and/or reduce placebo response

Several aspects of a double blind, placebo-control RCT study design impact placebo response and have been evaluated by researchers to address this problem. They include:

- Allocation Ratios: Unbalanced allocation ratios in favor of the active drug have been employed in several studies to facilitate recruitment. However, any deviation from balanced allocation ratios has been shown to result in an increase in placebo response rate for a trial and introduce allocation bias (Enck et al., 2011; Enck, Junne, Klosterhalfen, Zipfel, & Martens, 2010). A balanced randomization allocation ratio permits the most optimal assessment of treatment effects and ensures maximum power for the study (Mallinckrodt, Tamura, & Tanaka, 2011).
- Maintenance of blind: Effective maintenance of the study blind at the participant and the site level is another factor that substantially improves the ability to detect the efficacy signal in a trial. This can be assessed by collecting and evaluating a participants' and site personnel's belief of assigned intervention at the end of a study.
- Control of center effects: Standardizing recruitment and implementing a randomization approach that stratifies by study site could reduce placebo response.
- New study designs: Several alternatives and modifications to the RCT design to eliminate the placebo response during the study or permit robust estimation of placebo effects have been proposed (Table 1). They include enrichment designs to identify and exclude individuals who do not demonstrate a placebo response, and thereby reduce the placebo effect.

2.2 Crossover study design

Crossover designs assign each participant to the active drug arm and the placebo arm in a blinded manner with a washout period in between, with the order of the assignment being random to prevent order effects. A schema of this design is provided in Fig. 1. This design has several advantages over the parallel arm design. Since each participant serves as his/her own control,

Table 1 Common clinical trial designs to control placebo effects.

Designs to reduce placebo response
Crossover instead of parallel groups design
Enrichment design—Placebo run-in phase prior to randomization
Enrichment design—Randomized withdrawal or Active run-in phase prior to randomization
Enrichment design—Sequential Parallel Comparison Design (SPCD)
Two-way Enriched Design (TED)
Designs to estimate placebo effects
Delayed-start or randomized-start
Balanced placebo design and Balanced cross-over design
Hidden treatment
No treatment control arm

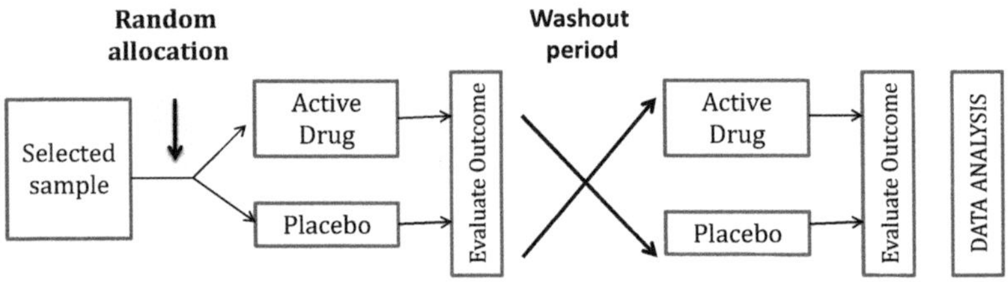

Fig. 1 Crossover study design: Participants are assigned to both the active drug and the placebo with a washout period in between. Order of assignment is randomized.

issues around confounding and placebo effects are eliminated. Sample sizes for such designs are smaller than parallel arm designs and recruitment is helped by the fact that each participant receives both interventions. The presence of carryover and learning effects as well as the loss of power even with minimal dropout make this a design that is not considered viable in many instances.

2.3 Placebo run-in and randomized withdrawal designs

A characteristic of an enrichment design is the use of a run-in period, which is a planned time period from formal patient enrollment in a trial to the point of randomization that enables exclusion of certain participants. In such trials, randomization takes place days or weeks after formal enrollment. During this post-inclusion pre-randomization period, all patients receive the same treatment, for example, a placebo (placebo run-in) or the experimental drug ("active run-in" or randomized withdrawal). The purpose of this run-in

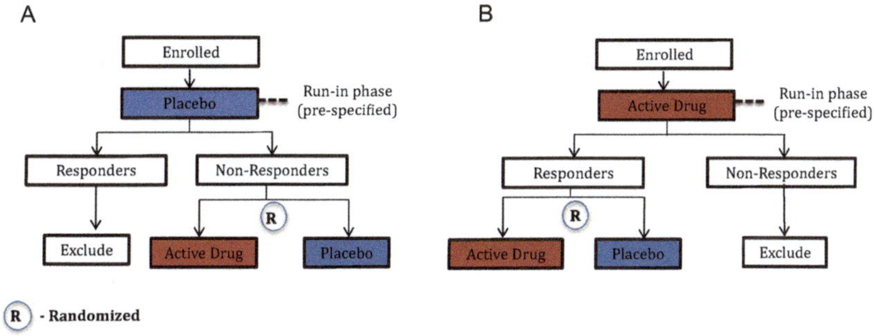

Fig. 2 Run-in study design. (A) Placebo run-in: Participants determined to be placebo responders are excluded from the study, while non-responders are randomized to receive drug or placebo. (B) Randomized withdrawal (active run-in): Participants determined to be responders are randomized to receive drug or placebo, while non-responders are excluded from the study.

phase is to identify the participants most likely to show a treatment effect. The schema for this design is shown in Fig. 2. Advantages of this design include potentially smaller trials with the potential for obtaining maximum treatment effect. However, these designs could result in the need for large number of pre-screens, possible ethical concerns over the binary definition of a responder, possible selection bias, as well as logistical issues if the outcome of interest is not accurately or easily measurable. Additionally, this design will not prevent the possibility of those initial non-responders receiving placebo who may respond at a later time in the study since a placebo response can occur at any stage of the trial. The power of the study is based on the enriched population randomized and may impact the generalizability of the results (Dunlop et al., 2012). Both the U.S. Department of Health and Human Services Food and Drug Administration Center for Drug Evaluation and Research (CDER), Center for Biologics Evaluation and Research (CBER) (2019) and the European Regulatory agency (Food and Drug Administration, HHS, 2001) in their new guidance document on enrichment designs favor the randomized withdrawal design.

2.4 Sequential parallel comparison design (SPCD)

This design, initially proposed by Fava, Evins, Dorer, & Schoenfeld (2003), is an enrichment study design, which attempts to combine the placebo run-in and the randomized withdrawal design. The Sequential Parallel Comparison Design (SPCD) consists of two phases with two points of randomization (Fig. 3). At the first stage, participants are randomized to

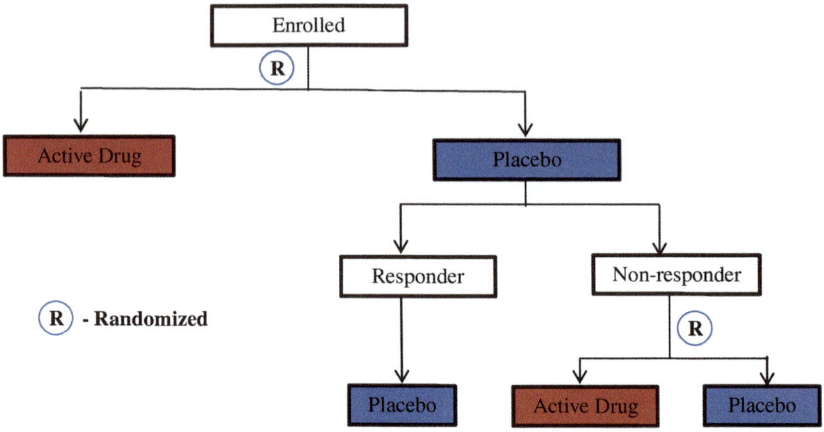

Fig. 3 Sequential parallel comparison design re-randomization format. Efficacy analysis is based on data from both stages. *Fava et al. Psychother Psychosom. 2003 May-Jun;72 (3):115–27.*

the active drug or placebo group. Individuals randomized to the placebo group are grouped into responders (based on the observed response) or non-responders. During the second stage of the design, the placebo non-responders are re-randomized to receive active drug or placebo. The data from the two phases are pooled and analyzed to estimate the average treatment effect and an overall *P*-value (Ivanova, Qaqish, & Schoenfeld, 2011), excluding the placebo non-responders and using weighing for the two phases. The authors of these designs believe that when compared to the conventional two arm clinical trial, the SPCD reduces the sample size by 20–50% under a wide range of parameters, while maintaining the same power and that the placebo response rates are lower in stage 2 than in stage 1 (Chen, Zhang, Tamura, & Chen, 2014; Ivanova & Tamura, 2015). The intellectual property rights to this design are now under the ownership of a private company, which could impact its use for a trial. Carryover effects are likely to play a role in these designs; however, authors believe that the assumption of no carryover effect is reasonable given that the re-randomization only occurs in those individuals who did not respond to placebo (Baer & Ivanova, 2013). One of the biggest challenges with this design is the use of Phase 1 data (a data-driven approach) to define placebo response as a binary response.

2.5 Two-way enriched design (TED)

This design (Fig. 4) combines the characteristics of the run-in design and the SPCD (Ivanova & Tamura, 2015). Participants are initially randomized to

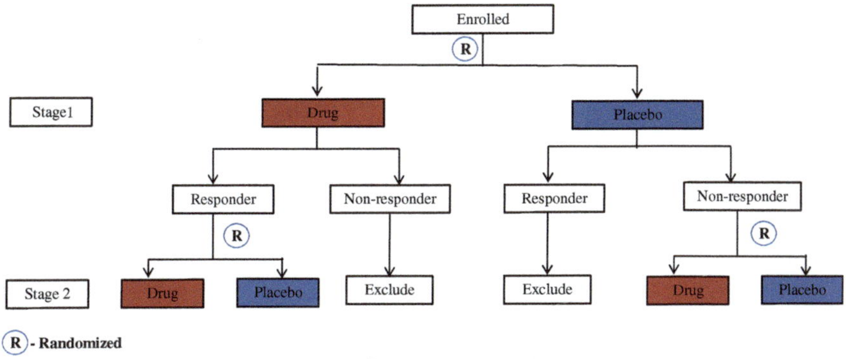

Fig. 4 Two-way enriched design. Participants are initially randomized to drug or placebo. During the second stage, placebo non-responders and drug responders are re-randomized to drug and placebo groups.

active drug or placebo. At the second stage of this design, active drug responders and placebo non-responders are re-randomized to active drug or placebo. The active drug non-responders and the placebo responders are not included in the final efficacy analysis. The authors posit that this design most closely resembles clinical practice since it maximizes participant benefit. This design has the potential to decrease sample sizes due to increased power. However, crossover effects continue to be an issue and this design could result in lengthy and expensive trials.

2.6 Designs to estimate placebo response

2.6.1 Delayed-start or randomized-start designs

The delayed start design is illustrated in Fig. 5. In this study design, participants are first randomized to investigational drug or placebo until the end of the double-blind phase (Stage 1). Then the participants in the placebo arm are rolled over to the active arm for an extended period of time called the open-label extension (OLE) phase (D'Agostino, 2009). The use of this study design may show whether an observed drug effect is disease modifying or symptomatic with positive results in both stages providing strong evidence of efficacy. Sample size estimation for this design requires assumptions regarding the observable treatment effect of time and therefore results in the need for larger sample sizes than a traditional trial design. Longer studies and adverse impact on estimates due to attrition are additional drawbacks to this design.

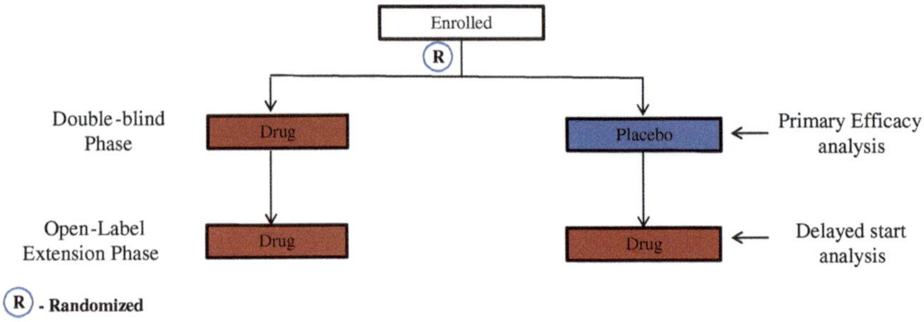

Fig. 5 Delayed start/randomized start design. The participants are randomly assigned to drug or placebo and followed for a period of time (double-blind phase). Then, the participants in the placebo group switch to receive the active drug for another period of time (open-label extension phase) until the end of the trial. Data from both phases are pooled for the delayed start analysis.

2.6.2 Balanced placebo design (BPD) and balanced cross-over design (BCD)

Unlike the enrichment designs that reduce placebo effects, these designs permit estimation of placebo effects by manipulating the information provided about the treatment assignment rather than withholding it. In the Balanced Placebo Design (Rohsenow & Marlatt, 1981), participants are randomized to active drug or placebo treatment. But this design orthogonally manipulates the instructions (told whether they have received active drug or placebo) and the intervention (receive active drug or placebo). Two major advantages of this design include (a) estimation of true placebo and active drug effect that is the based on the outcome of the placebo participants told that they have received active drug and outcome of active drug participants told that they have received placebo, (b) direct testing of the additivity assumption, namely, the treatment effect and placebo effect are entirely independent of each other. A major drawback of this design is the ability to truly unblind the trial. The Balanced Cross-Over Design (BCD) attempts to address this issue by including a cross-over element to the design, whereby each participant receives both active drug and placebo. The BCD suffers from the carryover effect observed in cross-over trials. Both study designs are not used as much due to ethical concerns due to deception of study participants.

2.6.3 Hidden treatment

This study design was proposed to address the ethical concerns raised by the balanced placebo design (Colloca, Lopiano, Lanotte, & Benedetti, 2004).

In this design, participants will be informed that they will receive both drug and placebo but will be not be informed of the timing and dose.

2.6.4 No treatment control arm

To estimate the true placebo effect in a trial, it is necessary to subtract all non-specific effects from the observed placebo response. Traditional RCT's do not permit the identification and quantification of the non-specific effect from the effect of simulation of the treatment. One proposed approach to permit the isolation of these effects is the use of third arm in a trial, namely, the no treatment control arm. The true placebo effect can then be calculated as the response between the no treatment control arm and the placebo arm. However, this design has been criticized for several practical and ethical issues it raises (Hróbjartsson, 2002), namely, the lack of blinding at the site and at the participant level and randomizing individuals to an intervention arm with no benefit. Alternative approaches include participant registries and trial-ready cohorts where participants are followed prospectively as part of a rigorous natural history study. The use of historical controls to replace a concurrent controls have also been proposed with the U. S. Food and Drug Administration (2019) issuing regulatory guidance for such designs in its guidance for rare diseases and The International Conference on Harmonization (ICH) E10 guidelines on choice of control group in clinical trials (2010).

3. Analytical approaches

Another area of research evaluating and addressing uncontrolled and varying levels of placebo response in RCT's are novel statistical methods. One set of methods attempt to estimate these placebo effects within a study and uses statistical models to control for these effects when estimating treatment effects. The other set of methods identify participants as responders/non-responders (binary classification) or assume that placebo response falls in a continuous continuum within an individual and should be treated as a latent variable. A detailed, excellent, discussion of these analytical approaches is provided in a recent article by Kessels, Mozer, & Bloemers (2019). A very brief summary is presented below.

3.1 Controlling for placebo effects

In RCT's, the use of blinding in study design reduces the influence of placebo effects. However, with some types of interventions, when a substantial placebo effect is expected, measuring and incorporating that effect in the statistical model when estimating the treatment effect should be considered.

3.1.1 Joint causal models

One approach suggested in the literature to separate the placebo and treatment effects when estimating the treatment effect in and RCT is the use of a joint causal model that includes the individual's randomized treatment (active, placebo) as well as their "treatmentality," which is a measure of the individual's belief about the treatment they have received (Zhang, Kotz, Wang, Ruan, & Ho, 2013). This method uses blinded assessment data collected in the trial, assumes that this assumption is binary (that is, the individual believe they have received active intervention or not) and argues that the causal treatment effect will be confounded by the individual's beliefs regarding the treatment received. As such, this approach seeks to deal with this measure of confounding by adapting the classic causal model (Rubin, 1974) to incorporate both placebo and treatment specific effects. Different approaches to measuring "treatmentality" were proposed by other authors. These include using treatment-related adverse events as a surrogate measure (Jamshidian, Hubbard, & Jewell, 2011) and randomization ratio (Roy, 2012). There are two main drawbacks to these approaches, namely, (a) the influence of site personnel's beliefs about participant's treatment assignment which could in turn influence the participant's own belief and the subjective nature of these belief's, and (b) the "don't know" response that do not provide clear information and could complicate the interpretation of the results.

3.1.2 Addressing site-specific placebo effects

In RCT's, investigator bias, especially if unintentional and unobserved, can affect the perceived response to treatment due to interactions between the participant and the site's study personnel (Rosenthal & Fode, 1963). In multi-site trials, such biases can lead to differential placebo responses across and within study sites when the outcome of interest is a subjective assessment such as pain or cognition. Optimistic staff and raters at a site could result in larger observed placebo effects compared to the other sites in a trial. To address this issue, Gomeni, Goyal, Bressolle, and Fava (2015), proposed a statistical approach that would incorporate site-specific weights corresponding to the observed placebo effects for a site. This model assumes that data from sites with high placebo response is less informative to assess overall treatment effect compared to the sites with low to no placebo effect.

This approach extends the traditional mixed-model repeated measures (MMRM) model used for longitudinal data to a non-linear longitudinal model (NLMMRM) that incorporates a performance level for each study site with a corresponding weight. Sites with higher levels of placebo response are given smaller weights to reduce the impact of the data from that site to the overall estimated treatment effect. This approach relies on the use of an appropriate pre-specified longitudinal regression model, rigorous and repeatable evaluation of performance level and corresponding weights and sufficiently large sample sizes at each study site.

3.1.3 Use of latent variable modeling to account for measurement error

Some researchers assume that a placebo response is simply a form of measurement error (Fava et al., 2003) and have used latent variable models to estimate the placebo and treatment effects (Eickhoff, 2008). Measurement error refers to the difference between the true value of an outcome and what is observed, while latent variable modeling is a multivariate technique to relate the observed endpoints to a set of under latent characteristics that cannot be observed directly. In this approach proposed by Eickhoff with Quality of Life (QoL) as the outcome of interest, QoL sub-scales are represented as latent variables and the placebo effect is incorporated as a misclassification parameter to obtain "placebo effect-adjusted" estimates.

3.2 Identify placebo responders (as a binary or continuous characteristic)

Another approach when attempting to control for placebo response is to characterize participants in a trial to their perceived placebo response. This can be done in two ways. One approach is to categorize individuals as placebo responders or non-responders and estimate the treatment effects within these two sub-groups. The second approach is to consider an individual's placebo response to be a continuous characteristic and model placebo response as a continuous distribution.

3.2.1 Deterministic models

In deterministic models, the output of the model is fully and uniquely determined by the parameters in the model and the initial inputs of these parameters. These models do not directly account for uncertainly or random variation and can therefore always be replicated for the same set of initial conditions.

This class of methods is based on a pre-specified rule using a cut-point to classify individuals randomized to the placebo group in an RCT as placebo responders or non-responders, thereby creating a binary classification of placebo response. Then, using covariate data from the placebo group, statistical classification methods such as logistic regression, Fisher's Linear Discriminant Analysis (LDA) or receiver operating characteristic (ROC) analyses are used to classify participants randomized to the treatment group or to use in a new trial. Examples of such rules include the use of a 50% reduction in symptoms as a criterion of a clinical response, a HAM-D score below 8 as the criterion of remission, and a change in a participant reported wellness outcomes such as quality of life (QoL) scores over time. This approach, while easy to understand and implement, has several major limitations. The decision rule used to define a placebo responder is subjective since it is determined by the study team and can be fairly arbitrary resulting in possible misclassification. Classification models also require extensive independent, external validation before one can be confident in their results.

3.2.2 Stochastic models

Stochastic models, unlike deterministic models, are models that estimate probability distributions of outcomes by incorporating inherent randomness in one or more inputs over time. As such, the same inputs and initial parameter values will result in different outputs.

This class of models assumes that the placebo response is a latent characteristic and can therefore be considered a random variable. Such statistical models, called mixture models, account for the heterogeneity observed in the trajectories for placebo responders and non-responders and allow for the assessment of the effect of treatment within these groups. Examples of these models include latent class models (Nagin, 1999), random effects mixture model (Xu & Hedeker, 2001), growth mixture models (Muthén & Asparouhov, 2009; Muthén, Brown, Leuchter, & Hunter, 2008; Muthén & Shedden, 1999) when considering placebo response as discrete and unobserved latent classes (responders/non-responders or responders/partial responders/non-responders) and infinite mixture models (Tarpey & Petkova, 2010) when modeling placebo response as a continuous latent variable. A drawback of these approaches is the assumptions and decisions made in the treatment effect estimation process within this mixture model framework since the most common practice estimates multiple models by specifying a different number of latent classes and then makes a decision on which model is "best," using commonly used criteria such as the Akaike information criterion (AIC) and the Bayes factor and/or the Bayesian information criterion (BIC).

4. Practical considerations and recommendations

Designing and conducting an RCT when anticipating higher than normal placebo response rates requires substantial upfront planning and critical collaboration among the subject matter experts (clinicians, neuropsychologists, biostatisticians) of the study scientific team. Each of the three facets of clinical trial development, namely, trial design, execution, and management, offer opportunities to improve study efficiency and minimize the impact of a placebo response. Sections 2 and 3 in this chapter provide an overview of some of the study design and/or statistical analysis approaches that can be considered. However, these considerations should ideally be made while the study is being designed and the final choice of design, sample size and analysis approach should account for feasibility (time, funding, resources, practical or ethical constraints, access to participants), statistical power (minimal clinically important effect size, variability and noise of outcome variables, amount and impact of non-compliance, missing data and attrition, and overall experiment wise Type I error rate) and approval from the appropriate regulatory agency (if applicable).

Table 2 provides additional details on contributors of placebo effect in RCT's and actionable recommendations to consider when designing and conducting such trials.

4.1 Study design

Many elements of the trial's study design play a big part as potential contributors to a placebo effect. Careful consideration of key elements of the study design with all the relevant subject matter experts is key to mitigating placebo effect. While there are statistical analysis approaches to deal with the issue once the trial is completed, the study has maximum power when determining the best statistical analysis procedure when designing the study. The use of randomized, double-blind study design will protect against selection and allocation bias. A recruitment and retention plan while developing the trial protocol will ensure that the right target population is being enrolled at appropriate sites. Additional recommendations include identifying a small number of rating scales as study outcomes, each with good psychometric

Table 2 Trial characteristics that possibly contribute to the placebo effect and actionable recommendations to mitigate it.

Topic	Issues to consider
Study design	• *Comparator treatment*: Standard of care or placebo. Randomizing to a no treatment arm is hard to justify for ethical reasons. • *Hypothesis*: Superiority trials with two-sided alternative hypothesis • *Schedule of assessments*: Focus on minimizing participant and site burden/exhaustion • *Study duration*: Long enough to eliminate learning effects and address hypothesis • *Appropriate endpoints*: Combine Participant Reported Outcomes (PRO's) with more objective outcomes; Instruments with good psychometric properties for the specific stage of disease • *Eligibility criteria*: Individuals at the right stage of the disease to reduce regression to the mean • *Sample size*: Adequate sample size based on endpoint sensitivity and variability in specific population of interest • *Randomization*: Use equal allocation ratios for confirmatory trials. Conduct randomization centrally via an automated system • *Blinding*: Keep investigators and participants blinded to key cutoff values used in entry criteria as well as randomization assignment
Study conduct	• *Expectation bias*: Consider participant and rater bias. Have different raters for different assessments and at sequential visits to minimize rater-participant relationships; Separate raters from rest of clinical staff • *Data and study management*: High-quality data as close to real time; visits conducted within visit window • *Rater training*: Rater training and certification to improve quality and consistency of test administration • *Statistical monitoring*: Centralized statistical monitoring to identify outlier and inlier sites with respect to recruitment strategies, rater training, variability in ratings, and data quality • *Attrition*: Monitor and minimize item-level and instrument level attrition • *Effectiveness of the blind*: Assess during and at the end of a trial
Statistical analysis	• *Statistical analysis plan (SAP)*: Comprehensive to minimize potential data driven analysis and maintain overall Type I error; Specify and finalize instrument scoring algorithm prior to study start to inform database and SAP development. Collaboration between trial biostatistician, clinician, and neuropsychologist • *Pre-specified subgroup analysis*: Consider all sub-groups of interest including responder analysis • *Interpretation of final results*: Clinical and Statistical importance; interpret results in totality (primary, secondary and sub-group); focus on estimates and confidence intervals instead of *P*-values
Data sharing	• *Open data sharing*: Share study data as openly and quickly as possible to facilitate meta-analysis and other sensitivity analysis; Use data standards • *Scientific collaborations*: Facilitates additional exploration and understanding of the role of placebo response with increased power

properties and ease of use. Developing a study schedule to minimize participant and site personnel fatigue is critical to reducing non-response and attrition through fatigue. Sample size estimation should be based on well-justified assumptions within the specific therapeutic area, account for anticipated placebo effect, and be scaled up for attrition, dropout and non-compliance.

4.2 Study conduct

This portion of a clinical trial, although frequently overlooked, is a crucial component for early identification of placebo effect. When identified early, adjustments to design and conduct can be made with minimal impact on study's statistical power. Data and study monitoring through centralized statistical monitoring, smart data collection and entry, protocol training, comprehensive quality assurance and rater certification and re-training (if needed) permits evaluation of study conduct across sites in real time and allows sites who are performing as inliers or outliers on any of these measures to be identified for remedial action.

4.3 Statistical analysis plan

This is one of the central documents in a trial, that, in conjunction with the trial protocol, permits transparency and reproducibility of a trial's statistical conduct, and, by extension, a trial's conclusions. A statistical analysis plan provides a detailed and comprehensive description of the trial's analysis and is finalized prior to final database lock. Assumptions of primary analysis approach along with pre-specified alternatives if assumptions are not met, along with pre-specified sub-group analysis (compliers analysis, responders analysis) are clearly described in the plan to prevent post-hoc analyses. Final interpretation of results should consider clinical and statistical relevance and be combined with other secondary outcomes of interest, trends in pre-specified sub-group analysis and safety signals. Guidelines to develop an SAP has recently been published by Gamble et al. (2017).

4.4 Data sharing and collaboration

Scientists in the United States are increasingly advocating responsible and open data sharing as a way to advance science through increased and improved scientific collaborations and improved transparency and rigor (Kuntz et al., 2019; Weninger et al., 2016). Use of therapeutic area data standards, data harmonization approaches and sharing of programs and approaches will permit robust meta-analysis, increase statistical power for post-hoc subgroup analysis and permit more precise assessment of placebo response in neurosciences clinical trials.

5. Summary

This chapter summarizes some of the common strategies being proposed to address the issue of high placebo response through novel approaches to study design and statistical methodologies to provide guidance in designing and analyzing clinical trials data in neurosciences in the presence of a larger than expected placebo effect. Every approach has its own advantages and disadvantages when addressing this issue and the choice of study design and statistical analysis for a trial should be made as part of a scientific collaborative effort among the subject matter experts, including but not limited to the lead trial clinician, statistician and neuropsychologist. In addition, particular focus and attention should be paid to study conduct to identify and potentially mitigate potential causes of placebo effect early in a trial. Additionally, results from all primary and secondary analyses pre-specified in a statistical analysis plan must be considered when interpreting the results and making conclusion about the effectiveness of an intervention. Finally, seeking guidance from the regulatory agencies on the study protocol and the statistical analysis plan will provide an understanding of the agency's current thinking about an intervention or approach.

References

Baer, L., & Ivanova, A. (2013). When should the sequential parallel comparison design be used in clinical trials? *Clinical Investigation*, *3*, 823–833.

Chen, Y. F., Zhang, X., Tamura, R. N., & Chen, C. M. (2014). A sequential enriched design for target patient population in psychiatric clinical trials. *Statistics in Medicine*, *33*, 2953–2967.

Colloca, L., Lopiano, L., Lanotte, M., & Benedetti, F. (2004). Overt versus covert treatment for pain, anxiety, and Parkinson's disease. *Lancet Neurology*, *3*, 679–684.

D'Agostino, R. B. (2009). The delayed-start design. *New England Journal of Medicine*, *361*, 1304–1306.

Dunlop, B. W., Thase, M. E., Wun, C. C., Favvad, R., Guico-Pabia, C. J., Musgnung, J., et al. (2012). A meta-analysis of factors impacting detection of antidepressant efficacy in clinical trials: The importance of academic sites. *Neuropsychopharmacology*, *37*, 2830–2836.

Eickhoff, J. C. (2008). Placebo effect-adjusted assessment of quality of life in placebo-controlled clinical trials. *Statistics in Medicine*, *3027*(9), 1387–1402.

Enck, P., Junne, F., Klosterhalfen, S., Zipfel, S., & Martens, U. (2010). Therapy options in irritable bowel syndrome. *European Journal of Gastroenterology and Hepatology*, *22*, 1402–1411.

Enck, P., Klosterhalfen, S., Weimer, K., Horing, B., & Zipfel, S. (2011). The placebo response in clinical trials: More questions than answers. *Philosophical Transactions of the Royal Society, B: Biological Sciences, 366*, 1889–1895.

Fava, M., Evins, A. E., Dorer, D. J., & Schoenfeld, D. A. (2003). The problem of the placebo response in clinical trials for psychiatric disorders: Culprits, possible remedies, and a novel study design approach. *Psychotherapy and Psychosomatics, 72*, 115–127.

Food and Drug Administration, HHS. (2001). International conference on harmonisation; choice of control group and related issues in clinical trials. *Federal Registery, 66*(93), 24390–24391.

Fournier, J. C., DeRubeis, R. J., Hollon, S. D., Dimidjian, S., Amsterdam, J. D., Shelton, R. C., et al. (2010). Antidepressant drug effects and depression severity: A patient level meta-analysis. *Journal of the American Medical Association, 303*, 47–53.

Freeman, T. B., Vawter, D. E., Leaverton, P. E., Godbold, J. H., Hauser, R. A., Goetz, C. G., et al. (1999). Use of placebo surgery in controlled trials of a cellular-based therapy for Parkinson's disease. *New England Journal of Medicine, 341*, 988–992.

Gamble, C., Krishan, A., Stocken, D., Lewis, S., Juszczak, E., Dore, C., et al. (2017). Guidelines for the content of statistical analysis plans in clinical trials. *Journal of the American Medical Association, 318*(23), 2337–2343.

Gomeni, R., Goyal, N., Bressolle, F., & Fava, M. (2015). A novel methodology to estimate the treatment effect in presence of highly variable placebo response. *Neuropsychopharmacology, 40*(11), 2588–2595.

Hróbjartsson, A. (2002). What are the main methodological problems in the estimation of placebo effects? *Journal of Clinical Epidemiology, 55*, 430–435.

Ivanova, A., Qaqish, B., & Schoenfeld, D. A. (2011). Optimality, sample size, and power calculations for the sequential parallel comparison design. *Statistics in Medicine, 30*, 2793–2803.

Ivanova, A., & Tamura, R. N. (2015). A two-way enriched clinical trial design: Combining advantages of placebo lead in and randomized withdrawal. *Statistical Methods in Medical Research, 24*, 871–890.

Jamshidian, F., Hubbard, A. E., & Jewell, N. P. (2011). Accounting for perception, placebo and unmasking effects in estimating treatment effects in randomized clinical trials. *Statistical Methods in Medical Research, 23*, 293–307.

Kessels, R., Mozer, R., & Bloemers, J. (2019). Methods for assessing and controlling placebo effects. *Statistical Methods in Medical Research, 28*(4), 1141–1156.

Khan, A., Schwartz, K., Kolts, R. L., Ridgway, D., & Lineberry, C. (2007). Relationship between depression severity entry criteria and antidepressant clinical trial outcomes. *Biological Psychiatry, 62*, 65–71.

Kuntz, R. E., Antman, E. M., Califf, R. M., Ingelfinger, J. R., Krumholz, H. M., Ommaya, A., et al. (2019). *Individual patient-level data sharing for continuous learning: A strategy for trial data sharing. NAM Perspectives*. Discussion Paper Washington, DC: National Academy of Medicine.

Mallinckrodt, C. H., Tamura, R. N., & Tanaka, Y. (2011). Recent developments in improving signal detection and reducing placebo response in psychiatric clinical trials. *Journal of Psychiatric Research, 45*, 1202–1207.

Muthén, B., & Asparouhov, T. (2009). Growth mixture modeling: Analysis with nongaussian random effects. In G. Fitzmaurice, M. Davidian, G. Verbeke, & G. Molenberghs (Eds.), *Longitudinal data analysis* (pp. 143–165). Boca Raton, FL: Chapman & Hall/CRC Press.

Muthén, B., Brown, H., Leuchter, A., & Hunter, A. (2008). General approaches to analysis of course: Applying growth mixture modeling to randomized trials of depression medication. In P. E. Shrout (Ed.), *Causality and psychopathology: Finding the determinants of disorders and their cures.* Washington, DC: American Psychiatric Publishing.

Muthén, B., & Shedden, K. (1999). Finite mixture modeling with mixture outcomes using the EM algorithm. *Biometrics, 55*(2), 463–469.

Nagin, D. S. (1999). Analyzing developmental trajectories: A semi-parametric, group-based approach. *Psychological Methods, 4*, 139–157.

Quessy, S. N., & Rowbotham, M. C. (2008). Placebo response in neuropathic pain trials. *Pain, 138*, 479–483.

Rohsenow, D. J., & Marlatt, G. A. (1981). The balanced placebo design: Methodological considerations. *Addictive Behaviors, 6*(2), 107–122.

Rosenthal, R., & Fode, K. L. (1963). The effect of experimenter bias on the performance of the albino rat. *Behavioral Science, 8*(3), 183–189.

Roy, J. (2012). Randomized treatment-belief trials. *Contemporary Clinical Trials, 33*, 172–177.

Rubin, D. B. (1974). Estimating causal effects of treatments in randomized and non-randomized studies. *Journal of Educational Psychology, 66*, 688–701.

Stolk, P., Ten Berg, M. J., Hemels, M. E. H., & Einarson, T. R. (2003). Meta-analysis of placebo rates in major depressive disorder trials. *Annals of Pharmacotherapy, 37*, 1891–1899.

Tarpey, T., & Petkova, E. (2010). Modelling placebo response via infinite mixtures. *JP Journal of Biostatistics, 4*(2), 161–179.

Turner, J. A., Deyo, R. A., Loeser, J. D., Von Korff, M., & Fordyce, W. E. (1994). The importance of placebo effects in pain treatment and research. *Journal of the American Medical Association, 271*, 1609–1614.

Tuttle, A. H., Tohyama, S., Ramsay, T., Kimmelman, J., Schweinhardt, P., Bennett, G. J., et al. (2015). Increasing placebo responses over time in U.S. clinical trials of neuropathic pain. *Pain, 156*(12), 2616–2626.

U.S. Department of Health and Human Services Food and Drug Administration Center for Drug Evaluation and Research (CDER) Center for Biologics Evaluation and Research (CBER) Guidance for Industry. (2019). *Enrichment strategies for clinical trials to support determination of effectiveness of human drugs and biological products guidance for industry. Retrieved from https://www.fda.gov/media/121320/download.*

Walsh, B. T., Seidman, S. N., Sysko, R., & Gould, M. (2002). Placebo response in studies of major depression: Variable, substantial, and growing. *Journal of the American Medical Association, 287*, 1840–1847.

Weninger, S., Carrillo, M. C., Dunn, B., Aisen, P. S., Bateman, R. J., Kotz, J. D., et al. (2016). Collaboration for Alzheimer's prevention: Principles to guide data and sample sharing in preclinical Alzheimer's disease trials. *Alzheimer's & Dementia, 12*(5), 631–632.

Xu, W., & Hedeker, D. (2001). A random-effects mixture model for classifying treatment response in longitudinal clinical trials. *Journal of Biopharmaceutical Statistics, 11*, 253–273.

Zhang, Z., Kotz, R. M., Wang, C., Ruan, S., & Ho, M. (2013). A causal model for joint evaluation of placebo and treatment-specific effects in clinical trials. *Biometrics, 69*(2), 318–327.

CHAPTER SIX

Nocebo and lessebo effects

Tiago A. Mestre[a,b,c,*]
[a]University of Ottawa Brain and Mind Research Institute, Ottawa, ON, Canada
[b]The Ottawa Hospital Research Institute, Ottawa, ON, Canada
[c]Division of Neurology, Department of Medicine, The Ottawa Hospital, Ottawa, ON, Canada
*Corresponding author: e-mail address: tmestre@toh.ca

Contents

Abstract

The power of placebos is commonly associated with the placebo effect. In contrast, detrimental effects related to the use of a placebo are little studied and less well recognized. This chapter covers the nocebo and lessebo effects defined, respectively, as expectation of harm in the form of adverse events in a placebo arm and reduction of therapeutic benefit due to the uncertainty of being allocated to placebo. The lessebo effect is a more recent concept and has been described only in depression, schizophrenia and Parkinson's disease. The nocebo response was evaluated in many neurological diseases, including epilepsy, multiple sclerosis, Parkinson's disease, Alzheimer's disease, restless leg syndrome, among others. Meta-analyses of randomized controlled trials in these conditions reveal a significant variability of the magnitude of the nocebo response and that factors related to study design, study participants or neurological disease can be associated with a nocebo response, although with the opposing findings across conditions. The knowledge about neurobiological mechanisms of the nocebo effect is poor for neurological diseases, and most of the information has been generated in pain. Functional neuroimaging suggests the existence of a distinct network for the anticipation and the experience of a hyperalgesia nocebo response. Different types of neurotransmitters have been involved, including cholecystokinin, dopamine and opioids. Recognizing the potential impact of nocebo and lessebo effects, mitigating strategies are in development with application to clinical research and clinical practice, such as a contextualized informed consent process, alternative study designs and enhancement of patient-physician communication.

International Review of Neurobiology, Volume 153
ISSN 0074-7742
https://doi.org/10.1016/bs.irn.2020.04.005

1. Introduction

The use of a placebo in randomized controlled clinical trials is the gold standard for the assessment of novel therapies. The comparison of efficacy results in the experimental and placebo arms allows differentiating the therapeutic benefit specific to the intervention and nonspecific effects. Nonspecific effects include the placebo effect associated with mechanisms of expectation of benefit and learning conditioning, but also other phenomena like regression to the mean, Hawthorne effect, and response (or survey) bias (Benedetti, Carlino, & Piedimonte, 2016).

The administration of a placebo is far from innocuous and can be associated with harm. In this chapter, the "dark side" of the use of placebos in randomized placebo-controlled trials is presented. The nocebo and lessebo effects are recognized as two negative phenomena associated with the use of placebos. The nocebo effect relates to the occurrence of adverse events (AEs) during exposure to placebo (Benedetti, Lanotte, Lopiano, & Colloca, 2007). The lessebo effect is a more recent concept defined as the reduction in the magnitude of the effect of an experimental intervention due to the possibility of being allocated to placebo (Sinyor et al., 2010).

In this chapter, the prevalence of nocebo and lessebo effects in neurological diseases is provided, followed by the current knowledge about the neurobiology of both nocebo and lessebo effects. Recognizing these deleterious effects, strategies designed to mitigate the nocebo and lessebo effects in clinical research and/or clinical practice are explored.

2. The nocebo effect

The term nocebo in Latin translates in English to "*I shall harm*" in contrast with the concept of placebo ("*I shall please*") effect (Benedetti et al., 2007). The knowledge about the nocebo effect is a first insight on how the use of a placebo in a clinical trial is associated with untoward events, as it is defined by the occurrence of "harmful" AEs in study participants allocated to a placebo arm during a randomized controlled trial. The monitoring of AEs in a placebo arm allows to control for non-specific AEs to the same extent the placebo arm is used to measure a nonspecific therapeutic benefit not attributable to the mechanism of action of an experimental intervention. The AEs documented in a placebo arm may also be associated with the natural history of the disease, comorbidities, random occurrence, or with a true nocebo effect. In fact, the term nocebo response is frequently used to denote the many contributing factors for the occurrence of AEs in a placebo arm. Later in this chapter, we introduce the concept of placebo-associated worsening.

Similar to the placebo effect, the nocebo effect has been explained by two models: expectancy and conditioning (Benedetti et al., 2007). Many studies have shown the role of expectation of harm in a nocebo response. The suggestion of a negative outcome in the form of new deleterious symptoms or clinical worsening of existing symptoms has been associated with AEs after administration of a placebo (Okusogu & Colloca, 2019). A systematic review of anti-migraine drug trials eloquently showed that the AE profile in a placebo arm was not the same across trials, and the observed differences mimicked the disclosed side effect profile of the antimigraine drug being evaluated in a given trial (Amanzio, Corazzini, Vase, & Benedetti, 2009), suggesting that the information about the safety of a drug provided prior to its administration translates into distinct expectations of harm and profiles of AEs in a placebo arm. A complementary observation comes from a systematic review of the nocebo response in anti-epileptic drug trials conducted in epilepsy, and other conditions such as pain, movement disorders, psychiatric disorders, substance abuse, obesity and binge eating disorders (Zaccara, Giovannelli, Giorgi, Franco, & Gasparini, 2016). AEs like ataxia and diplopia were more frequent in epilepsy trials, but other neurological symptoms such as dizziness, somnolence and headache were equally found in the epilepsy and nonepilepsy trials (Zaccara et al., 2016). Although the authors do not control for the distribution of individual antiepileptic drugs in epilepsy and nonepilepsy trials, the results allow to hypothesize that the condition being treated or associated co-morbidities may influence the nature of the nocebo response. Conditioning learning involves the establishment of an association between an inert stimulus (the placebo) and the active drug due to past learning experiences that predispose to the occurrence of an AE related to the active drug when exposed to a placebo (Tracey, 2010). More recently, research has explored the interactions of conditioning learning and expectancy, hypothesizing the role of conditioning in the generation of expectations (Okusogu & Colloca, 2019).

2.1 The nocebo effect in neurological diseases

The prevalence of the nocebo effect in neurological diseases has been evaluated through aggregated data meta-analyses of randomized placebo-controlled clinical trials reporting a pooled nocebo response across studies. Examples of neurological disease with meta-analyses of the nocebo response are epilepsy (Zaccara, Giovannelli, Cincotta, Loiacono, & Verrotti, 2015; Zaccara et al., 2014; Zis, Shafiq, & Mitsikostas, 2017), multiple sclerosis (Gklinos, Papadopoulos, & Mitsikostas, 2019; Papadopoulos & Mitsikostas, 2010), ataxia (Alam, Hadjivassiliou, & Zis, 2019), restless leg syndrome (Silva et al., 2017), Parkinson's disease (Leal Rato et al., 2019; Stathis, Smpiliris, Konitsiotis, & Mitsikostas, 2013), and Alzheimer's disease (Amanzio, Benedetti, & Vase, 2012; Zis & Mitsikostas, 2015). The nocebo effect has also been studied in neurological pain syndromes such as headache (Mitsikostas, Mantonakis, & Chalarakis, 2011), neuropathic pain (Papadopoulos & Mitsikostas, 2012), and in neuromuscular disorders like motor neuron disease (Shafiq, Mitsikostas, & Zis, 2017), myasthenia gravis (Varma & Zis, 2019), diabetic peripheral neuropathy (Hauser, Bartram, Bartram-Wunn, & Tolle, 2012), and chronic inflammatory demyelinating polyneuropathy (Zis, Hadjivassiliou, Sarrigiannis, Jenkins, & Mitsikostas, 2018) (Table 1).

In these studies, the outcome measure most commonly used to measure the nocebo response was the number of patients with the occurrence of at least one AE in a placebo arm (nocebo responder rate). Other outcome measures of a nocebo response include the discontinuation rate due to AEs in a placebo arm, used as *a proxy* for the severity of the nocebo response, and the occurrence of serious AEs. Additional measures of the nocebo response include a comparative evaluation of the profile of AEs and their prevalence in the active treatment groups expressed in the form of odds-ratio or proportion of symptoms not attributable to the pharmacological action (Leal Rato et al., 2019). Meta-analyses of the nocebo response commonly reported a significant degree of heterogeneity across studies which limits the validity of the results and authors caution for the existence of missing data due to inconsistent reporting of AEs. Overall, there is a substandard reporting quality of nocebo response in clinical trials. The access of raw data from clinical trials would permit a more comprehensive and less biased analyses of the nocebo response using meta-analytical approaches (Clarke, 2005).

In these meta-analyses, the number of participants that experienced at least one AEs in a placebo arm was as low as 13.8% in ataxia (Alam et al., 2019) and as high as 80% in neuromuscular conditions such as myasthenia gravis and motor neuron disease (Shafiq et al., 2017; Varma & Zis, 2019). The prevalence of discontinuation rate in a placebo arm due to the AEs

Table 1 Prevalence and magnitude of the nocebo response in neurology. Meta-analyses of randomized placebo-controlled trials are presented of various neurological diseases.

Neurological disease	Number of trials	Type of treatments	Nocebo response	
			Nocebo responder rate (95% CI)	Discontinuation rate due to AEs (95% CI)
Migraine (Mitsikostas et al., 2011)	59	Acute symptomatic	18.5 (14.9–22.2)	0.3 (0.2–0.5)
	31	Prevention	42.8 (34.7–51.4)	4.8 (3.3–6.5)
Cluster headache (Mitsikostas et al., 2011)	3	Acute symptomatic	18.7 (10.6–28.3)	–
Tension-type headache (Mitsikostas et al., 2011)	4	Prevention	24.0 (4.6–52.2)	5.4 (1.3–12.1)
Relapsing-remitting multiple sclerosis (Gklinos et al., 2019; Papadopoulos & Mitsikostas, 2010)	6	Oral DMT*	89 (88–90)	8 (5–12)
	3	Injectable DMT*	66 (51–80)	2 (0–2)
	56	All DMT*	74.4 (69.9–88.3)	2.1 (1.6–2.7)
	44	Symptomatic	25.3 (15.2–36.9)	2.3 (1.5–3.3)
Ataxia (Alam et al., 2019)	6	Pharmacological, various	13.8 (5.4–31.1)	4.8 (2.2–10.0)
Myasthenia gravis (Varma & Zis, 2019)	6	Pharmacological, various	80.1 (75.0–84.5)	2.4 (1.0–5.0)
Neuropathic pain (Papadopoulos & Mitsikostas, 2012)	12	Pharmacological, various	52.0 (35.7–67.9)	6.0 (4.5–8.0)
Diabetic neuropathy (Hauser et al., 2012)	62	Pharmacological, various	46.2 (36.5–56.1)	5.8 (5.1–6.6)
Chronic inflammatory demyelinating neuropathy (Zis et al., 2018)	3	Pharmacological, various	42	2.1 (0.3–7.3)

Continued

Table 1 Prevalence and magnitude of the nocebo response in neurology. Meta-analyses of randomized placebo-controlled trials are presented of various neurological diseases.—cont'd

			Nocebo response	
Neurological disease	**Number of trials**	**Type of treatments**	**Nocebo responder rate (95% CI)**	**Discontinuation rate due to AEs (95% CI)**
Motor Neuron Disease (Shafiq et al., 2017)	12	Pharmacological, various	78.3 (74.3–82.0) $n=4$	8.4 (6.7–10.4) ***
Refractory focal epilepsy	7 (children) (Zaccara et al., 2014) **	Pharmacological, various	81.3 (68.5–94.1)	3.6 (2.1–5.1)
	4 (pre-surgical) (Zis et al., 2017)	Pharmacological, various	60.8 ($n=3$)	3.2 (0.9–8.0)
	79 (Zaccara et al., 2015)	Pharmacological, various	60.3 (54.7–65.8)	3.9 (3.2–4.5)
Alzheimer's disease (Amanzio et al., 2012; Zis & Mitsikostas, 2015)	20	Pharmacological, various	57.8 (50.1–66.7)	6.6 (5.3–8.4)
Restless leg syndrome (Silva et al., 2017)	72	Pharmacological, various	45.4 (40.5–50.3)	2.1 (in 81 studies)
Parkinson's disease (Leal Rato et al., 2019; Stathis et al., 2013)	41	Pharmacological, various	64.7 (53.6–74.4)	8.8 (6.8–11.5)
	148	Pharmacological, various	56.0 (51.7–60.4)	5.7% (5.1–6.4)

*, disease-modifying treatments; **, placebo as add-on treatment; ***, AEs other than death.

ranged from 0.3% in migraine (Mitsikostas et al., 2011) to 12.6% in motor neuron disease (Shafiq et al., 2017). The differences in measures of nocebo response across neurological diseases remains to be established but may be associated with conditions being assessed in clinical trials, different profile of co-morbidities or co-medications, or study design factors.

The meta-analytical evaluation of the nocebo effect in neurological diseases explored the association of various factors with a nocebo response (Table 2). These factors have been divided in population-related (duration and severity of the study condition, age, and gender of participants), study-related (treatment duration, trial phase and study design), and intervention-related (route and frequency of administration). Other factors evaluated for an association with the nocebo response include the year of publication of the study and geographical location of clinical trial sites.

The presence and strength of the association of gender with the nocebo response vary across studies. A higher frequency of AEs in the placebo arms was associated with male gender in trials conducted for neuropathic pain (Papadopoulos & Mitsikostas, 2012), while the reverse association was reported in Alzheimer's disease trials (Zis & Mitsikostas, 2015). The role of age in a nocebo response is also unclear across different conditions. Older age has been correlated with a higher incidence of AEs and AE-related dropouts in placebo arms of Alzheimer's disease trials (Zis & Mitsikostas, 2015), while the opposite was found in diabetic neuropathic pain trials (Hauser et al., 2012).

The characteristics of the neurological disease being assessed were associated with different nocebo responses. In multiple sclerosis trials, trial participants with a progressive pattern, either primary or secondary, had a stronger nocebo response compared with those with a relapsing-remitting pattern (Papadopoulos & Mitsikostas, 2010). In cognitive impairment trials evaluating the acetylcholinesterase inhibitor donepezil, the rate of AEs was higher in participants with Alzheimer's disease compared with mild cognitive impairment. This finding can be justified by the greater frailty of Alzheimer's disease patients or a higher prevalence of somatic comorbidity (Amanzio et al., 2012). Additional factors related to the study population identified in meta-analyses have been as diverse as body mass index and cognitive function measured by the Mini-Mental State Examination (Table 2).

The characteristics of trial medications were associated with different magnitudes of a nocebo response. These data came mostly from multiple sclerosis trials of disease-modifying therapies. In these studies, a higher frequency of administration and an oral (vs. injectable) route of administration

Table 2 Factors associated with a nocebo response in various neurological diseases, using meta-analyses approach. Associations were tested using different measures of a nocebo response.

	Study population	Intervention	Study design	Other
Migraine		Drug class (Amanzio et al., 2009; Mitsikostas et al., 2011)		Year of publication (–) *
				Geographical location (clinical sites) (Loder, Goldstein, & Biondi, 2005)
Neuropathic pain (Papadopoulos & Mitsikostas, 2012)	Female (–)	–	–	Geographical location (clinical sites)
Multiple sclerosis (Papadopoulos & Mitsikostas, 2010)	Progressive forms > relapsing-remitting form	DMT: oral > injectable	Symptomatic treatment: phase II > phase III, parallel > cross-over	Year of publication (+)
		Administration frequency (+)		
Alzheimer's disease (Zis & Mitsikostas, 2015)	Age (+), disease duration (+), male (–), BMI (–), MMSE (–)	Not evaluated	Sample size (–)	Year of publication (–)
Diabetic neuropathy (Hauser et al., 2012)	Age (–), female (–)		Study duration	Year of publication (–)
Focal epilepsy (Zaccara et al., 2015)		Drug dose (+)	Sample size (+), study duration (+)	Year of publication (+) geographical location (clinical sites)

+, positive association or correlation; –, negative association or correlation; *, acute symptomatic; BMI, body mass index; MMSE, mini-mental status examination; DMT, disease-modifying therapy.

for disease-modifying therapies (Gklinos et al., 2019) were associated with a stronger nocebo response. In contrast, there was no difference in the nocebo dropout rate between subcutaneous and intravenous administration of immunoglobulins (Zis et al., 2018). In migraine trials, the drug class was associated with a different profile of nocebo responses (Mitsikostas et al., 2011).

Social and cultural factors have been associated with a nocebo response. In meta-analyses of neuropathic pain trials, the dropout rate was lower in North American studies compared with studies conducted in Denmark or Australia (Papadopoulos & Mitsikostas, 2012), although the reverse was observed between North American and European trials for oral triptans trials in migraine (Loder et al., 2005) and in focal epilepsy (Zaccara et al., 2015) trials. The role of social and cultural factors may be expressed through the patients but also the physicians.

An increase of the nocebo response (increased drop-out rate) with time measured by the year of publication was identified in multiple sclerosis trials (Papadopoulos & Mitsikostas, 2010), which may be related to a greater accessibility of information about medication on the internet, and a change in physician attitudes towards side effects with time (Papadopoulos & Mitsikostas, 2010). Year of publication was negatively associated with a nocebo response in Alzheimer's disease and diabetic neuropathic trials. In trials of symptomatic therapies for multiple sclerosis, a stronger nocebo response as found in parallel (vs. cross-over) design and phase II (vs. phase III) studies (Papadopoulos & Mitsikostas, 2010), but not in neuropathic pain trials (Papadopoulos & Mitsikostas, 2012). Smaller sample sizes were associated with a stronger nocebo response in the form of nocebo dropout rate in multiple sclerosis (Papadopoulos & Mitsikostas, 2010) and Alzheimer's disease trials (Zis & Mitsikostas, 2015). The nocebo response is also higher in open-label studies compared with double-blind controlled trials (Odinet, Day, Cruz, & Heindel, 2018). This evidence has been obtained, for the most part, in the context of biosimilar use of immune-mediated inflammatory diseases and may lead to undesired discontinuation of medications that have changed the management of these conditions dramatically (Odinet et al., 2018).

The strong correlation of the AEs in the placebo and active treatment arms in a randomized clinical trial is suggestive that the safety information provided in the informed consent is in itself a factor for a nocebo response (Amanzio et al., 2009). The nocebo response can positively correlate with a placebo response as demonstrated in Restless Leg syndrome (Silva et al., 2017) and

Alzheimer's disease (Zis & Mitsikostas, 2015) trials, which suggests that, in some circumstances, study participants may be equally predisposed to placebo and nocebo effects, and thus susceptible to both expectations of benefit and harm. This observation was not reported in other neurological diseases such as neuropathic pain (Papadopoulos & Mitsikostas, 2012).

2.2 Neurobiology of the nocebo effect

Compared with the knowledge of the mechanisms of the placebo effect, the study of the neurobiology of the nocebo effect is limited. Some authors (Benedetti et al., 2007) suggest that ethical challenges constitute the main barrier to assess the nocebo effect. The evaluation of the nocebo effect implicates a negative psychosocial context around the patient and the treatment (Benedetti et al., 2007), which is commonly operationalized with an exposure of study participants to an 'inert' substance together with a suggestion of harm or negative expectation, resulting in the occurrence of an untoward event.

Most of the insight into the biological mechanisms of a nocebo effect comes from the nocebo hyperalgesia and not from neurological diseases. Parkinson's disease is the exception. A few studies in Parkinson's disease patients with disabling motor complications or refractory tremor treated with deep brain stimulation provide some limited knowledge of the nocebo effect in Parkinson's disease. For example, a significant slowing of motor performance was observed after a verbal suggestion of motor worsening in a sham stimulation condition, in turn, overcome by an opposite verbal instruction of improvement (Pollo et al., 2002). In another study, slowing of motor performance was also observed in patients with uninterrupted neurostimulation after they were told of the deep brain stimulation was switched off when it was not (Benedetti et al., 2003). This study used deception and a double-blind (patient and examiner) paradigm (Benedetti et al., 2003) to control for patient and physician-related factors of a nocebo response. In Parkinson's disease, bradykinesia has been suggested to be more responsive to expectations compared with tremor or rigidity (Mercado et al., 2006). Others have not been able to replicate the association between verbal suggestions of harm and a nocebo response in patients with Parkinson's disease and deep brain stimulation (Keitel et al., 2013).

Outside neurology, the nocebo studies conducted in the field of pain provide insight into the brain mechanisms of the nocebo effect, thus deserving some attention. The normal processing of peripheral-induced pain involves the anterior cingulate, insular, prefrontal and somatosensory cortices, and the thalamus (Ploghaus et al., 1999). Pain processing has been grouped in two parallel systems: the lateral (sensory-discriminatory) and medial (affective-cognitive) pain systems (Tracey, 2010). In addition, painful stimuli conveyed through the dorsal horn of the spinal cord are under the modulatory action of supraspinal regions, such as the rostral anterior cingulate cortex, hypothalamus, amygdala, periaqueductal gray and the brainstem rostral ventromedial medulla (Tracey, 2010).

Neuroimaging studies using functional Magnetic Resonance Imaging (fMRI) have evaluated the neurological underpinnings of negative expectation, pain anticipation and nocebo hyperalgesia (Benedetti et al., 2007). In a study using a modified expectancy/conditioning manipulation model with administration of noxious stimuli and sham acupuncture, the authors documented the activation of various brain regions specifically during a hyperalgesia nocebo response that was distinct from a pain experience with a noxious stimulus. The hyperalgesia nocebo response network included the dorsal anterior cingulate cortex, insula, superior temporal gyrus bilaterally, the left frontal and parietal operculum, medial frontal gyrus, orbital prefrontal cortex, superior parietal lobule, and right claustrum/putamen, lateral prefrontal gyrus, and middle temporal gyrus (Kong et al., 2008). The authors concluded that the hyperalgesia nocebo response is associated with activation of the affective-cognitive pain pathway involving the left hippocampus and medial pain system (Kong et al., 2008).

Further, the anticipation and experience of pain are associated with distinct patterns of activation on fMRI. In a meta-analytic connectivity analysis of 19 fMRI studies, pain anticipation was associated with activation of the dorsolateral prefrontal cortex, the middle cingulate and anterior insula cortices, medial and inferior frontal gyri, inferior parietal lobule, middle and superior temporal gyrus, thalamus, and caudate (Palermo, Benedetti, Costa, & Amanzio, 2015). In contrast, reduced activity was found in the anterior cingulate, superior frontal gyrus, parahippocampal gyrus and in the claustrum (Palermo et al., 2015). The emotional components of a nocebo response have been explored: in a "smell study" paradigm, the activation of the left orbitofrontal cortex was associated with a disgust reaction by nocebo responders after the administration of distilled water preceded by a verbal suggestion of an unpleasant odor (Schienle, Hofler, Ubel, & Wabnegger, 2018). More recently, evidence of the spinal cord involvement in the processing of nocebo hyperalgesia before cortical involvement was documented in healthy volunteers using a nocebo heat pain paradigm and spinal fMRI (Geuter & Buchel, 2013).

The nocebo hyperalgesia response was also evaluated using neurophysiological studies. Enhanced low-alpha (8–10 Hz) activity was associated with pessimistic expectations of a more severe nocebo heat-induced pain experience (Albu & Meagher, 2016). Suggestions provided before the administration of a noxious LASER-stimuli modulated magnetoencephalographic evoked responses in the secondary somatosensory and anterior cingulate cortices: the intensity of the evoked response reduced when there was a suggestion of low-intensity pain and expanded when a high-intensity pain was anticipated (Lorenz et al., 2005).

The hyperalgesia nocebo response has been associated with different neurotransmitter systems, namely, dopamine (Scott et al., 2008), opioids (Scott et al., 2008), and cholecystokinin (Benedetti, Amanzio, Casadio, Oliaro, & Maggi, 1997), but also with other biochemical systems such as the cyclooxygenase pathway (Benedetti, Durando, & Vighetti, 2014). The known role of dopamine in expectancy and reward, and the involvement of opioids in the *endo*-analgesic system suggest a potential contribution to the nocebo response. In a [^{11}C] raclopride and [^{11}C] carfentanil PET imaging study, nocebo responders had a decrease in the μ-opioid neurotransmitter system bilaterally in the subgenual anterior cingulate, orbitofrontal cortex, anterior insular cortex, mediodorsal area of the thalamus, and nucleus accumbens and amygdala (Scott et al., 2008). Dopaminergic D2/D3 receptor activity was reduced in the right nucleus accumbens and left ventral putamen for the same nocebo responder group (Scott et al., 2008). For example, the involvement of the dopaminergic system in a hyperalgesia nocebo together with the knowledge that a placebo response in Parkinson's disease is associated with a greater release of dopamine (de la Fuente-Fernández et al., 2001), support further studies to evaluate the role of the dopaminergic system in a nocebo response in Parkinson's disease. In Parkinson's disease, it is possible that other neurotransmitter systems are involved considering the existence of non-dopaminergic symptoms that may be experienced as a nocebo response (Leal Rato et al., 2019).

In a model of postoperative pain, the mixed cholecystokinin 1 and 2 receptor antagonist proglumide inhibited a previously documented nocebo response associated with saline infusion (Benedetti et al., 1997), with some data suggesting a preferential action through cholecystokinin 2 receptors (Andre et al., 2005). The known role of the cholecystokinin system in anxiety led to hypothesize a link between brain neurochemistry and the role of anticipatory anxiety in the nocebo response. In a model of hypobaric hypoxia headache (Benedetti et al., 2014), a nocebo headache response was associated with an increase in salivary prostaglandins and thromboxane, suggesting the nocebo response can be a manifestation of the modulation of the cyclooxygenase pathway, similarly to commonly used non-steroidal inflammatory drugs, though in opposite directions.

There are no known genetic predictors of a nocebo response in neurological diseases. A single study used a conditioning model of immunosuppressive therapy and evaluated the catechol-*O*-methyltransferase Val158Met polymorphism. In this study (Wendt et al., 2014), homozygous carriers of the Val158 variant had a greater prevalence of drug-specific and general AEs side effects, and scored higher in somatosensory amplification scales. There were no differences in mood (anxiety) or biological variables such as immune and cardiovascular function between nocebo responders and non-responders (Wendt et al., 2014).

Overall, available experimental data demonstrates that there is a biological substrate for nocebo responses. Nevertheless, it remains to be established if the concept of a common modulatory for the nocebo response applies or not across conditions, namely neurological diseases.

3. The lessebo effect

When a placebo is portrayed as an inert substance (*sugar pill*) in an informed consent process, and study participants recognize the possibility of being blindly given placebo (and not the active 'beneficial' treatment), it is legitimate to conceptualize that this context may be perceived negatively by a study participant. There is scarce literature on how a patient entering a placebo-controlled randomized trial understands and reacts to the possibility of being blindly allocated to a placebo, and how that can artificially influence efficacy outcome measures. An unanswered issue is the full assessment of how patients see the use of a placebo in randomized controlled trials and how this perception may change their legitimate expectation of benefit.

In a survey conducted in family medicine practices, 40% of responders did not want to take part in placebo-controlled randomized trials; the majority were afraid of not receiving the study drug such that their condition could worsen. Among those willing to take part in a placebo-controlled randomized trial, as many as 66% did not think a placebo would be helpful (Chen & Johnson, 2009). In another survey with Parkinson's disease patients taking part in clinical trials, all of them hoped to have received the study drug rather than the placebo (Goetz, Janko, Blasucci, & Jaglin, 2003).

In fact, patients taking part in a randomized controlled trial and with a preference for an active treatment may either experience a positive "therapeutic effect" or a "resentful demoralization" depending on their belief or knowledge of being allocated to a preferred treatment or not, respectively, and that experience may have implications in the results of clinical trials (Adamson et al., 2008). Indirect but related evidence shows that patients' motivations for placebo-controlled randomized trials are highly variable and can range from self-benefit to sheer altruism (Kaptchuk et al., 2009; McCann, Campbell, & Entwistle, 2010). Even in those willing to participate in a trial for the benefit of helping others, it has been suggested that it is also important for these individuals to identify that participation will benefit them personally, in what has been labeled as conditional altruism (McCann et al., 2010).

In a sample of 50 consecutive Parkinson's disease patients from 14 different placebo-controlled randomized trials, not only all hoped to receive the study drug rather than the placebo, but 54% of these patients expressed surprise or shock by having been on placebo (Goetz et al., 2003). In the largest prospective study looking at patient's attitudes toward clinical trials in Parkinson's disease (Valadas et al., 2011), 63.9% of the 91 participants of Parkinson's disease trials responding to the survey reported that "placebo is a drug with no effect on Parkinson's disease," emphasizing the negative view patients with Parkinson's disease can have of the use of a placebo (Valadas et al., 2011). In addition, 41.6% of respondents would not take part in another placebo-controlled randomized trial (Valadas et al., 2011).

Based on these data, the hypothesis that using a placebo in randomized controlled trials may be deleterious to the experience of participants was put forward, with the corollary of a negative impact on patient's expectation of benefit and, ultimately, a reduction of the measured efficacy of an active treatment. This hypothesis is known as the lessebo effect, and was operationalized by comparing the magnitude of effect of active treatment arms from a placebo-controlled trial and from a clinical trial exclusively controlled with an active treatment (Sinyor et al., 2010).

3.1 The lessebo effect in neurological diseases

The lessebo effect has been assessed for the most part in psychiatry, namely, in depression and schizophrenia. More recently, the presence and magnitude of the lessebo effect were evaluated in Parkinson's disease (Mestre, Shah, Marras, Tomlinson, & Lang, 2014). The novel concept of the lessebo effect has not been reported in other conditions.

The landmark study that used the term lessebo effect for the first time was an aggregated-data meta-analysis that included 90 double-blinded randomized controlled trials of antidepressants for major depression. The authors found that the mean antidepressant response and remission rates were highest in those studies without a placebo, and in those placebo-controlled studies with a higher number of active treatment arms (Sinyor et al., 2010). These findings were replicated in other aggregated-data meta-analyses in depression studies (Henkel et al., 2012; Papakostas & Fava, 2009; Rutherford, Sneed, & Roose, 2009). The lessebo effect was also studied in schizophrenia. An aggregated data meta-analyses of 32 double-blinded randomized controlled trials of newer atypical antipsychotic medications showed that the improvement in the Brief Psychiatric Rating Scale estimated in active-controlled studies was almost double to the improvement documented in placebo-controlled studies (Woods, Gueorguieva, Baker, & Makuch, 2005).

Investigators have tried to test the lessebo effect prospectively in depression by randomly allocating study participants into a placebo-controlled study design or an active-controlled study without a placebo. In these studies, study participants were provided with the allocation to one of the study designs but blinded for final treatment allocation. Under these conditions, the expected magnitude of improvement after open randomization to study design was higher in the subjects allocated to the active-controlled study compared with the subjects in the placebo-controlled group (Rutherford, Sneed, Devanand, Eisenstadt, & Roose, 2010), and a higher baseline expectation of benefit of improvement was correlated with better lower final depression severity scores and greater improvement in depressive symptoms over the course of the study (Rutherford et al., 2013).

In Parkinson's disease, an aggregated data meta-analyses of 72 double-blind randomized controlled trials of dopamine agonists found a difference in the pooled mean score change from baseline to end of treatment between placebo-controlled trials and trials controlled solely with an active comparator using the most used motor efficacy outcome measures in Parkinson's disease, the motor Unified Parkinson's disease Rating Scale (mUPDRS) (Mestre et al., 2014). The difference of 1.75 units in this scale documented a greater improvement in active treatment arms of trials controlled solely with an active comparator and was considered clinically significant. Further, these meta-analyses identified that the lessebo effect in Parkinson's disease was higher in the early Parkinson's disease group without motor fluctuations (3.58 mUPDRS units) compared with a Parkinson's disease group with motor fluctuations, and for a study duration less than 3 months (4.21 mUPDRS units) (Mestre et al., 2014). The probability of placebo assignment was not predictive of a lessebo effect in the group of placebo-controlled studies (Mestre et al., 2014). These results were replicated in individual patient data meta-analyses, including a total of 815 Parkinson's disease patients without motor fluctuations. In this study, the presence of a placebo arm was associated with a smaller score change for both physician-rated (mUPDRS), and patient-reported (UPDRS part II) outcomes between the end of treatment and baseline in an active treatment arm compared with solely having an active treatment as a comparator (*personal communication*). For example, at 6 months of treatment, the lessebo effect corresponded to an adjusted mean score change difference of 6.0 (95% CI: 5.1–7.0) for the mUPDRS and of 1.8 (95% CI: 1.4–2.3) for the UPDRS part II between studies exclusively with an active comparator and no placebo and studies with a placebo arm, in favor of the former (*personal communication*). These results are well above what is considered a minimally clinically significant difference reported for the mUPDRS (2.0-unit points) and UPDRS part II (0.5-unit points) in active treatment arms of a placebo-controlled trial in Parkinson's disease (Hauser, Auinger, & Parkinson Study Group, 2011). The lessebo effect was also found as early as 3–4 weeks of a treatment period.

The lessebo effect was further evaluated in the context of disease-modifying therapies trials in Parkinson's disease, with the caveat that only placebo-controlled trials have been conducted to evaluate these therapies, and active-controlled trials without placebo do not exist since there are no known disease-modifying therapies in Parkinson's disease. Using a similar individual patient data meta-analyses approach, there was no association between the study factor *"placebo allocation probability"* and time to start dopaminergic treatment (*personal communication*). Various hypotheses stem from these results. It is possible that the lessebo effect is only present in symptomatic treatment trials, or that different probabilities of placebo allocation in trials with a placebo arm may not influence the expectation of benefit in patients with Parkinson's disease differently as observed previously (Mestre et al., 2014). In neurology, conditions such as Restless Leg syndrome (Silva et al., 2017) in which the positive placebo and negative nocebo effects seem to be correlated, are attractive disease models to evaluate the lessebo effect.

4. Nocebo and lessebo effects: Impact on clinical practice and research

A nocebo response has significant implications for clinical research since it may lead to an increase in the rate of drop-outs and a reduction of the statistical power of the trial due to nonspecific reasons not related to the experimental therapy being tested. In clinical practice, a nocebo response may lead to a lack of adherence and, ultimately, treatment failure with additional care costs and negative impact to general health of patients (Wells & Kaptchuk, 2012).

The lessebo effect may impact the interpretation of the results of placebo-controlled trials and affect the rigorous therapeutic evaluation of novel therapies (Mestre et al., 2014).

Various factors centered around the therapeutic act were associated with a nocebo response. These factors include the informed consent process with disclosure of a comprehensive list of AEs (Amanzio et al., 2009), the patient experience with previous treatments (Kessner, Wiech, Forkmann, Ploner, & Bingel, 2013) and erroneous association of symptoms to a new treatment. Further, psychological traits in patients such as somatization, anxiety, and depression may be associated with a nocebo response (Barsky, Saintfort, Rogers, & Borus, 2002). Some authors have noted the potential underrepresentation in clinical research of patients with anxiety or mistrust towards medications that may be more reluctant to take part in clinical trials, makings these psychological factors more relevant in clinical practice (Wells & Kaptchuk, 2012). Besides, the nature of the patient-physician relationship also modulates the nocebo response (Barsky et al., 2002; Enck, Benedetti, & Schedlowski, 2008), seen as a conduit for the interplay of the difference factors listed above. An interesting hypothesis is that the physicians' own negative beliefs for treatment may trigger a nocebo response in patients in what is called the Pygmalion or Rosenthal effect (Rosenthal, 1966). This effect has been studied in education whereas the children's performance seems to improve when a teacher is expecting an enhanced performance. The Rosenthal effect was studied little in therapeutics, and is not universal (Raudenbush, 1984).

The knowledge about factors that can modulate a nocebo response in a therapeutic context permits the delineation of mitigating strategies for the nocebo response. Some of these factors, such as the informed consent process, are exclusive of clinical trials. Other factors, such as the prior experience of AEs with treatments, may play a significant role both in clinical practice and research.

Clinical trials may warrant changes to minimize the occurrence of a nocebo response. The nocebo response has been heterogeneously defined across several studies and several interventions, creating confusion among researchers (Mestre, Lang, & Okun, 2016). The development of standards for the identification, reporting of AEs and establishment of a proof of causality has been put forward as means to provide a more comprehensive and transparent evaluation of the nocebo response (Leal Rato et al., 2019). The more widespread use of reporting guidelines is expected to contribute to a better evaluation of the nocebo response in the upcoming years (Moher, 1998). Similar to the placebo effect, there is a need to provide a conceptual framework for the occurrence of AEs in placebo treatment arms. Along the same lines as the concept of placebo-associated improvement (Diederich & Goetz, 2008), a concept of placebo-associated worsening applied systematically is warranted. A placebo-associated worsening may be associated with a nocebo effect rooted in negative expectancy and conditioning but also the natural history of the condition being treated, associated co-morbidities at a disease and individual level, side-effects from a placebo vehicle (injection, sham surgery), the Rosenthal effect (Rosenthal, 1966), and a potential survey bias due to the regular questioning of a standardized list of potential side effects. Alternative study designs may be considered to describe more accurately a placebo-associated worsening and permit the quantification of the nocebo effect. The inclusion of a "no-treatment" arm is thought to allow for the evaluation of the contribution of the natural history of the condition or co-morbidities to a nocebo response, but may raise ethical concerns if there is a well-established treatment (Benedetti et al., 2007). The open-hidden study design has been used to assess the nocebo effect in a paradigm of interruption of a treatment administration that is known (open condition) to or occulted (hidden condition) from the patient. The open-hidden study design was used in pain, anxiety and Parkinson's disease, and the difference in the worsening of symptoms in an open (earlier worsening) and hidden (delayed worsening) can be considered to be attributable to a nocebo effect (Colloca, Lopiano, Lanotte, & Benedetti, 2004).

Since negative expectations driven by the knowledge of side effects can promote a nocebo response, a modification to the traditional informed consent process to clinical research may be a valuable tool to reduce nocebo responses. This approach raises important ethical questions: (1) what is an accurate description of side effect, and (2) what is the balance between the right to informed consent, respect for person and transparency, the tenants of modern clinical research and clinical practice, and the principle of *primum non nocere* or non-maleficence, the foundation of clinical practice (Wells & Kaptchuk, 2012).

Alternative approaches such as the contextualized informed consent have been formulated to avoid the "nocebogenic" provision of a usually long list of common or severe AEs in an informed consent document (Wells & Kaptchuk, 2012). The contextualized informed consent approach advocates tailoring the information about medication side effects to the person being treated and the disease being studied (Bartels et al., 2017). In this novel approach, the information provided about AEs balances transparency with potential harm (Bartels et al., 2017) by proposing to list only more common drug-specific side effects and avoiding those AEs more commonly reported and deemed non-specific such as concentration difficulties, nausea, sedation, dizziness, fatigue, headache, and insomnia (Bartels et al., 2017). A contextualized informed consent would also take into consideration the severity of the condition being treated and those patients considered to be at higher risk of developing a nocebo response, to tailor the disclosure of AEs in an informed consent. Along the same lines, a balanced positive framing to disclose AEs of treatment was associated with less pain in labor epidural analgesia, and less sexual-related AEs with the use of medications for benign prostatic hyperplasia (Colloca & Finniss, 2012). Efforts are underway to develop further research in this area and build supportive evidence (Barnes et al., 2019).

The minimization of the nocebo response was tested experimentally in a randomized clinical trial using a conditioning strategy with a verbal suggestion (Bartels et al., 2017). In this experiment, a nocebo itch response was induced to all study participants, who were subsequently randomly allocated to a positive expectation condition or a group of maintenance of induction of a nocebo response. In the group provided with a positive expectation of lesser itch, the nocebo itch response was reversed (Bartels et al., 2017).

Another mitigating approach is the identification of those individuals more prone to a nocebo effect using a questionnaire before starting a treatment or being allocated in a clinical trial. The Q-No questionnaire is a 4-item questionnaire that predicted the occurrence of a nocebo response with a maximum 71.7% specificity, 67.5% sensitivity and 42.5% positive predictive value in an outpatient clinic setting (Mitsikostas & Deligianni, 2015). Due to the absence of placebo in this context, the authors used a non-validated criterion to identify a nocebo responder. Individual required to have at least three of the four following criteria: (1) at least two past discontinuations of previous treatments because of AEs; (2) self-discontinuation of treatment during the follow-up without communication with the treating physician; (3) experience of AEs not related to drug treatment; and (4) experience of AEs that were related to other concomitant conditions, or preexisted (Mitsikostas & Deligianni, 2015).

The same clinimetric approach has been considered for the lessebo effect, using an Expectancy questionnaire (EQ) developed specifically to capture expectations of benefit in a randomized placebo-controlled clinical trial. Despite expectation of benefit being of the utmost importance to understand the therapeutic effect of an intervention (Benedetti et al., 2016), expectation of benefit is not measured systematically in clinical trials (Benedetti et al., 2016). As a biological variable, expectation of benefit may vary from person to person, from trial to trial and more critically between treatment arms of the same trial. Although the process of randomization is expected to prevent bias for treatment allocation and produce unbiased estimates of treatment effects, randomization does not guarantee comparability between study arms (Yu, Chan, Hopewell, Deeks, & Altman, 2010). In fact, other baseline characteristics such as age, gender, disease severity are usually measured, and their comparability between study arms tested to document a balanced distribution after randomization. Adjusted analyses are critical when baseline variables are unbalanced across treatment groups and have a predictive value for the efficacy outcome (Saquib, Saquib, & Ioannidis, 2013). Since the patient expectation of benefit is not measured nor reported systematically in clinical trials, it is not possible to determine how balanced it is across treatment arms in a clinical trial, and consequently, it can be a potential threat to the internal validity of trials. The standardized use of the EQ may provide the ability to quantify the level of patient expectation of benefit across study arms and the opportunity to consider pre-specified adjusted analyses, with the expected outcome being a more rigorous ascertainment of therapeutic efficacy. Also, the baseline expectation of benefit measured by the EQ could be used for stratification at randomization. Overall, the possibility of adjusting efficacy results for the expectation of benefit in a randomized clinical trial could contribute substantially to more accurate estimates of treatment effects (Yu et al., 2010). The EQ is currently being tested in two placebo-controlled trials in Parkinson's disease (NCT02642393, NCT03205488) to test these hypotheses.

In clinical practice, recommendations were developed to help the clinician mitigate nocebo responses. For example, Barsky et al. (2002) suggest the need for the clinician to maintain a high index of suspicion for side effects that are vague, ambiguous, or prevalent, and identify patient factors predictive of a nocebo response such as a history of negative side effects to different classes of drug, and history of somatization, anxiety or depression (Barsky et al., 2002). Strategies that may help to mitigate the nocebo response in these circumstances include an effort of a therapeutic partnership between patient and physician developing an empathic and transparent communication style (Kleine-Borgmann & Bingel, 2018). The clinician needs to establish the patient's expectations about treatment and side effects, and present therapeutic benefits of treatment before addressing the potential side effects in a balanced fashion (Kleine-Borgmann & Bingel, 2018). When discussing side effects, the clinician will differentiate between drug-specific adverse events and nonspecific nocebo side effects The use of lay language and easy to read graphic illustrations could be used to convey this information (Kleine-Borgmann & Bingel, 2018). Additional strategies may involve starting at low doses and providing slower titrations, especially for those individuals more prone to a nocebo response (Barsky et al., 2002). Currently, there is no robust evidence to support these recommendations or on how to share clinical evidence for shared-decision making, such as when starting a medication (Epstein, Alper, & Quill, 2004).

5. Future directions

At present, the knowledge about the nocebo and the lessebo effects is very sparse. The meta-analyses of placebo-controlled trials provide the first insight on the significance of these effects in neurological disease, justifying further investigation. For example, the identification of the brain mechanisms underlying the nocebo and the lessebo effects using neuroimaging will inform their biological underpinnings. A critical advance in the field will be the identification of factors that consistently modulate the prevalence and magnitude of these effects to determine innovations in clinical trial design and contribute decisively to interpret and value the efficacy, safety and tolerability results of therapeutic interventions. The comprehensive knowledge of the nocebo and lessebo effects will justify a strong consideration of mitigating strategies for the design of clinical trials or the development of new ones. It remains to be defined how the knowledge nocebo and lessebo effects can be positively translated into clinical practice to be widely accepted by patients and clinicians, preserving patient autonomy and the ethics of the patient-physician relationship.

References

Adamson, S., Bland, J., Hay, E., Johnson, R., Jones, G., Kitchener, H., et al. (2008). Patients' preferences within randomised trials: Systematic review and patient level meta-analysis. *BMJ*, *337*, a1864. https://doi.org/10.1136/bmj.a1864.

Alam, J. M., Hadjivassiliou, M., & Zis, P. (2019). Nocebo in cerebellar ataxia: A systematic review and meta-analysis of placebo-controlled clinical trials. *Journal of the Neurological Sciences*, *401*, 112–117. https://doi.org/10.1016/j.jns.2019.04.039.

Albu, S., & Meagher, M. W. (2016). Expectation of nocebo hyperalgesia affects EEG alpha-activity. *International Journal of Psychophysiology*, *109*, 147–152. https://doi.org/10.1016/j.ijpsycho.2016.08.009.

Amanzio, M., Benedetti, F., & Vase, L. (2012). A systematic review of adverse events in the placebo arm of donepezil trials: The role of cognitive impairment. *International Psychogeriatrics*, *24*(5), 698–707. https://doi.org/10.1017/s1041610211002420.

Amanzio, M., Corazzini, L. L., Vase, L., & Benedetti, F. (2009). A systematic review of adverse events in placebo groups of anti-migraine clinical trials. *Pain*, *146*(3), 261–269. https://doi.org/10.1016/j.pain.2009.07.010.

Andre, J., Zeau, B., Pohl, M., Cesselin, F., Benoliel, J. J., & Becker, C. (2005). Involvement of cholecystokininergic systems in anxiety-induced hyperalgesia in male rats: Behavioral and biochemical studies. *The Journal of Neuroscience*, *25*(35), 7896–7904. https://doi.org/10.1523/JNEUROSCI.0743-05.2005.

Barnes, K., Faasse, K., Geers, A. L., Helfer, S. G., Sharpe, L., Colloca, L., et al. (2019). Can positive framing reduce nocebo side effects? Current evidence and recommendation for future research. *Frontiers in Pharmacology*, *10*, 167. https://doi.org/10.3389/fphar.2019.00167.

Barsky, A. J., Saintfort, R., Rogers, M. P., & Borus, J. F. (2002). Nonspecific medication side effects and the nocebo phenomenon. *JAMA*, *287*(5), 622–627. https://doi.org/10.1001/jama.287.5.622.

Bartels, D. J. P., van Laarhoven, A. I. M., Stroo, M., Hijne, K., Peerdeman, K. J., Donders, A. R. T., et al. (2017). Minimizing nocebo effects by conditioning with verbal suggestion: A randomized clinical trial in healthy humans. *PLoS One*, *12*(9), e0182959. https://doi.org/10.1371/journal.pone.0182959.

Benedetti, F., Amanzio, M., Casadio, C., Oliaro, A., & Maggi, G. (1997). Blockade of nocebo hyperalgesia by the cholecystokinin antagonist proglumide. *Pain*, *71*(2), 135–140. https://doi.org/10.1016/s0304-3959(97)03346-0.

Benedetti, F., Carlino, E., & Piedimonte, A. (2016). Increasing uncertainty in CNS clinical trials: The role of placebo, nocebo, and Hawthorne effects. *Lancet Neurology*, *15*(7), 736–747. https://doi.org/10.1016/S1474-4422(16)00066-1.

Benedetti, F., Durando, J., & Vighetti, S. (2014). Nocebo and placebo modulation of hypobaric hypoxia headache involves the cyclooxygenase-prostaglandins pathway. *Pain*, *155*(5), 921–928. https://doi.org/10.1016/j.pain.2014.01.016.

Benedetti, F., Lanotte, M., Lopiano, L., & Colloca, L. (2007). When words are painful: Unraveling the mechanisms of the nocebo effect. *Neuroscience*, *147*(2), 260–271. https://doi.org/10.1016/j.neuroscience.2007.02.020. S0306-4522(07)00181-9 [pii].

Benedetti, F., Pollo, A., Lopiano, L., Lanotte, M., Vighetti, S., & Rainero, I. (2003). Conscious expectation and unconscious conditioning in analgesic, motor, and hormonal placebo/nocebo responses. *The Journal of Neuroscience*, *23*(10), 4315–4323. https://doi.org/10.1523/JNEUROSCI.23-10-04315.2003. 23/10/4315 [pii].

Chen, G. F., & Johnson, M. H. (2009). Patients' attitudes to the use of placebos: Results from a New Zealand survey. *The New Zealand Medical Journal*, *122*(1296), 35–46.

Clarke, M. J. (2005). Individual patient data meta-analyses. *Best Practice & Research. Clinical Obstetrics & Gynaecology*, *19*(1), 47–55. https://doi.org/10.1016/j.bpobgyn.2004.10.011.

Colloca, L., & Finniss, D. (2012). Nocebo effects, patient-clinician communication, and therapeutic outcomes. *JAMA*, *307*(6), 567–568. https://doi.org/10.1001/jama.2012.115.

Colloca, L., Lopiano, L., Lanotte, M., & Benedetti, F. (2004). Overt versus covert treatment for pain, anxiety, and Parkinson's disease. *The Lancet Neurology*, *3*(11), 679–684. https://doi.org/10.1016/S1474-4422(04)00908-1. S1474442204009081 [pii].

de la Fuente-Fernández, R., Ruth, T. J., Sossi, V., Schulzer, M., Calne, D. B., & Stoessl, A. J. (2001). Expectation and dopamine release: Mechanism of the placebo effect in Parkinson's disease. *Science*, *293*(5532), 1164–1166. https://doi.org/10.1126/science.1060937. 293/5532/1164 [pii].

Diederich, N. J., & Goetz, C. G. (2008). The placebo treatments in neurosciences: New insights from clinical and neuroimaging studies. *Neurology*, *71*(9), 677–684. https://doi.org/10.1212/01.wnl.0000324635.49971.3d. 71/9/677 [pii].

Enck, P., Benedetti, F., & Schedlowski, M. (2008). New insights into the placebo and nocebo responses. *Neuron*, *59*(2), 195–206. https://doi.org/10.1016/j.neuron.2008.06.030.

Epstein, R. M., Alper, B. S., & Quill, T. E. (2004). Communicating evidence for participatory decision making. *JAMA*, *291*(19), 2359–2366. https://doi.org/10.1001/jama.291.19.2359.

Geuter, S., & Buchel, C. (2013). Facilitation of pain in the human spinal cord by nocebo treatment. *The Journal of Neuroscience*, *33*(34), 13784–13790. https://doi.org/10.1523/jneurosci.2191-13.2013.

Gklinos, P., Papadopoulos, D., & Mitsikostas, D. D. (2019). Nocebo in multiple sclerosis trials: A meta-analysis on oral and newer injectable disease-modifying treatments. *Multiple Sclerosis and Related Disorders*, *36*, 101389. https://doi.org/10.1016/j.msard.2019.101389.

Goetz, C. G., Janko, K., Blasucci, L., & Jaglin, J. A. (2003). Impact of placebo assignment in clinical trials of Parkinson's disease. *Movement Disorders*, *18*(10), 1146–1149. https://doi.org/10.1002/mds.10504.

Hauser, R. A., Auinger, P., & Parkinson Study Group (2011). Determination of minimal clinically important change in early and advanced Parkinson's disease. *Movement Disorders*, *26*(5), 813–818. https://doi.org/10.1002/mds.23638.

Hauser, W., Bartram, C., Bartram-Wunn, E., & Tolle, T. (2012). Adverse events attributable to nocebo in randomized controlled drug trials in fibromyalgia syndrome and painful diabetic peripheral neuropathy: Systematic review. *The Clinical Journal of Pain*, *28*(5), 437–451. https://doi.org/10.1097/AJP.0b013e3182321ad8.

Henkel, V., Casaulta, F., Seemuller, F., Krahenbuhl, S., Obermeier, M., Husler, J., et al. (2012). Study design features affecting outcome in antidepressant trials. *Journal of Affective Disorders*, *141*(2–3), 160–167. https://doi.org/10.1016/j.jad.2012.03.021.

Kaptchuk, T. J., Shaw, J., Kerr, C. E., Conboy, L. A., Kelley, J. M., Csordas, T. J., et al. (2009). "Maybe I made up the whole thing": Placebos and patients' experiences in a randomized controlled trial. *Culture, Medicine and Psychiatry*, *33*(3), 382–411. https://doi.org/10.1007/s11013-009-9141-7.

Keitel, A., Wojtecki, L., Hirschmann, J., Hartmann, C. J., Ferrea, S., Sudmeyer, M., et al. (2013). Motor and cognitive placebo –/nocebo-responses in Parkinson's disease patients with deep brain stimulation. *Behavioural Brain Research*, *250*, 199–205. https://doi.org/10.1016/j.bbr.2013.04.051.

Kessner, S., Wiech, K., Forkmann, K., Ploner, M., & Bingel, U. (2013). The effect of treatment history on therapeutic outcome: An experimental approach. *JAMA Internal Medicine*, *173*(15), 1468. https://doi.org/10.1001/jamainternmed.2013.6705.

Kleine-Borgmann, J., & Bingel, U. (2018). Nocebo effects: Neurobiological mechanisms and strategies for prevention and optimizing treatment. *International Review of Neurobiology*, *138*, 271–283. https://doi.org/10.1016/bs.irn.2018.02.005.

Kong, J., Gollub, R. L., Polich, G., Kirsch, I., Laviolette, P., Vangel, M., et al. (2008). A functional magnetic resonance imaging study on the neural mechanisms of hyperalgesic nocebo effect. *The Journal of Neuroscience*, *28*(49), 13354–13362. https://doi.org/10.1523/jneurosci.2944-08.2008.

Leal Rato, M., Duarte, G. S., Ferreira, A. N., Alves, M., Mainoli, B., Teodoro, T., et al. (2019). Nocebo response in Parkinson's disease: A systematic review and meta-analysis. *Parkinsonism & Related Disorders*. https://doi.org/10.1016/j.parkreldis.2019.04.015.

Loder, E., Goldstein, R., & Biondi, D. (2005). Placebo effects in oral triptan trials: The scientific and ethical rationale for continued use of placebo controls. *Cephalalgia*, *25*(2), 124–131. https://doi.org/10.1111/j.1468-2982.2004.00817.x.

Lorenz, J., Hauck, M., Paur, R. C., Nakamura, Y., Zimmermann, R., Bromm, B., et al. (2005). Cortical correlates of false expectations during pain intensity judgments—A possible manifestation of placebo/nocebo cognitions. *Brain, Behavior, and Immunity*, *19*(4), 283–295. https://doi.org/10.1016/j.bbi.2005.03.010.

McCann, S. K., Campbell, M. K., & Entwistle, V. A. (2010). Reasons for participating in randomised controlled trials: Conditional altruism and considerations for self. *Trials*, *11*, 31. https://doi.org/10.1186/1745-6215-11-31.

Mercado, R., Constantoyannis, C., Mandat, T., Kumar, A., Schulzer, M., Stoessl, A. J., et al. (2006). Expectation and the placebo effect in Parkinson's disease patients with subthalamic nucleus deep brain stimulation. *Movement Disorders*, *21*(9), 1457–1461. https://doi.org/10.1002/mds.20935.

Mestre, T. A., Lang, A. E., & Okun, M. S. (2016). Factors influencing the outcome of deep brain stimulation: Placebo, nocebo, lessebo, and lesion effects. *Movement Disorders*, *31*(3), 290–296. https://doi.org/10.1002/mds.26500.

Mestre, T. A., Shah, P., Marras, C., Tomlinson, G., & Lang, A. E. (2014). Another face of placebo: The lessebo effect in Parkinson disease meta-analyses. *Neurology*, *82*(16), 1402–1409. https://doi.org/10.1212/wnl.0000000000000340.

Mitsikostas, D. D., & Deligianni, C. I. (2015). Q-no: A questionnaire to predict nocebo in outpatients seeking neurological consultation. *Neurological Sciences*, *36*(3), 379–381. https://doi.org/10.1007/s10072-014-1959-2.

Mitsikostas, D. D., Mantonakis, L. I., & Chalarakis, N. G. (2011). Nocebo is the enemy, not placebo. A meta-analysis of reported side effects after placebo treatment in headaches. *Cephalalgia*, *31*(5), 550–561. https://doi.org/10.1177/0333102410391485.

Moher, D. (1998). CONSORT: An evolving tool to help improve the quality of reports of randomized controlled trials. Consolidated standards of reporting trials. *JAMA*, *279*(18), 1489–1491. https://doi.org/10.1001/jama.279.18.1489.

Odinet, J. S., Day, C. E., Cruz, J. L., & Heindel, G. A. (2018). The biosimilar nocebo effect? A systematic review of double-blinded versus open-label studies. *Journal of Managed Care & Specialty Pharmacy*, *24*(10), 952–959. https://doi.org/10.18553/jmcp.2018.24.10.952.

Okusogu, C., & Colloca, L. (2019). Placebo hypoalgesia: Above and beyond expectancy and conditioning. *Current Opinion in Behavioral Sciences*, *26*, 75–81. https://doi.org/10.1016/j.cobeha.2018.10.008.

Palermo, S., Benedetti, F., Costa, T., & Amanzio, M. (2015). Pain anticipation: An activation likelihood estimation meta-analysis of brain imaging studies. *Human Brain Mapping*, *36*(5), 1648–1661. https://doi.org/10.1002/hbm.22727.

Papadopoulos, D., & Mitsikostas, D. D. (2010). Nocebo effects in multiple sclerosis trials: A meta-analysis. *Multiple Sclerosis*, *16*(7), 816–828. https://doi.org/10.1177/1352458510370793.

Papadopoulos, D., & Mitsikostas, D. D. (2012). A meta-analytic approach to estimating nocebo effects in neuropathic pain trials. *Journal of Neurology, 259*(3), 436–447. https://doi.org/10.1007/s00415-011-6197-4.

Papakostas, G. I., & Fava, M. (2009). Does the probability of receiving placebo influence clinical trial outcome? A meta-regression of double-blind, randomized clinical trials in MDD. *European Neuropsychopharmacology, 19*(1), 34–40. https://doi.org/10.1016/j.euroneuro.2008.08.009. S0924-977X(08)00217-4 [pii].

Ploghaus, A., Tracey, I., Gati, J. S., Clare, S., Menon, R. S., Matthews, P. M., et al. (1999). Dissociating pain from its anticipation in the human brain. *Science, 284*(5422), 1979–1981. https://doi.org/10.1126/science.284.5422.1979.

Pollo, A., Torre, E., Lopiano, L., Rizzone, M., Lanotte, M., Cavanna, A., et al. (2002). Expectation modulates the response to subthalamic nucleus stimulation in Parkinsonian patients. *Neuroreport, 13*(11), 1383–1386.

Raudenbush, S. W. (1984). Magnitude of teacher expectancy effects on pupil IQ as a function of the credibility of expectancy induction: A synthesis of findings from 18 experiments. *Journal of Educational Psychology, 76*(1), 85–97. https://doi.org/10.1037/0022-0663.76.1.85.

Rosenthal, R. (1966). *Experimenter effects in behavioral research.* New York: Appleton Century Crofts.

Rutherford, B. R., Marcus, S. M., Wang, P., Sneed, J. R., Pelton, G., Devanand, D., et al. (2013). A randomized, prospective pilot study of patient expectancy and antidepressant outcome. *Psychological Medicine, 43*(5), 975–982. https://doi.org/10.1017/s003329 1712001882.

Rutherford, B., Sneed, J., Devanand, D., Eisenstadt, R., & Roose, S. (2010). Antidepressant study design affects patient expectancy: A pilot study. *Psychological Medicine, 40*(5), 781–788. https://doi.org/10.1017/s0033291709991085.

Rutherford, B. R., Sneed, J. R., & Roose, S. P. (2009). Does study design influence outcome?. The effects of placebo control and treatment duration in antidepressant trials. *Psychotherapy and Psychosomatics, 78*(3), 172–181. https://doi.org/10.1159/000209348.

Saquib, N., Saquib, J., & Ioannidis, J. P. (2013). Practices and impact of primary outcome adjustment in randomized controlled trials: Meta-epidemiologic study. *BMJ, 347*, f4313. https://doi.org/10.1136/bmj.f4313.

Schienle, A., Hofler, C., Ubel, S., & Wabnegger, A. (2018). Emotion-specific nocebo effects: An fMRI study. *Brain Imaging and Behavior, 12*(1), 180–187. https://doi.org/10.1007/s11682-017-9675-1.

Scott, D. J., Stohler, C. S., Egnatuk, C. M., Wang, H., Koeppe, R. A., & Zubieta, J. K. (2008). Placebo and nocebo effects are defined by opposite opioid and dopaminergic responses. *Archives of General Psychiatry, 65*(2), 220–231. https://doi.org/10.1001/archgenpsychiatry.2007.34.

Shafiq, F., Mitsikostas, D. D., & Zis, P. (2017). Nocebo in motor neuron disease: Systematic review and meta-analysis of placebo-controlled clinical trials. *Amyotrophic Lateral Sclerosis & Frontotemporal Degeneration, 18*(7–8), 576–582. https://doi.org/10.1080/21678421.2017.1335325.

Silva, M. A., Duarte, G. S., Camara, R., Rodrigues, F. B., Fernandes, R. M., Abreu, D., et al. (2017). Placebo and nocebo responses in restless legs syndrome: A systematic review and meta-analysis. *Neurology, 88*(23), 2216–2224. https://doi.org/10.1212/WNL.0000000 000004004.

Sinyor, M., Levitt, A. J., Cheung, A. H., Schaffer, A., Kiss, A., Dowlati, Y., et al. (2010). Does inclusion of a placebo arm influence response to active antidepressant treatment in randomized controlled trials? Results from pooled and meta-analyses. *The Journal of Clinical Psychiatry, 71*(3), 270–279. https://doi.org/10.4088/JCP.08r04516blu.

Stathis, P., Smpiliris, M., Konitsiotis, S., & Mitsikostas, D. D. (2013). Nocebo as a potential confounding factor in clinical trials for Parkinson's disease treatment: A meta-analysis. *European Journal of Neurology*, *20*(3), 527–533. https://doi.org/10.1111/ene.12014.

Tracey, I. (2010). Getting the pain you expect: Mechanisms of placebo, nocebo and reappraisal effects in humans. *Nature Medicine*, *16*(11), 1277–1283. https://doi.org/10.1038/nm.2229.

Valadas, A., Coelho, M., Mestre, T., Guedes, L. C., Finisterra, M., Noronha, A., et al. (2011). What motivates Parkinson's disease patients to enter clinical trials? *Parkinsonism & Related Disorders*, *17*(9), 667–671. https://doi.org/10.1016/j.parkreldis.2011.05.023. S1353-8020(11)00155-6 [pii].

Varma, A., & Zis, P. (2019). Nocebo effect in myasthenia gravis: Systematic review and meta-analysis of placebo-controlled clinical trials. *Acta Neurologica Belgica*, *119*(2), 257–264. https://doi.org/10.1007/s13760-019-01143-1.

Wells, R. E., & Kaptchuk, T. J. (2012). To tell the truth, the whole truth, may do patients harm: The problem of the nocebo effect for informed consent. *The American Journal of Bioethics*, *12*(3), 22–29. https://doi.org/10.1080/15265161.2011.652798.

Wendt, L., Albring, A., Benson, S., Engler, H., Engler, A., Hinney, A., et al. (2014). Catechol-O-methyltransferase Val158Met polymorphism is associated with somatosensory amplification and nocebo responses. *PLoS One*, *9*(9), e107665. https://doi.org/10.1371/journal.pone.0107665.

Woods, S. W., Gueorguieva, R. V., Baker, C. B., & Makuch, R. W. (2005). Control group bias in randomized atypical antipsychotic medication trials for schizophrenia. *Archives of General Psychiatry*, *62*(9), 961–970. https://doi.org/10.1001/archpsyc.62.9.961.

Yu, L. M., Chan, A. W., Hopewell, S., Deeks, J. J., & Altman, D. G. (2010). Reporting on covariate adjustment in randomised controlled trials before and after revision of the 2001 CONSORT statement: A literature review. *Trials*, *11*, 59. https://doi.org/10.1186/1745-6215-11-59.

Zaccara, G., Giovannelli, F., Cincotta, M., Loiacono, G., & Verrotti, A. (2015). Adverse events of placebo-treated, drug-resistant, focal epileptic patients in randomized controlled trials: A systematic review. *Journal of Neurology*, *262*(3), 501–515. https://doi.org/10.1007/s00415-014-7391-y.

Zaccara, G., Giovannelli, F., Franco, V., Cincotta, M., Tramacere, L., & Verrotti, A. (2014). Adverse events, placebo and nocebo effects in placebo-treated paediatric patients with refractory focal epilepsies. Analysis of double-blind studies. *Epilepsy Research*, *108*(10), 1685–1693. https://doi.org/10.1016/j.eplepsyres.2014.09.015.

Zaccara, G., Giovannelli, F., Giorgi, F. S., Franco, V., & Gasparini, S. (2016). Analysis of nocebo effects of antiepileptic drugs across different conditions. *Journal of Neurology*, *263*(7), 1274–1279. https://doi.org/10.1007/s00415-015-8018-7.

Zis, P., Hadjivassiliou, M., Sarrigiannis, P. G., Jenkins, T. M., & Mitsikostas, D. D. (2018). Nocebo in chronic inflammatory demyelinating polyneuropathy; a systematic review and meta-analysis of placebo-controlled clinical trials. *Journal of the Neurological Sciences*, *388*, 79–83. https://doi.org/10.1016/j.jns.2018.03.009.

Zis, P., & Mitsikostas, D. D. (2015). Nocebo in Alzheimer's disease; meta-analysis of placebo-controlled clinical trials. *Journal of the Neurological Sciences*, *355*(1–2), 94–100. https://doi.org/10.1016/j.jns.2015.05.029.

Zis, P., Shafiq, F., & Mitsikostas, D. D. (2017). Nocebo effect in refractory partial epilepsy during pre-surgical monitoring: Systematic review and meta-analysis of placebo-controlled clinical trials. *Seizure*, *45*, 95–99. https://doi.org/10.1016/j.seizure.2016.12.003.

CHAPTER SEVEN

Deception and the ethics of placebo

A. Jon Stoessl*
Pacific Parkinson's Research Centre and Djavad Mowafaghian Centre for Brain Health, University of British Columbia, Vancouver, BC, Canada
*Corresponding author: e-mail address: jstoessl@mail.ubc.ca

Contents

Abstract

The placebo effect in many areas of neurological therapeutics is common and prominent. The importance of the response means that any new treatment must account for the placebo effect, but this in turn raises major challenges as to how to conduct scientifically meaningful research in an ethically acceptable fashion. Basic principles that may be in tension with one another are those of beneficence and respect for autonomy. It may be challenging to respect autonomy if the scientific design of a study depends upon the use of deception, but this is often mitigated by the information provided to trial participants as part of the informed consent process. Deception is a particular challenge if placebos are to be used in a clinical/therapeutic setting, outside the context of a clinical trial. While practice-based placebo use may on occasion be both beneficial and ethically acceptable, close attention must then be paid to ensuring that basic ethical principles are respected and that placebos are either prescribed in an open and honest fashion, or that any deception is authorized.

International Review of Neurobiology, Volume 153
ISSN 0074-7742
https://doi.org/10.1016/bs.irn.2020.03.030

1. Introduction

The question of ethical use of placebos in neurological disorders is guided by the same basic principles that underlie medical ethics in general: autonomy, beneficence, non-maleficence and justice. While many of these principles have a much longer history, their modern-day enshrinement is typically traced back to the Nazi war crimes trials, culminating in the Nuremberg Code. Of 10 points, the very first was the essential nature of voluntary consent. While the crimes that necessitated the Code were quite different, this highlights the importance of autonomy in medical research and is one of the key principles relevant to a consideration of the use of placebos. While the absolute requirement for consent was relaxed somewhat in the ensuing Declaration of Helsinki (e.g., to permit consent by proxy where appropriate), the latter code also made clear that the investigator's over-riding duty is to the patient or volunteer and that the subject's welfare takes precedence over the interests of science and society. This has been interpreted to invalidate the use of placebos, particularly in situations where there is adequate demonstration of benefit from an existing treatment. This point is particularly relevant in disadvantaged countries where a placebo-controlled trial of a treatment readily available in the first world is conducted and participation in the trial may represent the only opportunity to access a potentially life-saving medication. Such a practice would be a violation of both the principles of autonomy (using a trial to allow patients to access medication is a form of coercion) and justice (the medication is more easily available elsewhere). However, the 4th and 5th revisions to the Declaration and subsequent notes of clarification attempted to explicitly address situations in which the use of placebo controls is ethically acceptable (i.e., where no proven intervention exists or where there is minimal harm inherent in their use). While the interpretation of these amendments and clarifications continues to be hotly debated, their practical application will be outlined below, using as the core neurological example, Parkinson's disease (PD).

The use of placebos in PD raises numerous ethical questions. While it must be admitted that there may not be consensus on the answers to these questions, they should always be taken into consideration in the conduct of clinical trials, as well as clinical practice.

1. Is it necessary to include a placebo/sham group?
2. Does the scientific return justify the potential risk of a placebo/sham therapy?
3. Is the use of deception justifiable?

4. Are there alternatives to the use of deception?
5. Are there additional risks associated with the use of placebo/sham?
6. Is the use of placebo/sham therapy justifiable in clinical practice, outside a research scenario?

Each of these questions is addressed in the rest of this chapter.

2. Is it necessary to include a placebo/sham group in PD research?

For the majority of this piece, the focus will be on the use of placebo or sham intervention within the context of a clinical trial, in which case (i) standard symptomatic therapy is deemed not yet necessary (i.e., in early disease) and the experimental therapy is being assessed for weak symptomatic effects or for benefits that are not demonstrably available from other treatments (i.e., disease modification) or (ii) the experimental therapy is being assessed in patients with more advanced disease as an adjunct to best available treatment, either alone or against an active comparator. While it would be ethically untenable to withhold active treatment of known benefit (i.e., levodopa) in a patient with moderately symptomatic PD, the scenarios described would satisfy the expectations embodied in the revised Declaration of Helsinki and in most cases meet the linked requirement for clinical equipoise (Touwen & Engberts, 2012; World Medical Association, 2013). The placebo effect in PD is common and may be of considerable magnitude (Goetz et al., 2008; Goetz, Leurgans, Raman, & Stebbins, 2000). The magnitude of the effect may be dictated by the perceived value of the intervention and may thus vary according to stated probability of benefit, stated magnitude of benefit, the invasiveness of the intervention, and even the price and appearance of the pills (Espay et al., 2015). There are several widely held misconceptions about placebo effects. One of these is that the effect is short-lived and that sustained benefit of a given therapy is therefore unlikely to be attributable to a placebo effect. This is untrue, as has now been demonstrated in a number of trials of surgical therapies for PD (Marks et al., 2010; McRae et al., 2004). While the mechanism(s) of sustained benefit attributable to placebo is(are) admittedly more complex than the case of an acute response, many factors may contribute, including halo and Hawthorne effects and repeated reinforcement of the expectation of benefit in a clinical trial setting. Irrespective of mechanism(s), placebo effects may indeed persist and this may be particularly true for invasive and potentially risky interventions.

A second misconception is that the placebo effect is subjective and therefore less likely to influence objective measures of disease. This too is untrue. Indeed, in one of the few double-blind, placebo-controlled trials of cell-based therapy (human fetal embryonic transplantation) for PD, the benefit was better predicted not by actual, but rather by perceived treatment allocation (active vs. sham) (McRae et al., 2004). This was true not only for "soft" or "subjective" outcome measures such as quality of life as reported by the patient, but also for "objective" measures of motor function as assessed by a blinded rater. Placebos have an impact on highly objective measures such as dopamine release as assessed by PET (de la Fuente-Fernandez et al., 2001; Strafella, Ko, & Monchi, 2006) or electrophysiological measures of subthalamic nucleus firing rate and pattern (Benedetti et al., 2004) in the short term; in the long term, sham intervention is associated with sustained changes in glucose metabolic networks, independent of changes associated with improvement in motor function (Ko et al., 2014).

These facts strongly suggest that it is not possible to truly assess the benefits of an active intervention without concurrently controlling for placebo effects.

3. Does the scientific return justify the potential risk of a placebo/sham therapy?

A number of authors have suggested that, particularly in the case of surgical therapies, there may be a substantial risk associated with the use of a sham intervention, especially if this involves penetrating the dura. In a survey of Parkinson Study Group investigators, almost all felt that a sham control study was scientifically preferable, but while 90% would support a sham procedure up to and including bilateral full burr holes, there was a precipitous decline (to 20% or less) in support for any procedure that involved brain penetration (Kim et al., 2005). These authors suggest that this risk is ethically unjustifiable when there is no likelihood of treatment-related benefit. Members of the public also recognize the importance of sham surgical procedures, provided that participants are adequately informed and provide consent (Pugh, Kahane, Maslen, & Savulescu, 2016). Not surprisingly, PD patients themselves are less likely to agree to participate in a blinded trial of surgery for PD, but are about evenly split on whether they would or would not agree to such a trial moving forward if they were members of an institutional review board. Higher educational level is associated with a greater likelihood to approve of the use of sham surgery controls in a hypothetical trial and to consider that the associate risk may be justified by the potential benefits to society (Frank et al., 2008). The position has been taken that the risk of studying invasive therapies in PD is more readily justified if there is strong preclinical evidence to support the treatment (Kimmelman et al., 2009). As well recognized by the authors, this position itself, while seeming inherently reasonable, is not free of difficulty, as animal models have to date been poor predictors of neuroprotective or neurorestorative capacity (Dawson, Golde, & Lagier-Tourenne, 2018; Lang & Espay, 2018). However, the authors also note that a factor mitigating risk is the strength of the evidence as seen from other perspectives, including the scientific rigor of the study, including the use of an appropriate control group.

Of course, in the presence of a strong (and sustained) placebo effect, the patient may indeed benefit, but most physicians would not accept this as adequate justification in its own right, although patients themselves might differ with respect to deriving benefit from a placebo effect (Köteles & Ferentzi, 2012). However, this view, while understandable, must be placed into the context of the risks of not adequately assessing benefit, and thereby potentially exposing large numbers of patients for many years to come to the risk of such invasive therapy, even when it may not work. The history of medicine is unfortunately replete with such examples.

Not all risks associated with placebo are ethically justifiable. It might be considered reasonable to test a novel acute therapy against placebo in a short-term trial conducted in patients with relatively early disease. It is standard to assess adjunctive therapies against placebo in patients with more advanced disease, but only when they remain on open-label optimized therapy, which almost always includes levodopa. It would not be deemed acceptable to withdraw standard medical therapy in order to test a novel therapy against placebo (i.e., novel therapy + nothing vs. placebo + nothing) in patients with moderately advanced disease without their knowledge and explicit consent. In reality, the latter is done on a routine basis in a research setting, when patients who have undergone deep brain stimulation surgery are taken off medication in order to assess the effects of surgery, but this is done in an open label fashion and for a short duration (e.g., Deuschl et al., 2006). This can also be done in clinical practice, in order to adjust stimulation parameters and to assess the need for ongoing medication.

4. Is the use of deception justifiable?

It is a basic tenet of medicine that lying to a patient is never acceptable (although some would argue that the true answer may be more nuanced—see below). This initial absolutism reflects a growing recognition of respect for patient autonomy and repudiation of a more traditional paternalistic approach to the practice of medicine (Annoni, 2018; Faden, Beauchamp, & King, 1986). However, respect for patient autonomy must be balanced against the primary purpose of the therapeutic interaction, which is to improve the patient's wellbeing—from that perspective, emphasizing the hope for positive outcome and reducing unnecessarily negative information may generally be considered as beneficial, as long as the information provided is consistent with truly informed consent (Alfano, 2015).

While overt deception is rarely if ever justifiable, an alternative perspective is that failure to allow patients to participate in a placebo-controlled study is itself paternalistic and disrespectful of their autonomy. While patients want to be made better, they also want hope—particularly in the face of a chronic disease without known cure (Groopman, 2004; Metzler & Just, 2018; Snyder, Adams, Crooks, Whitehurst, & Vallee, 2014). Hope and optimism are intrinsically linked to placebo effects, which are thought to be related to expectation of benefit. In more general terms, however, patients may derive hope from the very knowledge that new therapies are being developed and tested and their active participation in this process may foster a sense of autonomy. While levels of sophistication vary, most patients understand the rationale and need for testing new therapies with as much scientific rigor as possible and most are willing to run the "risk" of receiving a placebo if this will ultimately advance knowledge about their condition. Thus, in the case of agreeing to participate in a clinical trial that includes a placebo/sham arm, almost all deception is "authorized."

In attempting to respect patient autonomy, overt deception may not necessarily represent the biggest problem. Rather, even if one has explained the study procedure and the chance of being assigned to a placebo/sham arm, patients may want to believe that they are receiving active treatment and this may color their response to the intervention and their willingness to participate. Over time, consent forms have become increasingly long, detailed and complex, ironically in order to ensure that patients are fully informed, but with the unfortunate consequence that they are likely to confuse. Patients with PD may suffer from cognitive impairment that limits their capacity to understand and to make decisions. Patients may also be subject to pressure to participate from well-meaning family members or feel that they should participate in order to maintain a good relationship with their treating physician. These issues must all be thoroughly addressed, but they are a reason to pay close attention to the way in which a clinical trial is conducted, not a reason to avoid participating in properly controlled studies.

5. Are there alternatives to the use of deception?

In the setting of a randomized controlled trial, overt deception can never be recommended, as trial participants should be informed about their chances of receiving active vs. placebo/sham treatment (see more on this below). However, the purpose of most RCTs is to determine whether the active treatment performs significantly better than placebo. In many cases, when the purpose of the trial is to study the placebo effect itself, some degree of deception may be required. As an example, in our work on manipulating the strength of placebo-induced dopamine release in PD, we assessed dopamine release following administration of open label levodopa, as well as after varying stated probabilities of receiving levodopa (Lidstone et al., 2010). However, regardless of the stated probability, we administered a placebo and compared the degree of dopamine release to that obtained following open label drug. Not surprisingly, the degree of placebo-induced dopamine release was closely linked to the magnitude of release following open label levodopa. However, in order to conduct the experiment, subjects were deceived by telling them that they stood a particular probability (0%, 25%, 50% or 100%) of receiving verum ("true" or active as presented to patient and physician) medication at the time of the final scan. This may seem a relatively harmless untruth, considering the short-term nature of the particular study, as patients returned to their usual medication regimen as soon as the final scan was complete. Nonetheless, deception of any kind could be seen as disrespectful of the subjects' autonomy. In this case, we used "authorized deception," in which we advised subjects in advance of our need to deceive, coupled with a (true) promise to debrief them as soon as their own individual participation was complete. This approach is generally held to be ethically acceptable (Miller, Wendler, & Swartzman, 2005).

Another intriguing approach, championed by a modern-day pioneer in understanding the neurobiology of the placebo effect, is the use of an open vs. hidden design (Colloca, Lopiano, Lanotte, & Benedetti, 2004). In a series of fascinating and elegant experiments, Benedetti and his colleagues (2004) have demonstrated that the response to a given intervention is heavily dependent upon whether it is administered in an open (response heightened) or hidden (response diminished) fashion. A typical example would be to set up an infusion pump either behind a screen or fully visible to the individual. If people can see the drug being administered, their response is enhanced compared to hidden administration. Conversely, if the drug is withdrawn in an open fashion, the symptoms return faster and with greater severity compared to hidden withdrawal. This is true for several drug classes and conditions (opiates for pain, benzodiazepines for anxiety) and even for physiological responses (heart rate and blood pressure responses to beta blockers). The same approach can be taken with concealed vs. open manipulation of stimulation parameters in assessing the response to DBS. The difference in response between open and hidden states is thought to correspond to the placebo response. This may indeed represent an effective way to control for placebo effects in assessing the impact of acute symptomatic therapies for PD (including invasive surgical treatments such as DBS), but would not be applicable to treatments where a longer-term effect is under study.

Finally, for many years, it has been assumed that the placebo effect derives from expectation of therapeutic benefit, of which conditioning may represent a special case. Intriguingly, however, and entirely contrary to this assumption, is the demonstration that even subjects who are told they will be receiving placebo may derive benefit. In describing this seemingly unlikely observation, Kaptchuk differentiates between hope and expectation (Kaptchuk, 2018). Certain conditions in these open label placebo studies are worth noting. First, patients are prescribed treatment as usual, combined with either honestly prescribed placebo or nothing, but they know whether or not they are receiving a placebo or nothing. Secondly, while the treatment as usual is said to control for other effects, the time devoted to and the honesty of the discussion regarding the use of placebos, followed by their administration, may serve to substantially enhance the effect of treatment as usual. Thirdly, the conditions under study (irritable bowel syndrome, chronic low back pain, fatigue, migraine) are commonly associated with placebo responses and the patients were appropriately told this. The same is of course true for PD, but the very act of telling someone of possible benefit from a placebo may modify the response to therapy, whether one attributes this effect to expectation or hope. Fourth, the patients in the study were hoping for symptom relief, and from this perspective, the situation could be seen as analogous to a study in which PD patients are taking medication as a symptomatic adjunct to levodopa (treatment as usual), but is not comparable to the increasingly common scenario of medication for disease modification. It is also highly unlikely that patients or ethics boards would tolerate a scenario in which only sham invasive surgery, as opposed to the possibility of active vs. sham, was being offered. None of this is to take away from the fascinating observation that known placebos can be therapeutic, but it does not clearly indicate an alternative to placebo controls for clinical trials of novel therapies.

As pointed out by Justman (2013), there is an ethically important distinction between truly open disclosure of placebo use and the exploitation of ambiguous wording—the statement "I am going to give you something that has been shown to work for your condition in previous studies" when administering a placebo is technically truthful, but falls short of full disclosure. Others (Cohen & Shapiro, 2013) take a somewhat different view, however. These authors argue that the biological effect of the placebo (e.g., release of endogenous opioids or dopamine) may in some cases be comparable to those of verum therapy and in keeping with the desired therapeutic effect. The prescriber knows this and therefore believes there is a good chance of the desired biological and therapeutic outcomes. The degree of deception may under such circumstances therefore be morally defensible.

6. Are there additional risks associated with the use of placebo/sham?

While it may seem unlikely that a placebo could induce harm based on classical pathophysiological mechanisms in the absence of a surgical or other invasive intervention, just as placebos can induce anticipated benefits, they can also result in expected negative outcomes (the nocebo effect, see Mestre, 2020). While this may represent a useful control for assessing the effects of a novel therapy in a clinical trial setting, the harm induced may be very real. This raises a separate ethical concern related to the extent of information that is provided during an informed consent process and the need to balance appropriate disclosure of anticipated risks and benefits against the induction of nocebo-derived harm (Alfano, 2015; Colloca & Finniss, 2012). From this perspective, the increasing length and legalistic nature of most modern consent forms may lead to side effects and also defeat the purpose for which they were intended.

The very knowledge that one might receive a placebo may not only be sufficient to induce harm (nocebo effect) but may also detract from real benefits of verum therapy (so-called lessebo effect) (Mestre, Shah, Marras, Tomlinson, & Lang, 2014). Apart from the negative emotions presumably experienced by the affected trial participants, there is the additional very real risk of a Type II error, i.e., of failing to detect a truly beneficial response when one exists.

The use of deceptive placebo and consequent violation of patient autonomy may lead to a lack of trust, of the prescribing physician or researcher and a potentially more widespread mistrust of the profession in general, and unwillingness to participate in clinical trials. Finally, given the powerful linkage of placebo effects to activation of reward circuitry, it seems possible that placebos have the potential to be addictive in their own right.

7. Is the use of placebo/sham therapy justifiable in clinical practice, outside a research scenario?

While the focus of this manuscript has been on the ethics of using placebo/sham controls in experimental trials, many health care providers, and possibly even more patients, will ask whether the placebo effect can be harnessed for therapeutic benefit. Many aspects of the therapeutic relationship may result in benefit that is not directly attributable to a specific pharmacological action. Compassion and interest in the patient's condition combined with knowledge and an aura of authority may all contribute to "halo" effects, but these are typically (although not readily) differentiated from placebo effects. Physicians also knowingly (and to a shockingly high extent) prescribe or provide medication that is not associated with an anticipated specific effect on the condition they are treating (so-called "impure" placebos, e.g., vitamins, antibiotics for a viral illness) (Fässler, Meissner, Schneider, & Linde, 2010; Howick et al., 2013; Hróbjartsson & Norup, 2003; Tilburt, Emanuel, Kaptchuk, Curlin, & Miller, 2008). Note that the American Medical Association (AMA) definition of placebo includes a treatment or procedure that the prescribing physician believes to be devoid of physiological or pharmacological effect on the condition being treated—in other words, this definition encompasses the use of "impure" placebos and accordingly, AMA guidelines would explicitly condemn the unauthorized use of such treatments. In the not so distant past, some physicians would knowingly provide a completely inert substance, without disclosure, as a "therapeutic" agent, or would in some cases turn a response against the patient to "prove" that the patient's complaints had a functional basis. Apart from the disrespect for patient autonomy inherent in such an approach, the diagnostic validity is of course nullified by virtue of legitimate physiological mechanisms underlying placebo responses. Most (but not all) would agree that unauthorized deception of a patient is ethically untenable (Barnhill, 2011), unless there is an overwhelming benefit to the patient that justifies the use of deception (Annoni, 2018; Beauchamp & Childress, 2009). Some, however, believe that the therapeutic benefits derived from such an approach, coupled with the presumed lack of physiological harm, may justify the means. This has been the topic of considerable debate (Foddy, 2009; Miller & Colloca, 2009; Shah & Goold, 2009) and the prevailing view is that unauthorized deception is both ethically untenable and may, by damaging the therapeutic relationship, also be counterproductive. An intermediate position that would allow one to ethically determine for a given individual whether benefit derived from verum therapy is mostly if not entirely derived from placebo effects is that taken by the American Medical Association, in which one seeks the patient's permission to surreptitiously substitute a placebo for verum therapy, but without disclosing when this will be done (American Medical Association, 2019). Such an approach respects patient autonomy while allowing a valid comparison of placebo and verum responses; while this may be useful for avoiding the long-term use of unnecessary (and potentially harmful) therapies that are relatively ineffective, it may not be helpful as long-term therapy in its own right. There are additional harms that may arise from the unauthorized attempt to deceive patients in the clinical setting (see Miller & Colloca, 2009 for review). For instance, the unnecessary use of antibiotics as "impure placebos" may ultimately carry a societal risk in the form of promoting antibiotic resistance. Indiscriminate use of placebos may result in a widespread societal expectation that all problems can and should be fixed by a pill.

Members of the public themselves recognize that the use of a placebo may be acceptable in certain contexts, as long as there is partial disclosure, but unacceptable if overt lying is involved (Pugh et al., 2016). However, the acceptable partial disclosure may simply involve (truthfully) telling the patient that "I believe this pill will help you feel better because it has helped other people in your condition feel better," without explicitly indicating that it is a placebo. This is the position taken by Gold and Lichtenberg (2014), who also draw attention to the importance of the intent/beliefs of the prescriber—in other words, if the prescribing physician genuinely believes (as is entirely reasonable) that the prescription of a placebo may be helpful, it does not necessarily involve the use of deception, unless the physician actually implies that the treatment is in fact an agent with specific physiological or pharmacological properties related to the condition under consideration. However, Gold and Lichtenberg take this a step further, by suggesting that unless the physician believes that the patient would be upset by not being informed that the prescribed treatment was a placebo, withholding such information may be ethically acceptable.

It is in fact misguided to be overly sharp in demarcating the use of placebo vs. verum therapy, as the two effects may be additive (de la Fuente-Fernandez et al., 2001). Thus, placebo effects contribute to the beneficial effects of verum therapy, even when the latter is truly beneficial and can be ascribed to a well-defined physiological mechanism (Brody, 1982). The most obvious example of this may be the case of antidepressants, where meta-analysis has revealed that regardless of mechanism of action, the effect size of placebo treatment is always 75% of the effect size of the active therapy (Kirsch & Sapirstein, 1998). This means that we always use the placebo effect in clinical practice, although the extent may vary depending upon the way in which active therapy is offered—with diffidence vs. confidence, with warmth and empathy vs. in an emotionally neutral and/or detached fashion. Taking advantage of the additive benefits arising from the placebo effect would not be regarded as deception or disrespectful of patient autonomy—it is simply good clinical practice and maximizing one of the other major tenets of modern-day medical ethics, the need for beneficence.

The principle of beneficence and its counterpart, the avoidance of maleficence, may not uncommonly be in conflict with respect for patient autonomy (Ruddick, 1999). In treating Parkinson's, we know we are dealing with a progressive and incurable condition that is likely to result in reduced lifespan and that may progress to cognitive impairment, severe physical disability and the need for institutionalization. Yet no compassionate physician would consider it appropriate to outline such a trajectory until much later in the therapeutic relationship, and even then, only the context of assessing goals of care and advanced planning. In contrast, most physicians would emphasize all the positive things that the patient can do for him/herself and that may be achieved thanks to modern medicine and lifestyle considerations, such as exercise and a well-considered diet. Is this deception or is it providing hope? While technically somewhat different from the engagement of a placebo effect, there may be considerable overlap in both mechanistic and ethical considerations.

Brody (1982) also summarizes that the principles of beneficence and respect for autonomy need not be in conflict. Patients who are well educated about what to expect and how they can manage what is about to happen (post-operative pain), combined with reassurance that medication will be available if needed have lower analgesic requirements than those who are not provided this opportunity (Egbert, Battit, Welch, & Bartlett, 1964). This is a striking example of a placebo effect that involves no deception (and no placebos).

Traditionally, most investigators would attribute placebo responses to expectation of benefit and would accordingly argue that full disclosure would be doomed to failure, as there would be no realistic expectation of therapeutic benefit. Remarkably, as the work of Kaptchuk and others has shown, this is not the case (Charlesworth et al., 2017). While a full discussion of the potential mechanisms underlying benefit from Open Label Placebo is beyond the scope of this chapter, Kaptchuk highlights the importance of prediction-error processing (Kaptchuk, 2018). Given the importance of dopamine for signaling reward prediction error, this hypothesis is not incompatible with others, but may explain other features of the placebo response, such as the greater magnitude when baseline responses are more variable (Kaptchuk, 2018). Regardless of mechanism, this work supports the intriguing possibility that in some conditions, patients could be prescribed placebos with full disclosure of the intervention and yet have a realistic hope of deriving benefit. However, it should be remembered that not all placebo effects are the same—thus, the benefits of Open Label Placebo have to date been demonstrated in only a few conditions. Perhaps more importantly, the disclosure to patients generally includes a statement to the effect that prior studies have demonstrated that placebos may be effective. However, that is based on trials where patients were given placebos in the context of possibly receiving an active drug. As we have demonstrated in the case of placebo-evoked dopamine release in patients with Parkinson's, the magnitude of the placebo response is strongly affected by the degree of uncertainty as to whether or not one is receiving active drug, as well as the strength of the response to verum therapy (Lidstone et al., 2010). Thus, telling patients that placebos have been shown to work may itself be a statement that is only partially truthful (Blease, Colloca, & Kaptchuk, 2016).

Only a minority of patients surveyed believe that all clinical use of placebo is ethically unacceptable, but the degree of acceptability is linked to the prescriber's belief that the prescribed placebo therapy will be both safe and potentially beneficial (Hull et al., 2013). In that survey, 80% of respondents believed that a placebo could be effective only if the patients did not know they were receiving placebo and if they trusted the prescribing physician. However, most patients felt that physicians should be honest if asked what the patient had received (when effective). Not surprisingly, most people surveyed felt that the therapeutic relationship could be impacted if a patient discovered that s/he had been deliberately deceived—even if the placebo was effective. Most respondents would have been prepared to consider placebo therapy depending upon the circumstances. While it is difficult to generalize from the results of this survey (which presented a number of different possible theoretical scenarios), and similarly difficult to draw a unifying set of conclusions, the findings would seem to support the use of placebo therapy in a scenario where risk is low, where there is a reasonable expectation of benefit, and where it is prescribed in a fully open fashion, or in the context of authorized deception.

8. Conclusions

In summary, there is probably no single correct answer that would satisfy all positions on the use of placebo/sham controls. However, it is salutary to remember that while not all interesting and technically rigorous science is necessarily ethical, bad science is by definition unethical, as it exposes people to risk, be it high or low, with no conclusive outcome and it may serve to prolong or even increase the use of therapies that are at best unhelpful and may be associated with their own risks. This was indeed the second principle outlined in the Nuremberg Code. It therefore seems likely that placebos will be here to stay, at least in research into novel therapies for PD, and the placebo effect will likely continue to be exploited in clinical use of established treatments.

References

Alfano, M. (2015). Placebo effects and informed consent. *The American Journal of Bioethics*, *15*(10), 3–12. https://doi.org/10.1080/15265161.2015.1074302.

Annoni, M. (2018). The ethics of placebo effects in clinical practice and research. *International Review of Neurobiology*, *139*, 463–484. https://doi.org/10.1016/bs.irn.2018.07.031.

American Medical Association. (2019). *Code of medical ethics opinion 2.1.4: Use of placebo in clinical practice. Retrieved from* https://www.ama-assn.org/delivering-care/ethics/use-placebo-clinical-practice.

Barnhill, A. (2011). What it takes to defend deceptive placebo use. *Kennedy Institute of Ethics Journal*, *21*(3), 219–250. https://doi.org/10.1353/ken.2011.0015.

Beauchamp, T. L., & Childress, J. F. (2009). *Principles of biomedical ethics* (6th ed.). New York: Oxford University Press.

Benedetti, F., Colloca, L., Torre, E., Lanotte, M., Melcarne, A., Pesare, M., et al. (2004). Placebo-responsive Parkinson patients show decreased activity in single neurons of subthalamic nucleus. *Nature Neuroscience*, 7, 587–588.

Blease, C., Colloca, L., & Kaptchuk, T. J. (2016). Are open-label placebos ethical? Informed consent and ethical equivocations. *Bioethics*, *30*(6), 407–414. https://doi.org/10.1111/bioe.12245.

Brody, H. (1982). The lie that heals: The ethics of giving placebos. *Annals of Internal Medicine, 97*(1), 112–118. https://doi.org/10.7326/0003-4819-97-1-112.
Charlesworth, J. E. G., Petkovic, G., Kelley, J. M., Hunter, M., Onakpoya, I., Roberts, N., et al. (2017). Effects of placebos without deception compared with no treatment: A systematic review and meta-analysis. *Journal of Evidence-Based Medicine, 10*(2), 97–107. https://doi.org/10.1111/jebm.12251.
Cohen, S., & Shapiro, H. (2013). "Comparable placebo treatment" and the ethics of deception. *The Journal of Medicine and Philosophy, 38*(6), 696–709. https://doi.org/10.1093/jmp/jht052.
Colloca, L., & Finniss, D. (2012). Nocebo effects, patient-clinician communication, and therapeutic outcomes. *JAMA, 307*(6), 567–568. https://doi.org/10.1001/jama.2012.115.
Colloca, L., Lopiano, L., Lanotte, M., & Benedetti, F. (2004). Overt versus covert treatment for pain, anxiety, and Parkinson's disease. *Lancet Neurology, 3*(11), 679–684.
Dawson, T. M., Golde, T. E., & Lagier-Tourenne, C. (2018). Animal models of neurodegenerative diseases. *Nature Neuroscience, 21*(10), 1370–1379. https://doi.org/10.1038/s41593-018-0236-8.
de la Fuente-Fernandez, R., Ruth, T. J., Sossi, V., Schulzer, M., Calne, D. B., & Stoessl, A. J. (2001). Expectation and dopamine release: Mechanism of the placebo effect in Parkinson's disease. *Science, 293*(5532), 1164–1166.
Deuschl, G., Schade-Brittinger, C., Krack, P., Volkmann, J., Schafer, H., Botzel, K., et al. (2006). A randomized trial of deep-brain stimulation for Parkinson's disease. *The New England Journal of Medicine, 355*(9), 896–908.
Egbert, L. D., Battit, G. E., Welch, C. E., & Bartlett, M. K. (1964). Reduction of postoperative pain by encouragement and instruction of patients: A study of doctor-patient rapport. *The New England Journal of Medicine, 270*, 825–827. https://doi.org/10.1056/NEJM196404162701606.
Espay, A. J., Norris, M. M., Eliassen, J. C., Dwivedi, A., Smith, M. S., Banks, C., et al. (2015). Placebo effect of medication cost in Parkinson disease: A randomized double-blind study. *Neurology, 84*(8), 794–802. https://doi.org/10.1212/WNL.0000000000001282.
Faden, R. R., Beauchamp, T. L., & King, N. M. P. (1986). *A history and theory of informed consent*. New York: Oxford University Press.
Fässler, M., Meissner, K., Schneider, A., & Linde, K. (2010). Frequency and circumstances of placebo use in clinical practice—A systematic review of empirical studies. *BMC Medicine, 8*, 15. https://doi.org/10.1186/1741-7015-8-15.
Foddy, B. (2009). A duty to deceive: Placebos in clinical practice. *The American Journal of Bioethics, 9*(12), 4–12. https://doi.org/10.1080/15265160903318350.
Frank, S. A., Wilson, R., Holloway, R. G., Zimmerman, C., Peterson, D. R., Kieburtz, K., et al. (2008). Ethics of sham surgery: Perspective of patients. *Movement Disorders, 23*(1), 63–68. https://doi.org/10.1002/mds.21775.
Goetz, C. G., Leurgans, S., Raman, R., & Stebbins, G. T. (2000). Objective changes in motor function during placebo treatment in PD. *Neurology, 54*(3), 710–714.
Goetz, C. G., Wuu, J., McDermott, M. P., Adler, C. H., Fahn, S., Freed, C. R., et al. (2008). Placebo response in Parkinson's disease: Comparisons among 11 trials covering medical and surgical interventions. *Movement Disorders, 23*(5), 690–699.
Gold, A., & Lichtenberg, P. (2014). The moral case for the clinical placebo. *Journal of Medical Ethics, 40*(4), 219–224. https://doi.org/10.1136/medethics-2012-101314.
Groopman, J. E. (2004). *The anatomy of hope: How people prevail in the face of illness* (1st ed.). New York: Random House.
Howick, J., Bishop, F. L., Heneghan, C., Wolstenholme, J., Stevens, S., Hobbs, F. D., et al. (2013). Placebo use in the United Kingdom: Results from a national survey of primary care practitioners. *PLoS One, 8*(3), e58247. https://doi.org/10.1371/journal.pone.0058247.

Hróbjartsson, A., & Norup, M. (2003). The use of placebo interventions in medical practice—A national questionnaire survey of Danish clinicians. *Evaluation & the Health Professions*, *26*(2), 153–165. https://doi.org/10.1177/0163278703026002002.

Hull, S. C., Colloca, L., Avins, A., Gordon, N. P., Somkin, C. P., Kaptchuk, T. J., et al. (2013). Patients' attitudes about the use of placebo treatments: Telephone survey. *BMJ*, *347*, f3757. https://doi.org/10.1136/bmj.f3757.

Justman, S. (2013). Deceit and transparency in placebo research. *The Yale Journal of Biology and Medicine*, *86*(3), 323–331.

Kaptchuk, T. J. (2018). Open-label placebo: Reflections on a research agenda. *Perspectives in Biology and Medicine*, *61*(3), 311–334. https://doi.org/10.1353/pbm.2018.0045.

Kim, S. Y., Frank, S., Holloway, R., Zimmerman, C., Wilson, R., & Kieburtz, K. (2005). Science and ethics of sham surgery: A survey of Parkinson disease clinical researchers. *Archives of Neurology*, *62*(9), 1357–1360. https://doi.org/10.1001/archneur.62.9.1357.

Kimmelman, J., London, A. J., Ravina, B., Ramsay, T., Bernstein, M., Fine, A., et al. (2009). Launching invasive, first-in-human trials against Parkinson's disease: Ethical considerations. *Movement Disorders*, *24*(13), 1893–1901.

Kirsch, I., & Sapirstein, G. (1998). Listening to Prozac but hearing placebo: A meta-analysis of antidepressant medication. *Prevention & Treatment*, *1*(2). https://doi.org/10.1037/1522-3736.1.1.12a Article 0002a.

Ko, J. H., Feigin, A., Mattis, P. J., Tang, C. C., Ma, Y., Dhawan, V., et al. (2014). Network modulation following sham surgery in Parkinson's disease. *The Journal of Clinical Investigation*, *124*(8), 3656–3666. https://doi.org/10.1172/JCI75073.

Köteles, F., & Ferentzi, E. (2012). Ethical aspects of clinical placebo use: What do laypeople think? *Evaluation & the Health Professions*, *35*(4), 462–476. https://doi.org/10.1177/0163278712453993.

Lang, A. E., & Espay, A. J. (2018). Disease modification in Parkinson's disease: Current approaches, challenges, and future considerations. *Movement Disorders*, *33*(5), 660–677. https://doi.org/10.1002/mds.27360.

Lidstone, S. C., Schulzer, M., Dinelle, K., Mak, E., Sossi, V., Ruth, T. J., et al. (2010). Effects of expectation on placebo-induced dopamine release in Parkinson disease. *Archives of General Psychiatry*, *67*(8), 857–865. https://doi.org/10.1001/archgenpsychiatry.2010.88.

Marks, W. J., Jr., Bartus, R. T., Siffert, J., Davis, C. S., Lozano, A., Boulis, N., et al. (2010). Gene delivery of AAV2-neurturin for Parkinson's disease: A double-blind, randomised, controlled trial. *Lancet Neurology*, *9*(12), 1164–1172.

McRae, C., Cherin, E., Yamazaki, T. G., Diem, G., Vo, A. H., Russell, D., et al. (2004). Effects of perceived treatment on quality of life and medical outcomes in a double-blind placebo surgery trial. *Archives of General Psychiatry*, *61*(4), 412–420.

Mestre, T. A. (2020). Nocebo and lessebo effects. *International Review of Neurology*, *153*, 121–146.

Mestre, T. A., Shah, P., Marras, C., Tomlinson, G., & Lang, A. E. (2014). Another face of placebo: The lessebo effect in Parkinson disease: Meta-analyses. *Neurology*, *82*(16), 1402–1409. https://doi.org/10.1212/WNL.0000000000000340.

Metzler, I., & Just, P. (2018). "Think positively": Parkinson's disease, biomedicine, and hope in contemporary Germany. *Health (London, England)*, *22*(5), 483–499. https://doi.org/10.1177/1363459317715774.

Miller, F. G., & Colloca, L. (2009). The legitimacy of placebo treatments in clinical practice: Evidence and ethics. *The American Journal of Bioethics*, *9*(12), 39–47. https://doi.org/10.1080/15265160903316263.

Miller, F. G., Wendler, D., & Swartzman, L. C. (2005). Deception in research on the placebo effect. *PLoS Medicine*, *2*(9), e262.

Pugh, J., Kahane, G., Maslen, H., & Savulescu, J. (2016). Lay attitudes toward deception in medicine: Theoretical considerations and empirical evidence. *AJOB Empirical Bioethics*, 7(1), 31–38. https://doi.org/10.1080/23294515.2015.1021494.

Ruddick, W. (1999). Hope and deception. *Bioethics*, *13*(3–4), 343–357. https://doi.org/10.1111/1467-8519.00162.
Shah, K. R., & Goold, S. D. (2009). The primacy of autonomy, honesty, and disclosure—Council on Ethical and Judicial Affairs' placebo opinions. *The American Journal of Bioethics*, *9*(12), 15–17. https://doi.org/10.1080/15265160903316339.
Snyder, J., Adams, K., Crooks, V. A., Whitehurst, D., & Vallee, J. (2014). "I knew what was going to happen if I did nothing and so I was going to do something": Faith, hope, and trust in the decisions of Canadians with multiple sclerosis to seek unproven interventions abroad. *BMC Health Services Research*, *14*, 445. https://doi.org/10.1186/1472-6963-14-445.
Strafella, A. P., Ko, J. H., & Monchi, O. (2006). Therapeutic application of transcranial magnetic stimulation in Parkinson's disease: The contribution of expectation. *NeuroImage*, *31*(4), 1666–1672.
Tilburt, J. C., Emanuel, E. J., Kaptchuk, T. J., Curlin, F. A., & Miller, F. G. (2008). Prescribing "placebo treatments": Results of national survey of US internists and rheumatologists. *BMJ*, *337*, a1938. https://doi.org/10.1136/bmj.a1938.
Touwen, D. P., & Engberts, D. P. (2012). Those famous red pills—Deliberations and hesitations. Ethics of placebo use in therapeutic and research settings. *European Neuropsychopharmacology*, *22*(11), 775–781. https://doi.org/10.1016/j.euroneuro.2012.03.005.
World Medical Association. (2013). World medical association declaration of Helsinki: Ethical principles for medical research involving human subjects. *JAMA*, *310*(20), 2191–2194. https://doi.org/10.1001/jama.2013.281053.

PART 2

Placebo response in specific disease populations

CHAPTER EIGHT

Placebo effects in pain

Luana Colloca*
Department of Pain Translational Symptoms Science, School of Nursing, University of Maryland, Baltimore, MD, United States
Center to Advance Chronic Pain Research, University of Maryland, Baltimore, MD, United States
Departments of Anesthesiology and Psychiatry, School of Medicine, University of Maryland, Baltimore, MD, United States
*Corresponding author: e-mail address: colloca@umaryland.edu

Contents

Abstract

Placebo effects influence symptom perceptions and treatment outcomes. Placebo effects can be explored in laboratory settings controlling for natural history and expectations. Such a mechanistic approach to neurological disorders has been implemented in the domain of chronic clinical pain and other neurological disorders. This article therefore focuses on definitions and historical notes related to placebo effects and mechanisms of placebo effects in chronic pain. Knowledge on mechanisms of placebo effects could inform current clinical practice for the treatment of neurological disorders by focusing on patients (and providers) expectations for outcome optimization.

International Review of Neurobiology, Volume 153
ISSN 0074-7742
https://doi.org/10.1016/bs.irn.2020.04.001

1. Definitions and historical notes

The term "Placebo" is the Latin word for "I shall please" and appeared in the opening phrase, "Placebo domino in regione vivorum," of the Catholic Vesper for the Dead, psalm 116, 9th verse (Brown, 1998; de Craen, Kaptchuk, Tijssen, & Kleijnen, 1999; Wall, 1992). Despite some anomalies in both the origin and the translation of the term "placebo," it entered the medical lexicon to indicate sham interventions, treatments and substances (i.e. sugar pills, saline solutions, simulation of surgical interventions). Physicians have been deliberately given over the years to please or placate patients or to evoke a placebo response. However, the vestiges of this sort of consoling approach to cure are often used without any sort of evidence-based approach challenging the morale-boosting and pleasing use of remedies such as sugar pills, electuary or pharmaceutical syrups, all of which had no known pharmacodynamic actions (Brody, Colloca, & Miller, 2012; Colloca & Howick, 2018; Miller & Colloca, 2009). Those doctors who provided their patients with these sorts of morale-boosting remedies did so either to reassure their patients while the vis medicatrix naturæ (i.e. "the healing power of nature") that performed its normalizing restoring task, or to gratify their patients' need for being treated.

The New Medical Dictionary, published in 1785, described placebo as "a commonplace method or medicine" (de Craen et al., 1999). In 1811, placebo was defined as "an epithet given to any medicine adapted more to please than benefit the patients" (Hooper's Medical Dictionary, 1843). This time coincides with the advent of the clinic in Western medicine. By the mid-20th century, the technical term placebo was precisely applied within specialized medical domains of pharmacology, nosology, and etiology to denote the pharmacologically inert, dummy simulator of an active medication that serves as a control in clinical trials designed to validate the clinical efficacy of that particular medication. Thus, placebos have long been used in clinical (randomized) trials for the validation of new medication.

Within the 1950s, physicians and researchers began to acknowledge placebo as a confounding variable within experimental studies. Shortly after, placebos became defined by Gotzsche as "the difference in outcome between a placebo-treated group and an untreated control group in an unbiased experiment" (Gotzsche, 1994, 1995). It appears that the history of the placebos is not the history of the placebo effects, but rather is part of the history of medicine. Physicians have always been aware of the fact that patients may get better after taking inert drugs (Archer & Leier, 1992; Beecher, 1955; Benson & McCallie Jr., 1979; Brown, 1998; Turner, Deyo, Loeser, Von Korff, & Fordyce, 1994) raising a need to define the placebo effect as a distinct phenomenon.

Different definitions for placebo effects have been suggested over the years. According to Arthur K. Shapiro, "the placebo effect is the non-specific, psychological, or psychophysiological therapeutic effect produced by a placebo, or the effect of spontaneous improvement attributed to the placebo" (Shapiro & Shapiro, 1997). Therefore, a placebo effect appeared to be the response to placebos that is the response to an inert substance. However, it is important to point out that the effect is not due to the uselessness of the treatment per se. Daniel Moerman, a medical anthropologist, asserted many times that "placebos are inert and don't cause anything" (Hutchinson & Moerman, 2018; Moerman, 2000, 2002, 2013; Moerman & Jonas, 2000, 2002). Moerman suggested the use of the formulation "meaning response" rather than "placebo response." The meaning response is defined as the physiological or psychological effects of meaning on the treatment or illness. Conceptualizing the issue in terms of meaning may be important from an evolutionary perspective (Humphrey, 2002).

Brody indicated that the placebo effect is a change in the patient's health or bodily state attributable to the symbolic impact that one assigns to an event or object in the healing environment (Brody & Brody, 2000). In the same year, Vase, Riley 3rd, and Price (2002) proposed the definition of the placebo response as the reduction in a symptom that results from factors related to a subject's perception of the therapeutic intervention (Vase et al., 2002).

The term "placebo effect" has been also used to refer to any average improvement in the condition of a group of subjects that has received a placebo manipulation. In contrast, the term "placebo response" refers to the change in an *individual* caused by a placebo manipulation (Hoffman, Harrington, & Fields, 2005).

Benedetti (2002) emphasized the crucial role of context in triggering a placebo effect. In fact, any inert medical treatment is given within a "context" that surrounds the patients (Benedetti, 2002). Words, providers' behaviors and attitudes, and medical devices constitute the context that Balint (1955) called the "whole atmosphere around the treatment". Often the term "context effect" and "placebo effect" has been used, at least in part, interchangeably in order to overcome the negative connotations associated with the term placebo and to highlight the therapeutic nature of the healthcare context. Di Blasi, Harkness, Ernst, Georgiou, and Kleijnen (2001) examined the effect of health care interactions on the physical state of health across 25 identified randomized controlled trials. The authors found that there is consistency regarding the effectiveness of reassurance in the consultations in comparison with a formal patient-doctor interaction (Di Blasi et al., 2001). The fact that cognitive and emotional care can influence a patient's health outcomes is in line with research on patient-doctor relationships (Blasini, Peiris, Wright, & Colloca, 2018; Peiris, Blasini, Wright, & Colloca, 2018).

The *placebo effect* refers to "a powerful determinant of health across many different diseases and encounters; the placebo effect is due to expectancy of positive treatment outcomes" (Colloca, 2019; Colloca & Barsky, 2020). The *placebo response* refers to "outcome changes that are due to natural history, biases, regression to the mean, and other nonspecific effects" (Colloca, 2019). Similar definitions have been consensually accepted by the a team of experts representative of the research community that conducts placebo investigation who gathered together in Leiden, Netherlands (2017) to discuss this matter (Evers et al., 2018).

2. Scientific basis of placebo effects on pain

2.1 Neuro-anatomical bases of placebo effects in pain

Recently, functional imaging techniques, such as functional magnetic resonance imaging (fMRI), positron emission tomography (PET), electroencephalogragphy (EEG), and magnetoelectroencephalography (MEG) have given the opportunity to define the neuroanatomical bases of placebo effects in particular in the domain of experimental pain given the feasibility to manipulate expectancies around a placebo (Colloca, Benedetti, & Porro, 2008; Rainville & Duncan, 2006; Tracey, 2005).

The first study of placebo analgesia and brain imaging showed that a subset of brain regions is similarly affected by either a placebo or a μ-opioid agonist. In fact, by using positron emission tomography (Petrovic et al., 2005, 2010; Petrovic, Kalso, Petersson, & Ingvar, 2002), it was found that the very same regions in the brain are affected by both a placebo and the opioid agonist remifentanil, thus indicating a potential related mechanism in placebo-induced (psychodynamic effect) and opioid-induced analgesia (pharmacodynamic effect). In particular, Petrovic and colleagues reported that the administration of a placebo induces the activation of brain regions such as the rostral anterior cingulate cortex (rACC), the orbitofrontal cortex (OrbF), and the anterior insula (aINS) (Petrovic et al., 2002). Furthermore, they found that there was a significant co-variation in activity between rACC and the lower pons and medulla, and a co-variation between rACC and the periacqueductal gray (PAG). This suggests that a descending rACC-PAG-pons-medulla pain-modulating circuit is involved in placebo analgesia, as previously demonstrated and postulated by other authors (Fields, 2000, 2004, 2014, 2018).

An opioid neuronal network in the cerebral cortex and the brainstem has been described as a descending pain-modulating pathway that connects, either directly or indirectly, the cerebral cortex to the brainstem (Fields & Levine, 1981; Levine, Gordon, Bornstein, & Fields, 1978, 1979). In particular, ACC and OrbF project to the PAG which, in turn, modulates the activity of the rostral ventromedial medulla (RVM). ACC and PAG, together with some other nuclei in the brainstem (e.g. the parabrachial nuclei), are rich with opioid receptors. Therefore, this pain-modulating circuit is the same as that activated in placebo analgesia (Zubieta et al., 2005).

Other studies have revealed that the activity of brain areas related to pain, particularly the thalamus, aINS and caudal rACC, decreases by virtue of a placebo treatment and these changes have been interpreted as an effect that placebos reduce nociceptive transmission along the pain pathways (Lieberman et al., 2004; Price, Craggs, Verne, Perlstein, & Robinson, 2007; Wager et al., 2004).

Subcortical areas and their possible implications in placebo analgesia have also been considered (Bingel, Lorenz, Schoell, Weiller, & Buchel, 2006), as well as the differences between responders and non-responders in neurophysiological parameters such as laser-evoked event related potentials (LEP) (Colloca et al., 2008; Wager, Matre, & Casey, 2006). The evidence is that top-down mechanisms could inhibit or facilitate pain signals at the level of the spinal cord (Eippert, Finsterbusch, Bingel, & Buchel, 2009; Tinnermann, Geuter, Sprenger, Finsterbusch, & Buchel, 2017). This idea fits with a study that showed how expectancy induces a reduction of secondary hyperalgesia, which is known to involve spinal mechanisms (Matre, Casey, & Knardahl, 2006).

A recently published study indicated that price of medication can also impact the coupling with the PAG and rACC in the presence of negative placebo (nocebo) effects (Tinnermann et al., 2017). Healthy volunteers underwent two placebo creams labeled with two distinct prices and presented to the volunteers as creams that would have relieved itch with the side effect of inducing local pain. The cream labeled as more expensive induced larger nocebo hyperalgesic effects than the less expensive cream (Tinnermann et al., 2017). The authors used a combined cortico-spinal imaging technique and demonstrated that the price augmented the blood-oxygen-level-dependent (BOLD) fMRI imaging activity in the prefrontal cortex. The rACC and the PAG paralleled differential nocebo effects between the expensive and cheaper creams. These findings suggest that expectancies of higher pain-related side effects associated with the more expensive cream have most likely triggered a facilitation of nociception processes at the level of early subcortical areas and the spinal cord inducing a behavioral increase of pain perception.

Importantly, a recent article presented the results of 20 studies and 603 healthy subjects that involved the human brain functional neuroimaging with pain delivered under stimulus intensity-matched placebo and control conditions conducted between December 2015 and November 2017 (Zunhammer, Bingel, Wager, & Placebo Imaging, 2018). The main Neurological Pain Signature (NPS) a validated measure that tracks levels of nociceptive pain (Wager et al., 2013) was the main target of the meta-analysis. NPS changes following painful stimulations compared with baseline conditions were found in 575 participants, with a very large effect size ($g=2.30$ [95% CI, 1.92–2.69]). Placebo manipulations induced significant behavioral reductions on pain ratings in 17 of 20 studies ($g=-0.66$ [95% CI, −0.80 to −0.53]). Placebo effects on the bottom-up nociceptive processing via NPS response were significant in only 3 of 20 studies and were very small in the combined sample ($g=-0.08$ [95% CI, −0.15 to −0.01]). The small effects on NPS indicate that placebo treatments affect pain via brain mechanisms largely dependent on top-down cognitive processes (Zunhammer et al., 2018).

2.2 Anticipation of benefits and expectancies

Treatments are not merely given in a vacuum and when consciously active, humans (and animals) form anticipations of benefits (Colloca & Miller, 2011a, 2011b). During the anticipation phase of a placebo effect (e.g. pain relief), it has been reported an activation of the dorsolateral prefrontal cortex (DLPFC), OrbF, rostral medial and anterior prefrontal cortex (rmAPC and aAPC), superior parietal cortex (SPC) and PAG, suggesting the activation of decisional, cognitive-evaluative network just before the placebo effect. The increased activity of PAG also suggests the activation of endogenous opioids in the anticipatory phase of the placebo response (Wager et al., 2004; Wager & Nitschke, 2005; Wager, Scott, & Zubieta, 2007).

A recent activation likelihood estimation meta-analysis analyzed how anticipation of pain affects several brain regions including 19 functional magnetic resonance imaging human studies (Amanzio, Benedetti, Porro, Palermo, & Cauda, 2013). The authors reported that during anticipation there are active foci in the DLPC, midcingulate and AI cortices, medial and inferior frontal gyri, inferior parietal lobule, middle and superior temporal gyrus, thalamus, and caudate. Deactivated foci were identified in the ACC, superior frontal gyrus, parahippocampal gyrus and in the claustrum. Therefore, a highly distributed set of self-regulation mechanisms active emotion, action and perception during anticipation of experimental noxious pain.

Keltner et al. used a balanced experimental design with two levels of noxious thermal stimuli (high and low) and two corresponding levels of expectancy (high and low) produced by visual cues briefly conditioned with each of the two noxious stimuli (Keltner et al., 2006). Consistently with previous reports, significant fMRI blood-oxygen-level-dependent imaging, (BOLD) activation were found in nociceptive regions (thalamus, insular cortex, somatosensory cortex, and anterior cingulated cortex) when the high-intensity noxious stimulus was anticipated and accompanied by the high-intensity cue along with the activation of orbitofrontal cortex, amygdala, ventral striatum, and the nucleus cuneiformis (nCF) were found. The fMRI BOLD activation was significantly less when the high-intensity noxious stimulus was anticipated and accompanied by the low-intensity visual cue. These findings point to a significant contribution of expectancy to the magnitude of fMRI BOLD signals in response to cutaneous noxious thermal stimuli (Keltner et al., 2006).

By using an expectancy manipulation model combined with a sham acupuncture treatment, Kong et al. reported changes of fMRI BOLD signals in bilateral lateral/orbital prefrontal cortex, rACC, cerebellum, right fusiform, parahippocampus and, pons, suggesting that placebo analgesia may be configured through multiple brain pathways and mechanisms (Kong et al., 2006).

Expectations are often misaligned with experiential outcomes in clinical contexts and daily life circumstances. We modeled in a laboratory setting the lack of congruency between what is expected and what is actually experienced to determine the effects at the level of behavioral and neural placebo and nocebo responsiveness. We demonstrated that expectancies and outcomes that are irreconcilable result in a blockage of placebo effects and a mitigation of nocebo effects (Colloca, Schenk, Nathan, Robinson, & Grillon, 2019). This effect is reflected at the brain level by an activation of the left inferior parietal cortex. The left inferior parietal cortex has been reported as a brain region that may redirect attention in the presence of discrepancies between sensory and cognitive events (Zeidan, Lobanov, Kraft, & Coghill, 2015). These findings demonstrated for the first time how the lack of consistency between anticipation of outcomes (e.g. individual expectancies) and experiential intervention outcomes can be deleterious for the formation of therapeutic placebo effects. On the contrary, a discrepancy between expectancies and outcome can be exploited as a mechanism for intentionally reducing nocebo effects in clinical trials and practices. Future research efforts should be directed to translate these findings approach in studies with patients and then, real-world settings.

3. Clinical pain syndromes

3.1 Acute and clinical chronic pain disorders

The impact of placebo effects has been extensively explored in pain research mainly with healthy participants (Colloca, 2019). This abundance of studies has allowed us to understand the mechanisms underlying placebo effects. Recently, the trend and efforts have moved to translational research that can inform us about the nature of placebo effects in acute post-operative pain and chronic pain disorders with a focus on physiopathology of pain (e.g. neuropathic vs non-neuropathic pain), source of pain (post-surgical vs inflammatory pain), association with comorbidities (psychiatric vs other diseases) and psychological factors (catastrophizing vs somatization).

Responses to placebos affect the success of clinical trials. In 2011, Clinicaltrials.gov listed over 4000 pain-related trials, but the new approvals were limited to five existing drugs in new formulations or dosage forms (e.g. duloxetine, oxycodone, and fentanyl) (Bartfai & Lees, 2011). In clinical trials for new treatments of cancer and neuropathic pain, the failure rate has been over 90% in the past 10 years (Bartfai & Lees, 2011; Katz, Finnerup, & Dworkin, 2008). Therefore, knowledge about the mechanisms of placebos and their effects can be relevant to inform the designs of clinical trials.

Over the last 2 decades, placebo effects have been investigated in a variety of chronic pain disorders (Colloca, 2019). The studies described below have been conducted under controlled laboratory conditions and do not refer to placebo randomized control trials. Placebo effects and expectancies have been explored in neuropathic pain (Petersen et al., 2014, 2012; Vase, Petersen, & Lund, 2014), low back pain (Carvalho et al., 2016; Hashmi et al., 2012; Klinger, Kothe, Schmitz, Kamping, & Flor, 2017), knee osteoarthritis (Tetreault et al., 2016), irritable bowel syndrome (Kaptchuk et al., 2008; Vase, Robinson, Verne, & Price, 2003, 2005), fibromyalgia (Kosek et al., 2017), migraine (Kam-Hansen et al., 2014) and orofacial pain (Blasini, Movsas, & Colloca, 2018). Placebo analgesia appears to be present despite experiencing hyperalgesic chronic pain states. Expectations, medical history, duration of the disorder, concurrence of comorbidities and emotion regulation might play a role in engaging the endogenous modulatory systems underlying placebo effects in patients with chronic pain disorders.

3.2 Neuropathic pain

Placebo effects have been recently investigated in neuropathic pain patients (Petersen et al., 2014, 2012; Skyt et al., 2018a, 2018b; Vase et al., 2014; Vase, Skyt, & Hall, 2016) who suffer from nerve injury (Colloca et al., 2017). Peterson and colleagues exposed 19 patients who had developed neuropathic pain after thoracotomy to a placebo manipulation consisting of either open or hidden administrations of lidocaine (Petersen et al., 2012). Levels of spontaneous pain and expected pain were collected along with quantitative sensory testing of evoked pain including brush and cold allodynia, heat pain tolerance, area of pinprick hyperalgesia, and wind-up-like pain (that is increased *pain* intensity over time in response to a same-intensity stimulus is delivered repeatedly) after pinprick stimulation. The placebo manipulation significantly reduced the area of pinprick hyperalgesia, and this placebo effect was significantly related to low levels of negative affect. Following this first study (Petersen et al., 2012), the authors explored how emotional feelings via a quantitative/qualitative experiential method could affect the observed significant placebo effects on both ongoing and evoked neuropathic pain (Petersen et al., 2014). Expected pain levels accounted for significant amounts of the variance in ongoing (53.4%) and evoked pain (up to 34.5%) after the open lidocaine administration. High levels of positive and low levels of negative emotional feelings were observed in the placebo condition suggesting that patients' expectations may together with emotional feelings influence treatment outcomes.

Skyt and colleagues tested the hypothesis that patients with neuropathic pain may have large placebo effects that are mediated by the dopaminergic system (Skyt et al., 2018a). The authors investigated 19 patients with neuropathic pain related to thoracic surgery using an open/hidden administration of the pain-relieving agent lidocaine (2 mL) and no treatment. Additionally, patients with neuropathic pain received the dopamine antagonist haloperidol (2 mg) and the agonist levodopa/carbidopa (100/25 mg) to determine whether placebo effects would have been antagonized or enhanced, respectively. The outcomes include assessment of expected pain levels, desire for pain relief, and ongoing and evoked pain. The authors observed robust placebo effects on ongoing and evoked pain. Expectation ratings and desire of pain relief accounted for up to 41.2% and 71.5% of the variance in ongoing and evoked pain, respectively, following the open application of lidocaine. However, neither haloperidol nor levodopa/carbidopa affected placebo effects.

All these three studies are relevant. However, definitive conclusions are difficult to be drawn as the studies were underpowered due to their small sample size.

3.3 Low back pain

Low back pain represents a debilitative disorder (Hoy et al., 2014) and a variety of nonpharmacological treatments have been explored (Chou et al., 2017). A few studies have explored the possibility that placebo effects occur when psychological manipulation such as conditioning and verbal suggestions is used in laboratory settings. Klinger et al. completed a study in chronic back pain in which participants were randomized to two placebos—a sham opioid solution compared with an alleged neutral, completely inactive solution (same) solution (Klinger et al., 2017). When presented as an effective treatment, the sham "opioid" solution triggered placebo analgesia resulting in a low ratings of the patients' clinical back pain and acute experimental pain ratings. A change in expectations was also observed suggesting that can be clinically useful to explicitly integrate expectation manipulation in delivering any treatment for pain management (Klinger et al., 2017). Not only do patients with chronic low pain respond to saline solution given deceptively, but open-labeled placebos (patients deliberately informed about the use of sham interventions) can still induce a placebo therapeutic effect (Carvalho et al., 2016).

At the level of neural responses, it is likely that some brain changes may predict placebo responses. Hashmi et al. examined chronic back pain in patients who participated in a double-blind fMRI clinical trial and were treated, for 2 weeks, with topical analgesic or no drug patches (Hashmi et al., 2012). The baseline functional connectivity between left medial prefrontal cortex and bilateral insula accurately predicted post-treatment groups indicating that placebo responses can be potentially identified a priori (Hashmi et al., 2012). In another study with brain imaging of osteoarthritis patients, the right midfrontal gyrus connectivity identified placebo responders and the right parahippocampus gyrus connectivity predicted drug analgesia (Tetreault et al., 2016). More recently, Apkarian's team (Vachon-Presseau et al., 2018) demonstrated that using MRI and fMRI that the response to placebo pills is linked to the subcortical limbic volume asymmetry, sensorimotor cortical thickness, and functional coupling of prefrontal regions, anterior cingulate, and periaqueductal gray—regions that appear to be predictive of response.

3.4 Fibromyalgia and episodic migraine

Patients suffering from fibromyalgia who experience dysfunctions of central pain show that longer duration of the disease is negatively associated with placebo-induced analgesia (Kosek et al., 2017). Future studies are needed to understand whether the etiology of fibromyalgia may interfere with the activation of the endogenous pain modulatory system.

Episodic migraine has been also investigated for placebo effects and how information provided to patients influence placebo and drug effects in a prospective, within-subjects, repeated-measures study with medication labeling modified for placebo and drug effects. Participants received either placebo or Maxalt (10-mg rizatriptan) given under three information conditions—told placebo, told Maxalt or placebo, told Maxalt. Active drug Maxalt has a superior efficacy than placebo for pain relief. However, when was placebo labeled as (i) placebo, (ii) Maxalt or placebo, and (iii) Maxalt, the placebo effect increased progressively. Interestingly, when Maxalt was labeled as placebo and placebo labeled as Maxalt the efficacy of the two were similar and the benefit of placebo persisted even if placebo was open-labeled (no deception) (Kam-Hansen et al., 2014). These findings indicate the relevance of framing effects (Barnes et al., 2019) when both active and placebo treatments are given.

3.5 Orofacial pain

An emerging area of research in pain is orofacial pain (Blasini, Movsas, & Colloca, 2018; Wang et al., 2020). Wang et al. (2020) used the latent class analysis approach that uses a subset of structural equation modeling to find subtypes of cases in multivariate categorical data (Nylund, Asparouhov, & Muthén, 2007), to determine how learning patterns during classical conditioning may predict the formation of placebo analgesia (Wang et al., 2020). Classes characterized by persistently greater or increasing delta scores during conditioning displayed greater placebo analgesia during the test phase than those with persistently lower or decreasing delta scores. This study is the first to use latent class analysis approach to discern the relationship between prior experiences and learning patterns and the subsequent placebo analgesia in chronic orofacial pain patients (Wang et al., 2020). The possibility to model findings related to previous (therapeutic) responses to predict placebo effects is a nascent line of research that holds an interesting prospective. Other groups have theoretically postulated that the complex experience of pain is based on the predictive coding (Buchel, Geuter, Sprenger, & Eippert, 2014) and inferential processes in which prior information and self-healing experiences are merged to generate representations of future outcomes (Wiech, 2016).

3.6 Beyond pain disorders

The notion that placebo effects and therefore expectations modulate the therapeutic outcomes has been investigated in several studies targeting Parkinson's disease as a neurological disease (de la Fuente-Fernandez et al., 2001, 2002b; De La Fuente-Fernandez & Stoessl, 2002a; Dumitriu & Popescu, 2010; Quattrone, Barbagallo, Cerasa, & Stoessl, 2018).

In particular, it has been posited that expectations of a benefit (e.g. a positive reward) trigger the release of dopamine in the nucleus accumbens (Comings & Blum, 2000) and dopaminergic neurons respond to stimuli that predict a reward (Schultz, 1998). Experiments in Parkinsonian patients showed that expectations of a reward (e.g. reduction of Parkinsonian symptoms) trigger the release of dopamine in the nucleus accumbens and in the nigrostriatal pathway with a critical role of the dorsal striatum (de la Fuente-Fernandez et al., 2001).

The administration of a placebo given under distinct verbal suggestions (and distinct strength of expectations) of a clinical benefit influences placebo effects in patients with moderate Parkinson's disease. Patients were told that they had a 25%, 50%, 75%, or 100% probability of receiving active medication when they in fact received a placebo (Lidstone et al., 2010). A significant release of dopamine documented with [11C]raclopride radiotracer and positron emission tomography, was induced only by a 75%

probability of being treated with placebo-induced dopamine release in the striatum. Therefore, uncertainty and/or salience over and above a patient's prior treatment response may help trigger clinical placebo effects. Perception of treatment has been also reported in other studies with Parkinson patients. Parkinsonian patients undergone either human fetal cell transplant or a sham surgery. Post-surgery quality of life observed changes were depended on the patient's expectations. In fact, the overall clinical benefits of the surgery were modest in both groups, however the best predictor of improved quality of life and objective improvement in motor scores was the patient's perception of which surgery they had had rather than the type of surgery they actually received (McRae et al., 2004). The expectation of a therapeutic benefit has been documented in studies with repetitive transcranial magnetic stimulation (Strafella, Ko, & Monchi, 2006) and STN-DBS in Parkinsonian patients (Mercado et al., 2006). A remaining gap in the literature is the investigation of placebo (and nocebo) effects when Parkinson's and pain disorders co-exist.

4. Future directions

Research on placebo effects and neurological disorders is limited to a few disorders. Yet, expectancy and placebo effects can potentially influence responses to any treatment from neurological disorders. Therefore, it is important to emphasize the benefit of grasping the overall mechanisms of placebo and expectancy effects and use them in clinical practice and clinical trials. It is likely that placebo and expectancy effects may interaction with active treatments by an involvement of modulatory central nervous and peripheral mechanisms therefore biasing the results of clinical trials and challenging the concept that placebo and drug effects. In clinical settings, knowledge on placebo and expectancy effects should be used to enhance clinical outcomes, as chronic patients despite the neurological degenerations may overall benefit from expectancy empowerment and placebo effects.

Acknowledgments

The author has no conflicts of interest to declare.

This research is supported by National Institute Dental Craniofacial Research, NIDCR (R01 DE025946, L.C.). The funding agencies have no roles in the study. The views expressed here are the authors own and do not reflect the position or policy of the National Institutes of Health or any other part of the federal government.

The author would like thank John A. Melnicki for his continuous support and comment on the last version of this article.

References

Amanzio, M., Benedetti, F., Porro, C. A., Palermo, S., & Cauda, F. (2013). Activation likelihood estimation meta-analysis of brain correlates of placebo analgesia in human experimental pain. *Human Brain Mapping, 34*(3), 738–752. https://doi.org/10.1002/hbm.21471.

Archer, T. P., & Leier, C. V. (1992). Placebo treatment in congestive heart failure. *Cardiology, 81*(2–3), 125–133. https://doi.org/10.1159/000175787.

Balint, M. (1955). The doctor, his patient, and the illness. *The Lancet, 268*(6866), 683–688. https://doi.org/10.1016/s0140-6736(55)91061-8.

Barnes, K., Faasse, K., Geers, A. L., Helfer, S. G., Sharpe, L., Colloca, L., et al. (2019). Can positive framing reduce nocebo side effects? Current evidence and recommendation for future research. *Frontiers in Pharmacology, 10*, 167. https://doi.org/10.3389/fphar.2019.00167.

Bartfai, T., & Lees, G. V. (2011). Pharma TARP: A troubled asset relief program for novel, abandoned projects in the pharmaceutical industry. *ScientificWorldJournal, 11*, 454–457. https://doi.org/10.1100/tsw.2011.61.

Beecher, H. K. (1955). The powerful placebo. *Journal of the American Medical Association, 159*(17), 1602–1606.

Benedetti, F. (2002). How the doctor's words affect the patient's brain. *Evaluation & The Health Professions, 25*(4), 369–386. https://doi.org/10.1177/0163278702238051.

Benson, H., & McCallie, D. P., Jr. (1979). Angina pectoris and the placebo effect. *The New England Journal of Medicine, 300*(25), 1424–1429. https://doi.org/10.1056/NEJM197906213002508.

Bingel, U., Lorenz, J., Schoell, E., Weiller, C., & Buchel, C. (2006). Mechanisms of placebo analgesia: rACC recruitment of a subcortical antinociceptive network. *Pain, 120*(1–2), 8–15. https://doi.org/10.1016/j.pain.2005.08.027.

Blasini, M., Movsas, S., & Colloca, L. (2018). Placebo hypoalgesic effects in pain: Potential applications in dental and orofacial pain management. *Seminars in Orthodontics, 24*(2), 259–268. https://doi.org/10.1053/j.sodo.2018.04.001.

Blasini, M., Peiris, N., Wright, T., & Colloca, L. (2018). The role of patient-practitioner relationships in placebo and nocebo phenomena. *International Review of Neurobiology, 139*, 211–231. https://doi.org/10.1016/bs.irn.2018.07.033.

Brody, H. B., & Brody, D. (2000). Placebo and health—II. Three perspectives on the placebo response: Expectancy, conditioning, and meaning. *Advances in Mind-Body Medicine, 16*(3), 216–232. https://doi.org/10.1054/ambm.2000.0183.

Brody, H., Colloca, L., & Miller, F. G. (2012). The placebo phenomenon: Implications for the ethics of shared decision-making. *Journal of General Internal Medicine, 27*(6), 739–742. https://doi.org/10.1007/s11606-011-1977-1.

Brown, W. A. (1998). The placebo effect. *Scientific American, 278*(1), 90–95. https://doi.org/10.1038/scientificamerican0198-90.

Buchel, C., Geuter, S., Sprenger, C., & Eippert, F. (2014). Placebo analgesia: A predictive coding perspective. *Neuron, 81*(6), 1223–1239. https://doi.org/10.1016/j.neuron.2014.02.042.

Carvalho, C., Caetano, J. M., Cunha, L., Rebouta, P., Kaptchuk, T. J., & Kirsch, I. (2016). Open-label placebo treatment in chronic low back pain: A randomized controlled trial. *Pain, 157*(12), 2766–2772. https://doi.org/10.1097/j.pain.0000000000000700.

Chou, R., Deyo, R., Friedly, J., Skelly, A., Hashimoto, R., Weimer, M., et al. (2017). Nonpharmacologic therapies for low back pain: A systematic review for an American College of Physicians Clinical Practice Guideline. *Annals of Internal Medicine, 166*(7), 493–505. https://doi.org/10.7326/M16-2459.

Colloca, L. (2019). The placebo effect in pain therapies. *Annual Review of Pharmacology and Toxicology, 59*, 191–211. https://doi.org/10.1146/annurev-pharmtox-010818-021542.

Colloca, L., & Barsky, A. J. (2020). Placebo and nocebo effects. *The New England Journal of Medicine, 382*(6), 554–561. https://doi.org/10.1056/NEJMra1907805.

Colloca, L., Benedetti, F., & Porro, C. A. (2008). Experimental designs and brain mapping approaches for studying the placebo analgesic effect. *European Journal of Applied Physiology, 102*(4), 371–380. https://doi.org/10.1007/s00421-007-0593-6.

Colloca, L., & Howick, J. (2018). Placebos without deception: Outcomes, mechanisms, and ethics. *International Review of Neurobiology, 138*, 219–240. https://doi.org/10.1016/bs.irn.2018.01.005.

Colloca, L., Ludman, T., Bouhassira, D., Baron, R., Dickenson, A. H., Yarnitsky, D., et al. (2017). Neuropathic pain. *Nature Reviews Disease Primers, 3*, 17002. https://doi.org/10.1038/nrdp.2017.2.

Colloca, L., & Miller, F. G. (2011a). How placebo responses are formed: A learning perspective. *Philosophical Transactions of the Royal Society of London Series B, Biological Sciences, 366*(1572), 1859–1869. https://doi.org/10.1098/rstb.2010.0398.

Colloca, L., & Miller, F. G. (2011b). Role of expectations in health. *Current Opinion in Psychiatry, 24*(2), 149–155. https://doi.org/10.1097/YCO.0b013e328343803b.

Colloca, L., Schenk, L. A., Nathan, D. E., Robinson, O. J., & Grillon, C. (2019). When expectancies are violated: A functional magnetic resonance Imaging study. *Clinical Pharmacology and Therapeutics, 106*(6), 1246–1252. https://doi.org/10.1002/cpt.1587.

Colloca, L., Tinazzi, M., Recchia, S., Le Pera, D., Fiaschi, A., Benedetti, F., et al. (2008). Learning potentiates neurophysiological and behavioral placebo analgesic responses. *Pain, 139*(2), 306–314. https://doi.org/10.1016/j.pain.2008.04.021.

Comings, D. E., & Blum, K. (2000). Reward deficiency syndrome: Genetic aspects of behavioral disorders. *Progress in Brain Research, 126*, 325–341. https://doi.org/10.1016/S0079-6123(00)26022-6.

de Craen, A. J., Kaptchuk, T. J., Tijssen, J. G., & Kleijnen, J. (1999). Placebos and placebo effects in medicine: Historical overview. *Journal of the Royal Society of Medicine, 92*(10), 511–515. https://doi.org/10.1177/014107689909201005.

de la Fuente-Fernandez, R., Ruth, T. J., Sossi, V., Schulzer, M., Calne, D. B., & Stoessl, A. J. (2001). Expectation and dopamine release: Mechanism of the placebo effect in Parkinson's disease. *Science, 293*(5532), 1164–1166. https://doi.org/10.1126/science.1060937.

De La Fuente-Fernandez, R., & Stoessl, A. J. (2002a). The biochemical bases for reward. Implications for the placebo effect. *Evaluation & the Health Professions, 25*(4), 387–398. https://doi.org/10.1177/0163278702238052.

de la Fuente-Fernandez, R., & Stoessl, A. J. (2002b). The placebo effect in Parkinson's disease. *Trends in Neurosciences, 25*(6), 302–306. https://doi.org/10.1016/s0166-2236(02)02181-1.

Di Blasi, Z., Harkness, E., Ernst, E., Georgiou, A., & Kleijnen, J. (2001). Influence of context effects on health outcomes: A systematic review. *The Lancet, 357*(9258), 757–762.

Dumitriu, A., & Popescu, B. O. (2010). Placebo effects in neurological diseases. *Journal of Medicine and Life, 3*(2), 114–121.

Eippert, F., Finsterbusch, J., Bingel, U., & Buchel, C. (2009). Direct evidence for spinal cord involvement in placebo analgesia. *Science, 326*(5951), 404. https://doi.org/10.1126/science.1180142.

Evers, A. W. M., Colloca, L., Blease, C., Annoni, M., Atlas, L. Y., Benedetti, F., et al. (2018). Implications of placebo and nocebo effects for clinical practice: Expert consensus. *Psychotherapy and Psychosomatics, 87*(4), 204–210. https://doi.org/10.1159/000490354.

Fields, H. L. (2000). Pain modulation: Expectation, opioid analgesia and virtual pain. *Progress in Brain Research, 122*, 245–253. https://doi.org/10.1016/s0079-6123(08)62143-3.

Fields, H. (2004). State-dependent opioid control of pain. *Nature Reviews Neuroscience, 5*(7), 565–575. https://doi.org/10.1038/nrn1431.

Fields, H. L. (2014). Neuroscience. More pain; less gain. *Science*, *345*(6196), 513–514. https://doi.org/10.1126/science.1258477.

Fields, H. L. (2018). How expectations influence pain. *Pain*, *159*(Suppl. 1), S3–S10. https://doi.org/10.1097/j.pain.0000000000001272.

Fields, H. L., & Levine, J. D. (1981). Biology of placebo analgesia. *The American Journal of Medicine*, *70*(4), 745–746. https://doi.org/10.1016/0002-9343(81)90525-8.

Gotzsche, P. C. (1994). Is there logic in the placebo? *The Lancet*, *344*(8927), 925–926. https://doi.org/10.1016/s0140-6736(94)92273-x.

Gotzsche, P. C. (1995). Placebo effects. Concept of placebo should be discarded. *BMJ*, *311*(7020), 1640–1641. https://doi.org/10.1136/bmj.311.7020.1640b.

Hashmi, J. A., Baria, A. T., Baliki, M. N., Huang, L., Schnitzer, T. J., & Apkarian, A. V. (2012). Brain networks predicting placebo analgesia in a clinical trial for chronic back pain. *Pain*, *153*(12), 2393–2402. https://doi.org/10.1016/j.pain.2012.08.008.

Hoffman, G. A., Harrington, A., & Fields, H. L. (2005). Pain and the placebo: What we have learned. *Perspectives in Biology and Medicine*, *48*(2), 248–265. https://doi.org/10.1353/pbm.2005.0054.

Hooper's Medical Dictionary, 1843. Volume 2, K-Z, https://www.abebooks.com/servlet/BookDetailsPL?bi=30315360846&searchurl=an%3Drobert%2Bhooper%26sortby%3D17%26tn%3Dhoopers%2Bmedical%2Bdictionary%2B1843&cm_sp=snippet-_-srp1-_-title1.

Hoy, D., March, L., Brooks, P., Blyth, F., Woolf, A., Bain, C., et al. (2014). The global burden of low back pain: Estimates from the Global Burden of Disease 2010 study. *Annals of the Rheumatic Diseases*, *73*(6), 968–974. https://doi.org/10.1136/annrheumdis-2013-204428.

Humphrey, N. (2002). *The mind made flesh: Essays from frontiers of psychology and evolution.* Oxford: Oxford University Press.

Hutchinson, P., & Moerman, D. E. (2018). The meaning response, "Placebo," and methods. *Perspectives in Biology and Medicine*, *61*(3), 361–378. https://doi.org/10.1353/pbm.2018.0049.

Kam-Hansen, S., Jakubowski, M., Kelley, J. M., Kirsch, I., Hoaglin, D. C., Kaptchuk, T. J., et al. (2014). Altered placebo and drug labeling changes the outcome of episodic migraine attacks. *Science Translational Medicine*, *6*(218), 218ra215. https://doi.org/10.1126/scitranslmed.3006175.

Kaptchuk, T. J., Kelley, J. M., Conboy, L. A., Davis, R. B., Kerr, C. E., Jacobson, E. E., et al. (2008). Components of placebo effect: Randomised controlled trial in patients with irritable bowel syndrome. *BMJ*, *336*(7651), 999–1003. https://doi.org/10.1136/bmj.39524.439618.25.

Katz, J., Finnerup, N. B., & Dworkin, R. H. (2008). Clinical trial outcome in neuropathic pain: Relationship to study characteristics. *Neurology*, *70*(4), 263–272. https://doi.org/10.1212/01.WNL.0000275528.01263.6c.

Keltner, J. R., Furst, A., Fan, C., Redfern, R., Inglis, B., & Fields, H. L. (2006). Isolating the modulatory effect of expectation on pain transmission: A functional magnetic resonance imaging study. *The Journal of Neuroscience*, *26*(16), 4437–4443. [pii] 26/16/4437. https://doi.org/10.1523/JNEUROSCI.4463-05.2006.

Klinger, R., Kothe, R., Schmitz, J., Kamping, S., & Flor, H. (2017). Placebo effects of a sham opioid solution: A randomized controlled study in patients with chronic low back pain. *Pain*, *158*(10), 1893–1902. https://doi.org/10.1097/j.pain.0000000000000977.

Kong, J., Gollub, R. L., Rosman, I. S., Webb, J. M., Vangel, M. G., Kirsch, I., et al. (2006). Brain activity associated with expectancy-enhanced placebo analgesia as measured by functional magnetic resonance imaging. *The Journal of Neuroscience*, *26*(2), 381–388. https://doi.org/10.1523/JNEUROSCI.3556-05.2006.

Kosek, E., Rosen, A., Carville, S., Choy, E., Gracely, R. H., Marcus, H., et al. (2017). Lower placebo responses after long-term exposure to fibromyalgia pain. *The Journal of Pain, 18*(7), 835–843. https://doi.org/10.1016/j.jpain.2017.02.434.

Levine, J. D., Gordon, N. C., Bornstein, J. C., & Fields, H. L. (1979). Role of pain in placebo analgesia. *Proceedings of the National Academy of Sciences of the United States of America, 76*(7), 3528–3531. https://doi.org/10.1073/pnas.76.7.3528.

Levine, J. D., Gordon, N. C., & Fields, H. L. (1978). The mechanism of placebo analgesia. *The Lancet, 2*(8091), 654–657. https://doi.org/10.1016/s0140-6736(78)92762-9.

Lidstone, S. C., Schulzer, M., Dinelle, K., Mak, E., Sossi, V., Ruth, T. J., et al. (2010). Effects of expectation on placebo-induced dopamine release in Parkinson disease. *Archives of General Psychiatry, 67*(8), 857–865. https://doi.org/10.1001/archgenpsychiatry.2010.88.

Lieberman, M. D., Jarcho, J. M., Berman, S., Naliboff, B. D., Suyenobu, B. Y., Mandelkern, M., et al. (2004). The neural correlates of placebo effects: A disruption account. *NeuroImage, 22*(1), 447–455. https://doi.org/10.1016/j.neuroimage.2004.01.037.

Matre, D., Casey, K. L., & Knardahl, S. (2006). Placebo-induced changes in spinal cord pain processing. *The Journal of Neuroscience, 26*(2), 559–563. https://doi.org/10.1523/JNEUROSCI.4218-05.2006.

McRae, C., Cherin, E., Yamazaki, T. G., Diem, G., Vo, A. H., Russell, D., et al. (2004). Effects of perceived treatment on quality of life and medical outcomes in a double-blind placebo surgery trial. *Archives of General Psychiatry, 61*(4), 412–420. https://doi.org/10.1001/archpsyc.61.4.412.

Mercado, R., Constantoyannis, C., Mandat, T., Kumar, A., Schulzer, M., Stoessl, A. J., et al. (2006). Expectation and the placebo effect in Parkinson's disease patients with subthalamic nucleus deep brain stimulation. *Movement Disorders, 21*(9), 1457–1461. https://doi.org/10.1002/mds.20935.

Miller, F. G., & Colloca, L. (2009). The legitimacy of placebo treatments in clinical practice: Evidence and ethics. *The American Journal of Bioethics, 9*(12), 39–47. https://doi.org/10.1080/15265160903316263.

Moerman, D. E. (2000). Cultural variations in the placebo effect: Ulcers, anxiety, and blood pressure. *Medical Anthropology Quarterly, 14*(1), 51–72. https://doi.org/10.1525/maq.2000.14.1.51.

Moerman, D. E. (2002). *Meaning, medicine and the placebo effect.* Cambridge University Press.

Moerman, D. E. (2013). Against the "placebo effect": A personal point of view. *Complementary Therapies in Medicine, 21*(2), 125–130. https://doi.org/10.1016/j.ctim.2013.01.005.

Moerman, D. E., & Jonas, W. B. (2000). Toward a research agenda on placebo. *Advances in Mind-Body Medicine, 16*(1), 33–46. https://doi.org/10.1054/ambm.2000.0116.

Moerman, D. E., & Jonas, W. B. (2002). Deconstructing the placebo effect and finding the meaning response. *Annals of Internal Medicine, 136*(6), 471–476. doi:200203190-00011 [pii].

Nylund, K. L., Asparouhov, T., & Muthén, B. O. (2007). Deciding on the number of classes in latent class analysis and growth mixture modeling: A Monte Carlo simulation study. *Structural Equation Modeling: A Multidisciplinary Journal, 14*(4), 535–569. https://doi.org/10.1080/10705510701575396.

Peiris, N., Blasini, M., Wright, T., & Colloca, L. (2018). The placebo phenomenon: A narrow focus on psychological models. *Perspectives in Biology and Medicine, 61*(3), 388–400. https://doi.org/10.1353/pbm.2018.0051.

Petersen, G. L., Finnerup, N. B., Grosen, K., Pilegaard, H. K., Tracey, I., Benedetti, F., et al. (2014). Expectations and positive emotional feelings accompany reductions in ongoing and evoked neuropathic pain following placebo interventions. *Pain, 155*(12), 2687–2698. https://doi.org/10.1016/j.pain.2014.09.036.

Petersen, G. L., Finnerup, N. B., Norskov, K. N., Grosen, K., Pilegaard, H. K., Benedetti, F., et al. (2012). Placebo manipulations reduce hyperalgesia in neuropathic pain. *Pain, 153*(6), 1292–1300. https://doi.org/10.1016/j.pain.2012.03.011.

Petrovic, P., Dietrich, T., Fransson, P., Andersson, J., Carlsson, K., & Ingvar, M. (2005). Placebo in emotional processing—Induced expectations of anxiety relief activate a generalized modulatory network. *Neuron, 46*(6), 957–969. [pii] S0896-6273(05)00470-8. https://doi.org/10.1016/j.neuron.2005.05.023.

Petrovic, P., Kalso, E., Petersson, K. M., Andersson, J., Fransson, P., & Ingvar, M. (2010). A prefrontal non-opioid mechanism in placebo analgesia. *Pain, 150*(1), 59–65. https://doi.org/10.1016/j.pain.2010.03.011.

Petrovic, P., Kalso, E., Petersson, K. M., & Ingvar, M. (2002). Placebo and opioid analgesia—Imaging a shared neuronal network. *Science, 295*(5560), 1737–1740. https://doi.org/10.1126/science.1067176.

Price, D. D., Craggs, J., Verne, G. N., Perlstein, W. M., & Robinson, M. E. (2007). Placebo analgesia is accompanied by large reductions in pain-related brain activity in irritable bowel syndrome patients. *Pain, 127*(1–2), 63–72. https://doi.org/10.1016/j.pain.2006.08.001.

Quattrone, A., Barbagallo, G., Cerasa, A., & Stoessl, A. J. (2018). Neurobiology of placebo effect in Parkinson's disease: What we have learned and where we are going. *Movement Disorders, 33*(8), 1213–1227. https://doi.org/10.1002/mds.27438.

Rainville, P., & Duncan, G. H. (2006). Functional brain imaging of placebo analgesia: Methodological challenges and recommendations. *Pain, 121*(3), 177–180. https://doi.org/10.1016/j.pain.2006.01.011.

Schultz, W. (1998). Predictive reward signal of dopamine neurons. *Journal of Neurophysiology, 80*(1), 1–27. https://doi.org/10.1152/jn.1998.80.1.1.

Shapiro, A. K., & Shapiro, E. (1997). The placebo: Is it much ado about nothing. In A. Harrington (Ed.), *An interdisciplinary exploration* (pp. 12–36). Harvard University Press.

Skyt, I., Moslemi, K., Baastrup, C., Grosen, K., Benedetti, F., Petersen, G. L., et al. (2018a). Dopaminergic tone does not influence pain levels during placebo interventions in patients with chronic neuropathic pain. *Pain, 159*(2), 261–272. https://doi.org/10.1097/j.pain.0000000000001089.

Skyt, I., Moslemi, K., Baastrup, C., Grosen, K., Svensson, P., Jensen, T. S., et al. (2018b). Does conditioned pain modulation predict the magnitude of placebo effects in patients with neuropathic pain? *European Journal of Pain, 22*(4), 784–792. https://doi.org/10.1002/ejp.1164.

Strafella, A. P., Ko, J. H., & Monchi, O. (2006). Therapeutic application of transcranial magnetic stimulation in Parkinson's disease: The contribution of expectation. *NeuroImage, 31*(4), 1666–1672. https://doi.org/10.1016/j.neuroimage.2006.02.005.

Tetreault, P., Mansour, A., Vachon-Presseau, E., Schnitzer, T. J., Apkarian, A. V., & Baliki, M. N. (2016). Brain connectivity predicts placebo response across chronic pain clinical trials. *PLoS Biology, 14*(10), e1002570. https://doi.org/10.1371/journal.pbio.1002570.

Tinnermann, A., Geuter, S., Sprenger, C., Finsterbusch, J., & Buchel, C. (2017). Interactions between brain and spinal cord mediate value effects in nocebo hyperalgesia. *Science, 358*(6359), 105–108. https://doi.org/10.1126/science.aan1221.

Tracey, I. (2005). Nociceptive processing in the human brain. *Current Opinion in Neurobiology, 15*(4), 478–487. https://doi.org/10.1016/j.conb.2005.06.010.

Turner, J. A., Deyo, R. A., Loeser, J. D., Von Korff, M., & Fordyce, W. E. (1994). The importance of placebo effects in pain treatment and research. *JAMA, 271*(20), 1609–1614.

Vachon-Presseau, E., Berger, S. E., Abdullah, T. B., Huang, L., Cecchi, G. A., Griffith, J. W., et al. (2018). Brain and psychological determinants of placebo pill response in chronic pain patients. *Nature Communications*, *9*(1), 3397. https://doi.org/10.1038/s41467-018-05859-1.

Vase, L., Petersen, G. L., & Lund, K. (2014). Placebo effects in idiopathic and neuropathic pain conditions. *Handbook of Experimental Pharmacology*, *225*, 121–136. https://doi.org/10.1007/978-3-662-44519-8_7.

Vase, L., Riley, J. L., 3rd, & Price, D. D. (2002). A comparison of placebo effects in clinical analgesic trials versus studies of placebo analgesia. *Pain*, *99*(3), 443–452doi: S0304395902002051 [pii].

Vase, L., Robinson, M. E., Verne, G. N., & Price, D. D. (2003). The contributions of suggestion, desire, and expectation to placebo effects in irritable bowel syndrome patients. An empirical investigation. *Pain*, *105*(1–2), 17–25.

Vase, L., Robinson, M. E., Verne, G. N., & Price, D. D. (2005). Increased placebo analgesia over time in irritable bowel syndrome (IBS) patients is associated with desire and expectation but not endogenous opioid mechanisms. *Pain*, *115*(3), 338–347. https://doi.org/10.1016/j.pain.2005.03.014.

Vase, L., Skyt, I., & Hall, K. T. (2016). Placebo, nocebo, and neuropathic pain. *Pain*, *157*(Suppl. 1), S98–105. https://doi.org/10.1097/j.pain.0000000000000445.

Wager, T. D., Atlas, L. Y., Lindquist, M. A., Roy, M., Woo, C. W., & Kross, E. (2013). An fMRI-based neurologic signature of physical pain. *The New England Journal of Medicine*, *368*(15), 1388–1397. https://doi.org/10.1056/NEJMoa1204471.

Wager, T. D., Matre, D., & Casey, K. L. (2006). Placebo effects in laser-evoked pain potentials. *Brain, Behavior, and Immunity*, *20*(3), 219–230. https://doi.org/10.1016/j.bbi.2006.01.007.

Wager, T. D., & Nitschke, J. B. (2005). Placebo effects in the brain: Linking mental and physiological processes. *Brain, Behavior, and Immunity*, *19*(4), 281–282. https://doi.org/10.1016/j.bbi.2005.04.009.

Wager, T. D., Rilling, J. K., Smith, E. E., Sokolik, A., Casey, K. L., Davidson, R. J., et al. (2004). Placebo-induced changes in FMRI in the anticipation and experience of pain. *Science*, *303*(5661), 1162–1167. https://doi.org/10.1126/science.1093065303/5661/1162. [pii].

Wager, T. D., Scott, D. J., & Zubieta, J. K. (2007). Placebo effects on human mu-opioid activity during pain. *Proceedings of the National Academy of Sciences of the United States of America*, *104*(26), 11056–11061. https://doi.org/10.1073/pnas.0702413104.

Wall, P. D. (1992). The placebo effect: An unpopular topic. *Pain*, *51*(1), 1–3. https://doi.org/10.1016/0304-3959(92)90002-s.

Wang, Y., Trichou, C., Raghuraman, N., Akintola, A., Haycock, N. R., Blasini, M., et al. (2020). Modeling learning patterns to predict placebo analgesic effects in healthy and chronic orofacial pain participants. *Frontiers in Psychiatry*, *11*, 39. https://doi.org/10.3389/fpsyt.2020.00039. eCollection 2020.

Wiech, K. (2016). Deconstructing the sensation of pain: The influence of cognitive processes on pain perception. *Science*, *354*(6312), 584–587. https://doi.org/10.1126/science.aaf8934.

Zeidan, F., Lobanov, O. V., Kraft, R. A., & Coghill, R. C. (2015). Brain mechanisms supporting violated expectations of pain. *Pain*, *156*(9), 1772–1785. https://doi.org/10.1097/j.pain.0000000000000231.

Zubieta, J. K., Bueller, J. A., Jackson, L. R., Scott, D. J., Xu, Y., Koeppe, R. A., et al. (2005). Placebo effects mediated by endogenous opioid activity on mu-opioid receptors. *The Journal of Neuroscience*, *25*(34), 7754–7762. https://doi.org/10.1523/JNEUROSCI.0439-05.2005.

Zunhammer, M., Bingel, U., Wager, T. D., & Placebo Imaging, C. (2018). Placebo effects on the neurologic pain signature: A meta-analysis of individual participant functional magnetic resonance Imaging data. *JAMA Neurology*, *75*(11), 1321–1330. https://doi.org/10.1001/jamaneurol.2018.2017.

CHAPTER NINE

Placebo responses in Parkinson's disease

Jau-Shin Lou*
Roger Gilbertson Endowed Chair in Neurology, Department of Neurology, University of North Dakota School of Medicine & Health Science, and Sanford Health, Fargo, ND, United States
*Corresponding author: e-mail address: jaushin.lou@und.edu

Contents

Abstract

Parkinson's disease (PD) patients exhibit strong placebo responses in clinical trials. Patient characteristics that affect placebo include patients' expectations of good outcomes, genetic variants, and personality. The presence of motor fluctuation and high baseline UPDRS motor scores predicted placebo response. However, gender, age, duration of PD, religion, or level of education do not correlate with placebo response. PD patients who are preconditioned with active treatment such as apomorphine have more robust placebo effects. Studies that focused on patients with motor fluctuations, surgical intervention, or higher probability of placebo assignment had higher rates of placebo response. Patients view participating in placebo-controlled trials positively.

Placebo effect can be measured objectively using neuroimaging and neurophysiological techniques. PET studies show that placebo-induced improvement is associated

International Review of Neurobiology, Volume 153
ISSN 0074-7742
https://doi.org/10.1016/bs.irn.2020.03.031

with dopamine release in the dorsal striatum and that the expectation of receiving the reward, not the reward itself, increased dopamine release in the ventral striatum. Expectations of benefitting from repetitive transcranial magnetic stimulation also induced dopamine release. Expectations of receiving a dopaminergic drug induced changes in fMRI in a reward-learning task. Single cell recordings demonstrate that placebo response is associated with changes of single neuronal activities in the basal ganglia circuit. These studies demonstrate that placebo effects are genuine biological responses to the administration of placebo.

In clinical trials, we can use several approaches to minimize placebo responses. In clinical practice, we can use approaches to harness the power of placebo and minimize nocebo effects to improve patients' outcome.

1. History: Double-blind placebo-controlled trials in Parkinson's disease (PD)

Double-blind placebo-controlled (DBPC) trials have played a critical role in developing new medical and surgical treatments for PD (Connolly & Lang, 2014; Ramirez-Zamora & Ostrem, 2018; Rughani et al., 2018; Zhang & Tan, 2016) since Yahr, Duvoisin, Schear, Barrett, and Hoehn (1969) conducted the first DBPC trial of levodopa 50 years ago. In these DBPC trials, researchers use placebo arms to evaluate the comparative efficacy of medical treatments or surgical interventions. Researchers then use the differences of the response rates between the treatment arm and the placebo arm to determine if the tested drug or intervention is effective.

Ideally, placebos such as pharmacologically inactive agents or sham surgery should not have any significant effects on subjects during clinical trials. Regardless, strong placebo effects have been reported in many neurological disorders such as pain, depression, and PD (Benedetti & Amanzio, 2011). The presence of the placebo effect reduces the difference of the effects between the study drug and the placebo in DBPC trials. One example is gene therapy with neurturin in PD (Olanow, Bartus, Baumann, et al., 2015). This trial was a 15- to 24-month, multicenter, double-blind trial with 51 patients with advanced PD who were randomly assigned to receive AAV2-neurturin injected bilaterally into the substantia nigra and putamen or sham surgery. The primary endpoint was change from baseline to final visit in the motor scores of the UPDRS in the practically defined off state. The AAV2-neurturin group met the criterion of a clinically significant improvement in the UPDRS motor score (Shulman et al., 2010) with −7.0 points over 2 years. Remarkably, the sham surgery group also experienced the similar magnitude of clinically significant improvement in the UPDRS motor score with −5.2 points. Therefore, the trial failed. The results of this trial demonstrated powerful placebo-related improvements even in a neurodegenerative disorder such as PD typically characterized by progressive decline.

Placebo-associated clinical improvement can have several contributors (Benedetti, Carlino, & Piedimonte, 2016): (1) Hawthorne effect: Participants can have changes in behavioral, clinical, or physiological measurements simply because of their awareness of being observed and lead to improvement; (2) Spontaneous remission: improvement due to the nature history of the diseases; (3) Regression to the mean: Repeated measures of a variable phenomenon tend to be closer to the mean; (4) Methodological biases such as report biases by patients or measurement biases by investigators.

Placebo effects have often been viewed as an annoyance in the past because a robust placebo effect has the potential of negating the efficacy estimates of the study drug as discussed above. However, with the groundbreaking advancement in neuroimaging and neurophysiology researches over last 20 years (Quattrone, Barbagallo, Cerasa, & Stoessl, 2018), we now view placebo effects not merely as Hawthorne effects, spontaneous remission, regression to the mean, or methodological biases, but as genuine biological responses to psychosocial cues to the administration of inactive medical treatments or sham surgery.

This review will focus on the current understanding of the placebo response in PD. Several more detailed reviews of the topics have been published as a special mini-series in *Movement Disorders* (Ferreira, Trenkwalder, & Mestre, 2018; Mestre & Ferreira, 2018; Mestre, Shamy, Benedetti, & Lang, 2018; Quattrone et al., 2018; Witek, Stebbins, & Goetz, 2018).

2. Determinants of placebo

2.1 Patient characteristics

2.1.1 Expectations

Expectations, defined as subjects' anticipation that they know which treatment they will receive or the outcome of such treatment, may play a role in the placebo effect. Several studies have examined the effect of expectations on clinical outcomes in PD patients who have DBS as the stimulator's intensity can be adjusted on command to induce prompt clinical responses.

One study showed that expectations of good outcomes induced by verbal suggestion improved movement velocity. Pollo et al. (2002) analyzed how expectations of good or bad outcomes influenced the effects of subthalamic stimulation on the velocity of movement of the right hand in seven PD patients who had bilateral subthalamic stimulators. The patients were tested in two different conditions in random orders in which opposite expectations were induced. In the first condition designed to induce bad expectations, they were told the truth about what was going to happen. When the stimulus intensity was reduced, they were told that motor performance was going to worsen. When the stimulus was increased, they were told that a small improvement of motor performance was going to occur. The second condition was designed to induce good expectations. When the stimulus was reduced, the patients were told that nothing was going to change in their motor performance, so that no worsening was expected. When the stimulus was increased, the patients were told that a big improvement in motor performance was going to occur. When patients expected a good motor performance, they moved faster than when they expected a bad motor performance even when the stimulators were set at the same intensity. The expectation of good performance was induced by verbal suggestions and all these effects occurred within minutes, suggesting that expectations induce neural changes very quickly.

Another study showed that patients who believed that they received dopamine neuron transplantation had better quality of life regardless of whether they received real or sham transplants. McRae et al. (2004) did a study in 30 PD patients who were part of a large double-blind sham surgery–controlled trial designed to determine the effectiveness of transplantation of human embryonic dopamine neurons into the brains of persons with advanced PD. They wanted to determine how quality of life improvement was affected by the actual intervention (true or sham) they received or by the perceived treatment (the type of surgery patients thought they received). They compared quality of life and medical outcomes between the actual transplant and sham surgery groups and the perceived treatment groups. The results showed that patients who thought they received transplantation of embryonic dopamine neurons (hence expected a good outcome) reported more improvement in quality of life than those who believed they did not receive the transplant (hence did not expect a good outcome), regardless of the actual intervention they received.

Awareness of the deep brain stimulator status (ON or OFF) affects PD patients' motor scores on standard rating measures such as the Unified Parkinson's Disease Rating Scale (UPDRS). Mercado et al. (2006) examined whether the degree to which a patient with PD was expecting therapeutic benefit from subthalamic nucleus-deep brain stimulation (STN-DBS) influenced the magnitude of motor improvement in 10 PD patients. They obtained four consecutive UPDRS motor examination scores in four different conditions: stimulation OFF/patient aware; stimulation OFF/patient blind; stimulation ON/patient aware; and stimulation ON/patient blind. They found when the stimulation was OFF, patients aware of this condition had higher (worse) UPDRS motor scores than when they were blinded. When the stimulation was ON, UPDRS motor scores were lower (better) when the patients were aware of the stimulation compared with when they were blinded. The authors concluded that the results supported the role of expectations and placebo effects in patients with STN-DBS.

Expectations induced by verbal suggestions also affected tremor severity and verbal fluency. Keitel, Stefano Ferrea, Südmeyer, Schnitzler, and Wojtecki (2013) investigated how expectations modulated the effect of STN-DBS on resting tremor and its interaction with verbal fluency. Although STN-DBS improves cardinal PD motor symptoms including tremor, impairment of verbal fluency (VF) has been observed as a potential side-effect (Saez-Zea, Escamilla-Sevilla, Katati, & Minguez-Castellanos, 2012). They manipulated the expectations of 24 tremor-dominant PD patients regarding the impact of STN-DBS on motor symptoms using verbal suggestions (positive [placebo], negative [nocebo], or neutral [control]). Patients participated with (MedON) and without (MedOFF) antiparkinsonian medication. They used accelerometry to measure resting tremor, finger tapping to measure bradykinesia, and 3D ultrasound motion detection to assess diadochokinesia. VF was quantified by lexical and semantic tests. They used prespecified criteria to determine a placebo response (tremor decreased by 10%) or a nocebo response (tremor increased by 10%). They found that in MedON, 8 out of 20 patients showed a placebo response with tremor reduction and 5 patients displayed a nocebo response. In MedOFF, seven patients were characterized by a placebo response and two patients showed a nocebo response. Interestingly, nocebo responders in MedON were additionally characterized by significant impairment in semantic verbal fluency. In contrast, bradykinesia was not affected by expectations in this study. They concluded that the therapeutic effect of STN-DBS on tremor can be modulated by expectations in a subgroup of patients and suggested that tremor was also among the parkinsonian symptoms responsive to placebo and nocebo interventions. In summary, these studies demonstrated that expectations play an important role in placebo effect. Expectations induced by verbal suggestions, awareness of the stimulator status, or the perception of receiving a real tissue transplant can affect motor function such as motor scores on the UPDRS, movement velocity or tremor severity as well as verbal fluency.

2.1.2 Preconditioning

PD patients who are preconditioned with active treatment such as apomorphine have more robust placebo effects. Benedetti, Frisaldi, Carlino, et al. (2016) examined the placebo response at the single-neuron level (see the section on single cell recording, p24) in the thalamus of 42 advanced Parkinson patients who underwent DBS procedure to see the differences between first time administration of placebo and administration after pharmacological preconditioning with apomorphine. They found that when the placebo was given for the first time, it induced neither clinical improvement, as assessed through muscle rigidity reduction at the wrist, nor neuronal changes in thalamic neurons. However, if placebo was given after two, three or four prior administrations of an anti-Parkinson drug, apomorphine, it produced both reduction in rigidity clinically and increased neuronal responses in the thalamus. Most interestingly, they found the greater the number of previous apomorphine administrations, the larger the magnitude and the longer the duration of the clinical and neuronal placebo responses. In a subsequent study (Frisaldi et al., 2017) investigated the effects of prior exposure to apomorphine on the placebo response of another cardinal symptom of Parkinson's disease, bradykinesia, by a movement time analyzer. They found no placebo response if the placebo was given for the first time, whereas the placebo response was substantial after prior pharmacological conditioning with apomorphine. These two studies (Benedetti, Frisaldi, et al., 2016; Frisaldi et al., 2017) indicate that prior exposure to drugs is a critical factor in the occurrence and magnitude of placebo effects. These learning effects should be carefully assessed in clinical trials in which patients receive the active treatment first and then are randomized.

2.1.3 Genetic variants

Recent advances in neurophysiological studies have increased our understanding of neurotransmitter pathways involved in the placebo effects (Benedetti, 2014). Dysfunction in the dopamine system is associated with diseases such as PD, schizophrenia, ADHD and addiction. Dopamine mediates the placebo response in PD, opioid in pain, and serotonin in depression (Benedetti, 2014). There is evidence that genetic variations in these neurotransmitters can modify placebo effects. The study of the genomic effect on placebo response has recently been named "Placebome" (Hall, Loscalzo, & Kaptchuk, 2015).

Genetic variations in dopamine metabolism and signaling pathway gene are prime candidates to serve as biomarkers for placebo response in PD because dopamine is involved in the placebo effect in PD (de la Fuente-Fernandez et al., 2002, 2001). Rs4680 is the most studied polymorphism in the gene for catechol-O- methyltransferase (COMT), an enzyme that metabolizes dopamine and other catecholamines in dopamine metabolism (Lachman et al., 1996). Rs4680 encoding a valine (Val)-to-methionine (Met) change at codon 158 (Val158Met) results in a three to fourfold reduction in enzymatic activity. Homozygotes of the less-active met allele have been associated with higher levels of dopamine in the prefrontal cortex, a region implicated in the placebo response pathway (Yavich et al., 2007).

Although the effect of rs4680 on placebo effect has not been examined in PD, it has been examined in other conditions which demonstrated patients with rs4680 low activity Met allele have more placebo responses (Hall et al., 2012; Yu et al., 2014). A DBPC trial in irritable bowel syndrome (IBS) (Hall et al., 2012) showed that patients homozygous for the rs4680 low activity Met allele (Met/Met), known to have the high levels of dopamine, had the greatest placebo response. The high activity Val allele homozygous (Val/Val) patients had the lowest placebo response. The Val/Met heterozygotes had an intermediate response. A small neuroimaging study in healthy volunteers (Yu et al., 2014) looked at genetic variations in *COMT* in relation to brain activity in the reward system. The study showed that in healthy volunteers, the suppression of pain by placebo was linearly correlated with the number of rs4680 Met alleles.

COMT Val158Met polymorphism affects the effect of entacapone in PD patients. Entacapone increases ON time in PD by the inhibition of L-dopa metabolism through the COMT pathway, leading to an increase in the bioavailability of L-DOPA (Bonifati & Meco, 1999). Corvol et al. (2011) examined the effects of COMT Val158Met polymorphism on the response to entacapone in PD. They randomized 17 PD patients who were homozygous for the COMT alleles COMT(Val/Val) ($n = 17$) and 16 patients with COMT(Met/Met) ($n = 16$) in a double-blind crossover trial consisting of 2 successive acute levodopa challenges associated with 200 mg entacapone or placebo. They found the gain in the best ON time was higher in COMT(Val/Val) than in COMT(Met/Met) patients. The COMT(Val/Val) genotype in PD patients enhanced the effect of entacapone on the pharmacodynamics and pharmacokinetics of levodopa. Although this study examined the effect of COMT Val158Met polymorphism on the effect of entacapone, they did not examine its effect on placebo.

The COMT Val158Met polymorphism seems to be involved in the development of nocebo response as well. Wendt et al. (2014) used an established paradigm of behaviorally conditioned immunosuppressive effects to examine if COMT Val158Met polymorphism could predict nocebo responses. They analyzed drug specific and general side effects before and after immunosuppressive medication and subsequent placebo intake in 62 healthy male subjects. They found more side effects were reported from homozygous carriers of the Val158 variant during medication as well as placebo treatment compared to the other genotype groups. They suggested that COMT Val158Met polymorphism has the potential of predicting nocebo responses after drug or placebo intake, which might be utilized to minimize nocebo effects in clinical trials and medical practice.

In summary, high activity Val allele homozygous (Val/Val) patients had the lowest placebo response in IBS. However, normal subjects with the high activity Val allele homozygous (Val/Val) had higher nocebo response in a behaviorally conditioned immunosuppressive paradigm. No studies had been done on the effect of this polymorphism on placebo or nocebo responses in PD.

2.1.4 Personality

The impact of personality on placebo has been well studied in pain. In particular, optimism, somatic focus (attention toward the body), and empathic resonance are related to higher placebo analgesic effects (Colloca & Benedetti, 2009; Geers, Helfer, Kosbab, Weiland, & Landry, 2005; Geers, Wellman, Fowler, Helfer, & France, 2010; Hunter, Siess, & Colloca, 2014). Hypnotic susceptibility and responsiveness to verbal suggestions influence placebo analgesia (Huber, Lui, & Porro, 2013). Other factors such as Neuroticism-Extraversion-Openness to experience (NEO), NEO Altruism, NEO Straightforwardness, and Ego-Resiliency, also influence placebo responses (Peciña et al., 2013). Conversely, anxiety (Staats, Staats, & Hekmat, 2001), harm avoidance, and persistence (Corsi, Emadi Andani, Tinazzi, & Fiorio, 2016) are associated with nocebo effects.

In terms of PD itself, several studies illuminate personality patterns in placebo responders. PD patients who were placebo responders tended to be more open to experience. Lou et al. (2013) examined if personality affects placebo effect in PD by asking subjects to complete the NEO-Five Factor

Inventory, General Perceived Self- Efficacy Scale, and Perceived Stress Scale. NEO-FFI measures five personality domains: neuroticism, extraversion, openness, agreeableness, and conscientiousness (Costa & McCrae, 1992). *General Perceived* Self-Efficacy Scale assesses the presence of optimistic self-beliefs to cope with a variety of difficult demands in life (Schwarzer & Jerusalem, 1995). *Perceived Stress Scale* evaluates the degree to which situations in one's life are appraised as stressful (Cohen, Kamarck, & Mermelstein, 1983). They found responders tended to be more open to experience than non-responders.

2.1.5 Patient demographics

An early meta-analysis demonstrated that patient demographics did not consistently correlate with placebo response. Shetty, Friedman, Kieburtz, Marshall, and Oakes (1999) used strict criteria to select studies of PD from 1969 to 1996 including the Parkinson Study Group database for Deprenyl & Tocopherol Antioxidative Therapy of Parkinsonism (DATATOP) study. They compared demographic factors between placebo responders and placebo non-responders. Their review demonstrated that although there is clearly a placebo response in PD, the variation in placebo response does not correlate with demographic factors such as gender, age, duration of PD, religion, or level of education.

Patient demographics also did not predict placebo response in a ropinirole monotherapy trial. Goetz, Leurgans, Raman, and Stebbins (2000) examined the frequency, temporal development, and stability of objective motor changes during placebo treatment in a 6-month DBPC trial of ropinirole monotherapy. They evaluated placebo-associated effects on the motor section of the UPDRS including tremor, bradykinesia, rigidity, and gait/balance/midline functions over a 6-month study. They applied a rigorous definition of placebo-associated improvement as an improvement over baseline score in motor UPDRS of at least 50% or a change in at least two motor items at any one visit by $\geq$2 points. The results showed that 16% of subjects improved on placebo treatment. The prevalence of response was steady (8–9%) at any one visit without a predominance of an early effect. Like the study by Shetty et al. (1999), gender, age, disease duration, and baseline disability score did not influence the likelihood of placebo effects.

The presence of motor fluctuation and high baseline UPDRS motor scores predicted placebo response. Shin et al. (2016) used meta-analysis to examine the predictors of the placebo response in randomized placebo-controlled trials on PD. The placebo response was defined as the mean change in the UPDRS part III score from baseline to the primary efficacy endpoint in the placebo group. They included 48 studies (consisting of 5618 participants on placebo). They found motor fluctuation and baseline UPDRS part III score were significant predictors, but not age, duration of treatment from baseline to primary endpoint or baseline daily levodopa dose. The high baseline UPDRS part III score significantly increased the magnitude of the positive placebo response.

In summary, motor fluctuation and high baseline motor UPDRS scores are good predictors for placebo response, but not patient age, gender, or disease duration.

2.2 Study designs

In addition to patient factors discussed above, study design factors, such as the methods of intervention (surgical versus medical interventions), duration of the studies, probability of placebo assignment, and type of patients (early vs late PD; with or without dyskinesia) may also affect rates of placebo response.

Studies that focused on patients with motor fluctuations, surgical intervention, or higher probability of placebo assignment had higher rates of placebo response. Goetz et al. (2008) analyzed the rates of placebo responses from 11 double-blind, placebo-controls trials published between 1993 and 2005, including eight with medical interventions and three with surgical interventions. They defined a positive placebo response at each visit as at least 50% improvement in total UPDRS motor score or a positive change by at least two points on at least two UPDRS in motor items compared to baseline. In the event that 50% overall improvement occurred, but two items worsened by two or more points, the patient was not considered to be a placebo responder. Study-dependent factors included medical vs surgical interventions, likelihood of placebo assignment <50% vs 50%, and disease severity inclusion criteria (PD patients without need for symptomatic treatment, PD patients without motor fluctuations but in need of symptomatic treatment, and PD patients with motor fluctuations in need of symptomatic treatment). The overall placebo response rate was 16% (range: 0–55%). Patients with higher baseline UPDRS scores and studies that focused on PD with motor fluctuations, surgical interventions, or those with a higher probability of placebo assignment showed increased odds of positive placebo response. Placebo responses were temporally distributed similarly during early (3–7 weeks), mid (8–18 weeks), and late phases

(23–35 weeks) of follow-up. Placebo-related improvements occur in most PD clinical trials and are similarly distributed across all 6 months of follow-up.

2.3 Medication cost

Perceptions of cost are capable of altering placebo response. Espay et al. (2015) examined the effect of medication cost on placebo response. They randomized 12 patients with moderate to severe PD and motor fluctuation to a "cheap" or "expensive" subcutaneous novel injectable dopamine agonist placebo (normal saline). Patients were crossed over to the alternate arm approximately 4 h later. They found that although both placebos improved motor function, benefit was greater when patients were randomized first to expensive placebo, with a magnitude halfway between that of cheap placebo and levodopa. They concluded that expensive placebo significantly improved motor function in a direction and magnitude comparable to levodopa. This study was conducted in the United States where patients largely pay for their medications, so that their price is readily apparent. In cultures and medical systems where public insurance underwrites medication costs, the importance of this factor to placebo effects may not be pertinent.

2.4 Placebo effects viewed from the patient's perspective

Improved motor function in placebo responders did not predict a parallel improvement in patient-based ratings of their activities of daily living (ADL). Goetz et al. (2002) compared the objective motor function improvement in the UPDRS motor section and the subjective ADL assessment in 199 PD patients who were assigned to the placebo group in the 26-week DATATOP trial. Similar to their previous study (Goetz et al., 2000), they found 17% of the patients met the strict criteria for placebo-associated improved motor function and the group prevalence of response was steady (7–10%) at any one visit during the trial. However, ADL improvements did not identify the subjects with objectively defined placebo-associated improvement. The authors pointed out that although the notion of placebo effect often implies patient-based perceptions, they found subjective changes to be infrequent in placebo-treated PD patients.

Most PD patients have a positive view about participating in DBPC trials. Goetz, Janko, Blasucci, and Jaglin (2003) examined how placebo assignment affects PD patients in clinical trials. They interviewed 50 consecutive placebo-treated patients from 14 placebo-controlled clinical trials of

PD after completion of their involvement. The study showed PD patients clearly understand the concept of placebo-controlled trials they participated in but are inaccurate in assessing whether they received placebo treatment. Although most wished they received the active compound being tested, most PD patients have a very positive view of participating in placebo-controlled trials.

3. Nocebo and lessebo responses in PD

Nocebo response refers to adverse events (AEs) generated by patients' negative expectations that medical treatment will likely be harmful instead of healing. Nocebo effect can be assessed in DBPC trials by examining AEs in the placebo arms in DBPCs. It is discussed in full detail elsewhere in this volume (Chapter "Nocebo and lessebo effects" by Mestre).

Whereas placebo effects are generally considered positive, a second consideration is the potential that placebo treatment can elicit a negative expectation of not receiving the active treatment. This negative expectation is named "lessebo effect", again discussed in full detail in another section of this volume (Chapter "Nocebo and lessebo effects" by Mestre). In both instances, PD has been the focus of much research, and the examples in Chapter "Nocebo and lessebo effects" by Mestre will draw from those studies.

4. Objective measures and markers of placebo (imaging, biologic, physiologic biomarkers)

4.1 PET studies

PET studies gave the first insights that the dopaminergic system plays a role in placebo effect by showing that placebo-induced improvement is associated with dopamine release in the dorsal striatum. In 2001, De la Fuente-Fernandez el al. used positron emission tomography (PET) to estimate pharmacologically or behaviorally induced dopamine release based on the competition between endogenous dopamine and [11C]raclopride (RAC) for binding to dopamine D2/D3 receptors. They examined the striatal RAC binding potential of six patients with PD under two conditions: Condition one, a placebo-controlled, blinded study in which the patients did not know when they were receiving placebo or active drug (apomorphine)—all patients received both placebo and active drug; and condition two, an open study in the same patients without placebo. They found a significant decrease in striatal RAC binding potential in the dorsal striatum (the caudate nucleus and the putamen) when the patients received placebo compared with open baseline observations. The magnitude of the placebo response was comparable to that of therapeutic doses of levodopa, or apomorphine. Interestingly, those patients who perceived a clinical benefit in response to placebo released more dopamine than those who did not. They concluded that there must be a causal link between the increase in the striatal levels of dopamine and clinical improvement. They suggested that this placebo-induced dopamine release is most likely related to the expectation of therapeutic effect.

Another PET study further demonstrated that the expectation of receiving the reward, not the reward itself, increased dopamine release in the ventral striatum (Nucleus Accumbens). de la Fuente-Fernandez et al. (2002) investigated the relationship between placebo effect and dopamine release in the ventral striatum. By considering the clinical benefit in response to placebo as the reward, they hoped to compare the contribution of expectation of reward and the reward itself to dopamine release. They studied 12 PD patients who were distributed in two groups of six: a placebo group, which was studied in a blind fashion under placebo intervention (as described below); and an open group, which was studied in an open fashion without placebo intervention. Patients of the placebo group were told 1 month in advance of the study that they would receive four subcutaneous injections: three different doses of apomorphine (0.03, 0.06, or 0.12 mg/kg) and a placebo injection (saline)/administered in an unspecified order. Patients of the open group were informed in advance there were three different doses of apomorphine (0.03, 0.06 or 0.12 mg/kg) or no injection. Scans were obtained after each injection. Patients were explicitly asked whether or not they perceived an improvement in the symptoms of their PD after each injection. They found that three patients in the placebo group reported clinical benefit after placebo injection, and the other three did not perceive any clinical effect. All six patients in the placebo group showed a decrease in [11C]raclopride binding potential in the ventral striatum after placebo injection as compared to open baseline values. However, there were no differences in [11C]raclopride binding potential changes between patients who experienced the reward (those who reported placebo-induced clinical benefit) and those who did not in the dorsal striatum. They conclude that the release of dopamine in the ventral striatum (nucleus accumbent) is related to the expectation of reward and not to the reward itself.

Expectations of benefitting from repetitive transcranial magnetic stimulation (rTMS) also induced dopamine release in PD even though the effect of rTMS in improving PD symptoms is not universally established (Pascual-Leone et al., 1994). Strafella, Ko, and Monch (2006) enrolled seven PD patients who were told that rTMS had potential of providing transitory clinical improvement of PD motor symptoms. They were informed of the possibility of receiving a sham rTMS (placebo-rTMS) and that they had a 50% chance of receiving active vs placebo-rTMS, while in fact, they all received only the placebo-rTMS. Placebo-rTMS induced a significant bilateral reduction in [11C]raclopride binding potential in the dorsal and ventral striatum as compared to the baseline condition. This reduction of the binding potential is indicative of an increase in dopamine neurotransmission. The changes in [11C]raclopride binding were more evident in the hemisphere contralateral to the more affected side suggesting that the more severe the symptoms, the greater the drive for symptom relief, and therefore, the placebo response.

Simple verbal manipulation of expectations of having potential benefits can also induce dopamine release in the striatum. Lidstone et al. (2010) enrolled 35 PD patients and applied verbal manipulation to modulate the expectations of patients, who were told that they had a particular probability (25%, 50%, 75%, or 100%) of receiving active medication when they, in fact, received placebo. Significant dopamine release occurred when the declared probability of receiving active medication was 75%, but not at other probabilities. Placebo-induced dopamine release in all regions of the striatum was also highly correlated with the dopaminergic response to open administration of active medication, whereas response to prior medication was the major determinant of placebo-induced dopamine release in the motor striatum. Expectation of clinical improvement was additionally required to drive dopamine release in the ventral striatum.

In summary, PET studies in PD demonstrated that expectations of improvements evoked by apomorphine, repetitive transcranial magnetic stimulation or verbal manipulation induce dopamine release in the striatum.

4.2 Functional MRI (fMRI)

Functional MRI (fMRI) is also useful in studying the neurophysiological basis of placebo in Parkinson's disease. Using fMRI, Hare, O'Doherty, Camerer, Schultz, and Rangel (2008) showed the ability of reward-learning depends on midbrain dopamine neurons and their striatal and prefrontal targets. Functional MRI studies in healthy participants have shown that the learning of rewards correlated with blood oxygen level–dependent (BOLD) activity in the prefrontal cortex, whereas errors of reward prediction correlate with activity in the striatum (Hare et al., 2008).

Expectations of receiving a dopaminergic drug induced changes in fMRI in a reward-learning task. Schmidt, Braun, Wager, and Shohamy (2014) combined a reward-learning task with fMRI to test how expectations of receiving dopaminergic drug modulated learning-related activity in the brain. They enrolled 18 PD patients and used a within-patient design to selectively test the effects of dopaminergic medication (comparing on versus off drug) and whether or not placebo had a similar effect (comparing placebo versus off drug). Each patient was tested under three conditions: off drug (no treatment), placebo (sham treatment), and on drug (standard dopaminergic medication). In the off-drug condition, participants were withdrawn overnight from dopaminergic medication. In the on-drug condition, participants received their standard dopaminergic medication crushed into orange juice before scanning. In the placebo condition, participants watched their sham medication being crushed and dissolved in a glass of orange juice. Unbeknownst to them, they were administered a glass of orange juice containing crushed placebo (starch) pills. While being scanned with fMRI, PD patients performed an instrumental learning task. On each trial, participants chose between two shapes and subsequently observed an outcome. Participants were instructed to maximize their winnings by learning the optimal shapes. In the gain condition, the optimal choice led to a gain ($10) 75% of the time and to no reward ($0) 25% of the time. In the loss condition, the optimal choice led to no reward ($0) 75% of the time and to a loss (−$10) 25% of the time. Gain and loss trials, distinguished by unique pairs of symbols, were randomly intermixed in a learning session. They found that the mere expectation of receiving a dopaminergic drug enhanced reward-learning that was associated with reward-learning-related signals in the striatum and the ventromedial prefrontal cortex in fMRI. These effects were selective to learning by reward; neither medication nor placebo had an effect on learning to avoid monetary loss. These findings suggest a neurobiological mechanism by which expectations shape learning.

Espay et al. (2015), in their study examining the effect of medication cost on placebo response (Section 2.3), also performed fMRI to measure brain activity using a feedback-based visual-motor associative learning fMRI task. Order effect was examined using stratified analysis. They found that although both placebos improved motor function, benefit was greater when patients were randomized first to the expensive placebo which improved motor function but decreased brain activation (in a manner qualitatively similar to levodopa), whereas cheap placebo exerted a robust level of brain activation with only modest effect on motor improvement. They concluded that expensive placebo significantly improved motor function in a direction and magnitude comparable to levodopa.

In summary, studies demonstrated that placebo responses are associated with brain activity changes in fMRI, providing further evidence for the neurobiological evidence of the placebo responses.

4.3 Single cell recordings

Placebo response is associated with changes of single-neuronal activities in the basal ganglia circuit.

In a neurophysiological study, Benedetti, Colloca, Torre, et al. (2004) showed that placebo response in PD was associated with reduced activity in subthalamic neurons. In that double-blind study, they recorded single neuron activities in the subthalamic nuclei in 11 PD patients before and after placebo to examine if changes in neuronal activity was associated with clinical placebo response. To enhance placebo response, they preconditioned these patients with apomorphine, a powerful antiparkinsonian drug before the implantation of DBS. Apomorphine has been shown to reduce neuronal activity in subthalamic nuclei (Lafreniere-Roula et al., 2010). Benedetti et al. (2004) recorded the activity of 100 subthalamic neurons before the implantation of the electrode as a control. The patients then received the placebo treatment, which was the injection of normal saline along with the verbal suggestion of a motor improvement. Neuronal activity was then recorded from110 neurons before implantation of the second electrode into the other subthalamic nuclei. They defined the placebo response as the decrease of arm rigidity of at least one point on the UPDRS. Those patients who showed a clear-cut clinical placebo response had a significant decrease of neuronal discharges compared to the pre-placebo control. Most interestingly, they observed no significant difference between pre-placebo and post-placebo neuronal discharges in those patients who did not show any placebo improvement.

In addition to the neuronal activity changes in the subthalamic nucleus, there are also changes of neuronal activities in the thalamus and the substantia nuclei associated with placebo responses. Given that changed activity of subthalamic neurons is associated with the placebo response

(Benedetti et al., 2004), Benedetti et al. (2009) reasoned that there might be changes of the neuronal activity in other parts of the basal ganglia circuits such as the substantia nigra, the internal globus pallidus, and the motor thalamus (Delong & Wichmann, 2007). To address these questions, they recorded the neuronal activities from subthalamus, substantia nigra pars reticulata and ventral anterior and anterior lateral thalamus in two groups of PD patients: the first group of 12 patients did not receive any treatment, while the second group of 12 patients received an intraoperative placebo treatment along with verbal suggestions of motor improvement. In those patients who showed a clinical placebo response, there was a decrease in firing rate in subthalamic neurons that was associated with a decrease in the substantia nigra pars reticulata and an increase in the ventral anterior and anterior ventral lateral thalamus. By contrast, placebo non-responders showed no changes of neuronal activity in this circuit. They concluded that a placebo treatment, which was saline injection followed by verbal suggestions of benefit, can reverse the malfunction of a complex neuronal circuit, although these placebo-associated neuronal changes are short-lasting and occur only in some patients but not in others.

In a subsequent study, Benedetti, Carlino, and Piedimonte (2016) and Benedetti, Frisaldi, et al. (2016) further demonstrated that the short-lasting neuronal activity changes in response to placebo can be significantly lengthened by preconditioning with apomorphine. In this study, they divided 42 PD patients into six groups randomly: the first group did not receive any treatment (the control group); the second group received a subcutaneous placebo treatment intraoperatively with suggestions of motor improvement, but without apomorphine preconditioning (Placebo 0 group). The other four groups received an intraoperative subcutaneous placebo after either one injection of 2 mg apomorphine, 1 day before surgery (Placebo 1 group), one injection of apomorphine per day for 2 days before surgery (Placebo 2 group), one injection of apomorphine per day for 3 days before surgery (Placebo 3 group) or one injection of apomorphine per day for 4 days before surgery (Placebo 4 group). They analyzed the placebo response at the single-neuron level in the thalamus in these six groups of PD patients to see the differences between first time administration of placebo and administration after pharmacological preconditioning. They found that when the placebo was given for the first time, it induced neither clinical improvement (muscle rigidity reduction at the wrist), nor neuronal changes in thalamic neurons. However, if placebo was given after two, three or four prior administrations of apomorphine, it produced both clinical and neuronal responses. Both the magnitude and the duration of these clinical and placebo responses depended on the number of prior exposures to apomorphine. The greater the number of previous apomorphine administrations, the larger the magnitude and the longer the duration of the clinical and neuronal placebo responses. They concluded that learning plays a crucial role in the placebo response and suggest that placebo non-responders can be turned into placebo responders, with important clinical implications.

In summary, single cell recordings demonstrated that the placebo response is associated with changes in neuronal activities in the basal ganglion circuit and that learning can enhance the changes in neuronal activity associated with the placebo response.

4.4 Transcranial magnetic stimulation

Transcranial magnetic stimulation (TMS) is useful in measuring cortical excitability. One of the parameters used to measure cortical excitability is the amplitude of motor evoked potential (MEP) evoked by TMS. The cortical excitability is increased if the stimulation of the same intensity evokes higher MEP amplitude and decreased if the stimulation of the same intensity evokes lower amplitude. Lou, Benice, Kearns, Sexton, and Nutt (2003) showed cortical excitability was markedly increased in PD patients while they were off PD medicine compared to normal controls and that this increased cortical excitability was reduced by the administration of levodopa.

Placebo response is associated with changes in cortical excitability in PD (Lou et al., 2013). Because expectation induces dopamine release in dorsal and ventral striatum (de la Fuente-Fernandez et al., 2002, 2001) and dopamine reduces cortical excitability in PD (Lou et al., 2013), Lou et al. (2013) carried out a study to examine if expectation of receiving dopamine could reduce cortical excitability in PD patients. They randomized 26 PD patients to one of three groups: 0%, 50% and 100% expectancy of receiving levodopa. However, all subjects received placebo regardless of expectancy group. Degree of expectancy had a significant effect on motor evoked potential (MEP) elicited by transcranial magnetic stimulation (TMS) ($P < 0.05$). Subjects in the 50% and 100% expectancy groups responded with a decrease in MEP, while those in the 0% expectancy group responded with an increase in MEP ($P < 0.05$). Therefore, expectations exert a similar effect on the amplitude of MEP to levodopa.

5. Placebo response: Where do we go from here?

We have made great progress in understanding the neuropsychophysiological mechanisms of the placebo response in PD over the last 20 years. Rather than treating the placebo responses as nuisance, we can now take advantage of our knowledge of the placebo response and apply it to the design of the clinical trials and daily clinical practice (Mestre et al., 2018).

5.1 Minimizing placebo response in clinical trials

One strategy to increase the likelihood of a successful DBPC trial is to minimize the placebo effect in the placebo arm. Often, a DBPC trial fails because of the robust placebo effects. Due to robust placebo effects, the difference between the active treatment arm and the placebo arm is reduced and thus the trial fails to reach the statistical significance. Therefore, minimizing the placebo effect may make a DBPC trial more likely to be successful.

The following approaches may be useful to minimize the placebo response rate when conducting a DBPC trial (Mestre et al., 2018):

1. Using a neutral tone in recruiting a subject: A clinical investigator should use a neutral tone while presenting a study to a potential subject and to emphasize that it is not clear the study drug will be beneficial because placebo effect is associated with expectations of potential benefits in participating in a trial.
2. Designing a longer trial: A longer trial is associated with lower placebo response.
3. Using 50:50 randomization.

One novel approach that had been used in designing a DBPC trial to minimize placebo responses is excluding placebo responders by including a run-in period with the administration of placebo only to identify placebo responders (Parkinson Study Group, 2017). However, it may be unethical to include only placebo non-responders in a trial. Furthermore, it is unknown how the results from such a trial can be applied to general population. Most importantly, it will be a challenging conversation for a clinical investigator to inform placebo responders that they will be dropped out of the study because they respond to placebo.

5.2 Optimizing placebo response in clinical practice

Traditionally, clinical researchers in PD have focused on reducing placebo effects in clinical trials. However, with increasing understanding of the neurophysiological mechanisms of placebo effects, there is growing interest in optimizing placebo effects to improve available treatments and in finding ways to minimize nocebo effects to improve clinical outcome (Mestre et al., 2018; Petrie & Rief, 2019).

Managing expectations appropriately can improve the outcomes of clinical encounters. Expectations are among the major contributing factors to placebo effects. Expectations can be influenced by previous experience (self-learning), observational learning from other people, instructions or information from physicians and other clinical staff, or information from online resources or personal cultural beliefs. Expectations are also strongly triggered by sensory cues of the clinical setting, such as stimuli from the white coats of physicians, friendly or comforting manners of a nurse, the feeling of receiving an injection, the shape or color of a pill, and the act of swallowing a medication (Mestre et al., 2018; Petrie & Rief, 2019).

Empathy in clinician-patient interaction can affect patients' expectations of good outcomes and influence placebo effects. Katptchuk et al. (2008) studied how a clinician's empathy influenced placebo responses in patients with irritable bowel syndrome (IBS). One group of the clinicians (treatment group) were instructed to use a very empathetic way (using a warm and supportive manner in interacting with patients) while giving a placebo treatment. The other groups of the clinicians (the control group) administered the same placebo treatment but were asked to act formally and without warmth toward the participants. Although both groups improved more than the observation treatment group (no placebo is given), the patients whose clinicians had empathy showed the greatest improvement, demonstrating the positive interaction in the clinician-patient relationship can increase placebo effects.

Competence of the providers, in addition to empathy, also influences patients' expectations of having good outcomes. Howe, Goyer, and Crum (2017) used a novel design to examine how providers' characteristics affect the placebo effect. After inducing an allergic reaction in participants through a histamine skin prick test, a health care provider administered a cream with no active ingredients and set either positive expectations (cream will reduce reaction) or negative expectations (cream will increase reaction). The provider demonstrated either high or low warmth, and either high or low competence. They found that the impact of expectations on allergic response was enhanced when the provider acted both warmer and more competent and negated when the provider acted colder and less competent. They concluded that placebo effects should be viewed as a psychological phenomenon that can be understood and harnessed to improve treatment outcomes.

In addition to clinician-patient relationship, other factors can also modulate the expectations of good outcomes. Certain physical characteristics of treatment can enhance placebo response. For example, capsules (rather than tablets), larger pills, and strong-tasting medicines seems to have more powerful placebo effects (Buckalew & Coffield, 1982). Espay et al. (2015) showed that more expensive medicine elicits stronger placebo responses.

5.3 Minimizing nocebo response in clinical practice

Negative expectations are the main contributor of nocebo effects (Petrie & Rief, 2019). Negative expectations can be generated by warnings of side effects that may occur from a drug or procedure, or even by increases in general awareness about the potential adverse effects of a treatment or procedure. The creation of negative expectations by warning of side effects was demonstrated in a clinical trial for unstable angina (Myers, Cairns, & Singer, 1987). In a multicenter trial of aspirin or sulfinpyrazone in the treatment of unstable angina, the authors (Myers et al., 1987) examined the importance of mentioning potential side effects in the consent form. Inclusion of a statement outlining possible gastrointestinal side effects in two of the three centers resulted in a sixfold increase in the number of subjects in these centers withdrawing from the study because of subjective gastrointestinal symptoms. The study concluded that the inclusion of gastrointestinal side effects in the consent form may have increased the likelihood of patients attributing gastrointestinal symptoms to drug therapy, leading to subsequent withdrawal from the study. Similarly, among patients who were being treated with beta blockers, those warned of sexual dysfunction side effects reported these side effects three to four times more frequently than a control group who did not receive this information (Cocco, 2009).

Clinicians can minimize nocebo effects by balancing presentation of adverse effects with expected benefit information (Petrie & Rief, 2019). Physicians can optimize treatment expectation and expectation of adverse effects. They can also teach and train patients to cope with adverse effects in advance. Patients can be referred to web-based services that provide evidence-based information, instead of unproven, anxiety-increasing comments usually seen in blogs (Bingel, 2014)

References

Benedetti, F. (2014). Placebo effects: From the neurobiological paradigm to translational implications. *Neuron, 84*, 623–637.

Benedetti, F., & Amanzio, M. (2011). The placebo response: How words and rituals change the patient's brain. *Patient Education and Counseling, 84*, 413–419.

Benedetti, F., Carlino, E., & Piedimonte, A. (2016). Increasing uncertainty in CNS clinical trials: the role of placebo, nocebo, and Hawthorne effects. *Lancet Neurology, 15*, 736–747.

Benedetti, F., Colloca, L., Torre, E., et al. (2004). Placebo-responsive Parkinson patients show decreased activity in single neurons of subthalamic nucleus. *Nature Neuroscience*, 7, 587–588.

Benedetti, F., Frisaldi, E., Carlino, E., et al. (2016). Teaching neurons to respond to placebos. *The Journal of Physiology, 594*, 5647–5660.

Benedetti, F., Lanotte, M., Colloca, L., Ducati, A., Zibetti, M., & Lopiano, L. (2009). Electrophysiological properties of thalamic, subthalamic and nigral neurons during the anti-parkinsonian placebo response. *The Journal of Physiology, 587*, 3869–3883.

Bingel, U. (2014). Avoiding nocebo effects to optimize treatment outcome. *JAMA, 312*, 693–694.

Bonifati, V., & Meco, G. (1999). New, selective catechol-O-methyltransferase inhibitors as therapeutic agents in Parkinson's disease. *Pharmacology & Therapeutics, 81*, 1–36.

Buckalew, L. W., & Coffield, K. E. (1982). An investigation of drug expectancy as a function of capsule color and size and preparation form. *J Clin Psychopharcol, 1982*(2), 245–248.

Cocco, G. (2009). Erectile dysfunction after therapy with metoprolol: The Hawthorne effect. *Cardiology, 112*, 174–177.

Cohen, S., Kamarck, T., & Mermelstein, R. (1983). A global measure of perceived stress. *Journal of Health and Social Behavior, 24*, 385–396.

Colloca, L., & Benedetti, F. (2009). Placebo analgesia induced by social observational learning. *Pain, 144*, 28–34.

Connolly, B. S., & Lang, A. E. (2014). Pharmacological treatment of Parkinson disease: A review. *JAMA, 311*, 1670–1683.

Corsi, N., Emadi Andani, M., Tinazzi, M., & Fiorio, M. (2016). Changes in perception of treatment efficacy are associated to the magnitude of the nocebo effect and to personality traits. *Scientific Reports, 6*, 30671.

Corvol, J. C., Bonnet, C., Charbonnier-Beaupel, F., Bonnet, A. M., Fiévet, M. H., Bellanger, A., et al. (2011). The COMT Val158Met polymorphism affects the response to entacapone in Parkinson's disease: A randomized crossover clinical trial. *Annals of Neurology, 69*, 111–118.

Costa, P. T., & McCrae, R. R. (Eds.), (1992). *Revised NEO personality inventory and NEO five-factor inventory: Professional manual.* Lutz, FL: Psychological assessment resources, Inc. 1992. p. 101.

de la Fuente-Fernandez, R., Phillips, A. G., Zamburlini, M., Sossi, V., Calne, D. B., Ruth, T. J., et al. (2002). Dopamine release in human ventral striatum and expectation of reward. *Behavioural Brain Research, 136*, 359–363.

de la Fuente-Fernandez, R., Ruth, T. J., Sossi, V., Schulzer, M., Calne, D. B., & Stoessl, A. J. (2001). Expectation and dopamine release: Mechanism of the placebo effect in Parkinson's disease. *Science, 293*, 1164–1166.

Espay, A. J., Norris, M. M., Eliassen, J. C., Dwivedi, A., Smith, M. S., Banks, C., et al. (2015). Placebo effect of medication cost in Parkinson disease: A randomized double-blind study. *Neurology, 84*, 794–802.

Ferreira, J. J., Trenkwalder, C., & Mestre, T. A. (2018). Placebo and nocebo responses in other movement disorders besides Parkinson's disease: How much do we know? *Movement Disorders, 33*, 1228–1235.

Frisaldi, E., Carlino, E., Zibetti, M., Barbiani, D., Dematteis, F., Lanotte, M., et al. (2017). The placebo effect on bradykinesia in Parkinson's disease with and without prior drug conditioning. *Movement Disorders, 32*, 1474–1478.

Geers, A. L., Helfer, S. G., Kosbab, K., Weiland, P. E., & Landry, S. J. (2005). Reconsidering the role of personality in placebo effects: Dispositional optimism, situational expectations, and the placebo response. *Journal of Psychosomatic Research, 58*, 121–127.

Geers, A. L., Wellman, J. A., Fowler, S. L., Helfer, S. G., & France, C. R. (2010). Dispositional optimism predicts placebo analgesia. *The Journal of Pain, 11*, 1165–1171.

Goetz, C. G., Janko, K., Blasucci, L., & Jaglin, J. A. (2003). Impact of placebo assignment in clinical trials of Parkinson's disease. *Movement Disorders, 18*, 1146–1149.

Goetz, C. G., Leurgans, S., Raman, R., & Parkinson Study Group (2002). Placebo-associated improvements in motor function: Comparison of subjective and objective sections of the UPDRS in early Parkinson's disease. *Movement Disorders, 17*, 283–288.

Goetz, C. G., Leurgans, S., Raman, R., & Stebbins, G. T. (2000). Objective changes in motor function during placebo treatment in PD. *Neurology, 54*, 710–714.

Goetz, C. G., Wuu, J., McDermott, M. P., Adler, C. H., Fahn, S., Freed, C. R., et al. (2008). Placebo response in Parkinson's disease: Comparisons among 11 trials covering medical and surgical interventions. *Movement Disorders, 23*, 690–699.

Hall, K. T., Lembo, A. J., Kirsch, I., Ziogas, D. C., Douaiher, J., Jensen, K. B., et al. (2012). Catechol-O-methyltransferase val158met polymorphism predicts placebo effect in irritable bowel syndrome. *PLoS One*, 7e48135.

Hall, K., Loscalzo, J., & Kaptchuk, T. J. (2015). Genetic and the placebo effect: The placebome. *Trends in Molecular Medicine, 21*, 285–294.

Hare, T. A., O'Doherty, J., Camerer, C. F., Schultz, W., & Rangel, A. (2008). Dissociating the role of the orbitofrontal cortex and the striatum in the computation of goal values and prediction errors. *The Journal of Neuroscience, 28*, 5623–5630.

Howe, L. C., Goyer, J. P., & Crum, A. J. (2017). Harnessing the placebo effect: Exploring the influence of physician characteristics on placebo response. *Health Psychology, 36*, 1074–1082.

Huber, A., Lui, F., & Porro, C. A. (2013). Hypnotic susceptibility modulates brain activity related to experimental placebo analgesia. *Pain, 154*, 1509–1518.

Hunter, T., Siess, F., & Colloca, L. (2014). Socially induced placebo analgesia: A comparison of a pre-recorded versus live face-to-face observation. *European Journal of Pain, 18*, 914–922.

Kaptchuk, T. J., Kelly, J. M., Conboy, L. A., Davis, R. B., Kerr, C. E., Jacobson, E. E., et al. (2008). Component of placebo effect: Randomized controlled trial in patients with irritable bowel syndrome. *BMJ (Clinical Research Ed), 336*, 999–1003.

Keitel, A., Stefano Ferrea, S., Südmeyer, M., Schnitzler, A., & Wojtecki, L. (2013). Expectation modulates the effect of deep brain stimulation on motor and cognitive function in tremor-dominant Parkinson's disease. *PLoS One*, 8e81878.

Lachman, H. M., Papolos, D. F., Saito, T., Yu, Y. M., Szumlanski, C. L., & Weinshilboum, R. M. (1996). Human catechol-O-methyltransferase pharmacogenetics: Description of a functional polymorphism and its potential application to neuropsychiatric disorders. *Pharmacogenetics, 6*, 243–250.

Lafreniere-Roula, M., Darbin, O., Hutchison, W. D., Wichmann, T., Lozano, A. M., & Dostrovsky, J. O. (2010). Apomorphine reduces subthalamic neuronal entropy in parkinsonian patients. *Experimental Neurology, 225*, 455–458.

Lidstone, S. C., Schulzer, M., Dinelle, K., Mak, E., Sossi, V., Ruth, T. J., et al. (2010). Effects of expectation on placebo-induced dopamine release in Parkinson disease. *Archives of General Psychiatry*, *67*, 857–865.

Lou, J. S., Benice, T., Kearns, G., Sexton, G., & Nutt, J. (2003). Levodopa normalizes exercise related cortico- motoneuron excitability abnormalities in Parkinson's disease. *Clinical Neurophysiology*, *114*, 930–937.

Lou, J. S., Dimitrov, D. M., Hammerschlag, R., Nutt, J., Hunt, E. A., Eaton, R. W., et al. (2013). The effect of expectancy and personality on cortical excitability in Parkinson's disease. *Movement Disorders*, *28*, 1257–1262.

McRae, C., Cherin, E., Yamazaki, T., Diem, G., Vo, A. H., Russel, D., et al. (2004). Effects of perceived treatment on quality of life and medical outcomes in a double-blind placebo surgery trial. *Archives of General Psychiatry*, *61*, 412–420.

Mercado, R., Constantoyannis, C., Mandat, T., Kumar, A., Schulzer, M., Stoessl, A. J., et al. (2006). Expectation and the placebo effect in Parkinson's disease patients with subthalamic nucleus deep brain stimulation. *Movement Disorders*, *21*, 1457–1461.

Mestre, T. A., & Ferreira, J. J. (2018). The placebo response in Parkinson's disease and other movement disorders. *Movement Disorders*, *33*, 1193–1194.

Mestre, T. A., Shamy, M., Benedetti, F., & Lang, A. E. (2018). Harnessing the power of placebos in movement disorders: Insights from Parkinson's disease in clinical research and practice. *Movement Disorders*, *33*, 1196–1203.

DeLong, M. R., & Wichmann, T. (2007). Circuits and circuit disorders of the basal ganglia. *Archives of Neurology*, *64*, 20–24.

Myers, M. G., Cairns, J. A., & Singer, J. (1987). The consent form as a possible cause of side effects. *Clinical Pharmacology and Therapeutics*, *42*, 250–253.

Olanow, C. W. R., Bartus, R. T., Baumann, T. L., et al. (2015). Gene delivery of neurturin to putamen and substantia nigra in Parkinson disease: A double-blind, randomized, controlled trial. *Annals of Neurology*, *78*, 248–257.

Parkinson Study Group. (2017). A randomized trial of relamorelin for constipation in Parkinson's disease (MOVE-PD): Trial results and lessons learned. *Parkinsonism & Related Disorders*, *37*, 101–105.

Pascual-Leone, A., Valls-Sole´, J., Brasil-Neto, J. P., Cammarota, A., Grafman, J., & Hallett, M. (1994). Akinesia in Parkinson's disease: II shortening of choice reaction time and movement time with subthreshold repetitive transcranial motor cortex stimulation. *Neurology*, *44*, 892–898.

Peciña, M., Azhar, H., Love, T. M., Lu, T., Fredrickson, B. L., Stohler, C. S., et al. (2013). Personality trait predictors of placebo analgesia and neurobiological correlates. *Neuropsychopharmacology*, *38*, 639–646.

Petrie, K. J., & Rief, W. (2019). Psychological mechanisms of placebo and nocebo effects: Pathways to improve treatments and reduce side effects. *Annual Review of Psychology*, *70*, 599–625.

Pollo, A., Torre, E., Lopiano, L., Rizzone, M., Lanotte, M., Cavanna, A., et al. (2002). Expectation modulates the response to subthalamic nucleus stimulation in parkinsonian patients. *Neuroreport*, *13*, 1383–1386.

Quattrone, A., Barbagallo, G., Cerasa, A., & Stoessl, A. J. (2018). Neurobiology of placebo effect in Parkinson's disease: What we have learned and where we are going. *Movement Disorders*, *33*, 1213–1227.

Ramirez-Zamora, A., & Ostrem, J. L. (2018). Globus pallidus internal or subthalamic nucleus deep brain stimulation for Parkinson disease: A review. *JAMA Neurology*, *75*, 367–372.

Rughani, A., Schwalb, J. M., Sidiropoulos, C., Pilitsis, J., Ramirez-Zamora, A., Sweet, J. A., et al. (2018). Congress of neurological surgeons systematic review and evidence-based guideline on subthalamic nucleus and globus pallidus internus deep brain stimulation for the treatment of patients with Parkinson's disease: Executive summary. *Neurosurgery*, *82*, 753–756.

Saez-Zea, C., Escamilla-Sevilla, F., Katati, M. J., & Minguez-Castellanos, A. (2012). Cognitive effects of subthalamic nucleus stimulation in Parkinson's disease: A controlled study. *European Neurology*, *68*, 361–366.

Schmidt, L., Braun, E. K., Wager, T. D., & Shohamy, D. (2014). Mind matters: Placebo enhances reward learning in Parkinson's disease. *Nature Neuroscience*, *17*, 1793–1797.

Schwarzer, R., & Jerusalem, M. (1995). In J. Weinman, S. Wright, & M. Johnston (Eds.), *Measures in health psychology: A user's portfolio* (pp. 35–37). Windsor, England: NFER-NELSON.

Shetty, N., Friedman, J. H., Kieburtz, K., Marshall, F. J., & Oakes, D. (1999). The placebo response in Parkinson's disease. Parkinson Study Group. *Clinical Neuropharmacology*, *22*, 207–212.

Shin, C. W., Hahn, S., Park, B. J., Kim, J. M., Park, E. O., & Jeon, B. (2016). Predictors of the placebo response in clinical trials on Parkinson's disease: A meta-analysis. *Parkinsonism & Related Disorders*, *29*, 83–89.

Shulman, L. M., Gruber-Baldini, A. L., Anderson, K. E., et al. (2010). The clinically important difference on the unified Parkinson's disease rating scale. *Archives of Neurology*, *67*, 64–70.

Staats, P. S., Staats, A., & Hekmat, H. (2001). The additive impact of anxiety and a placebo on pain. *Pain Medicine*, *2*, 267–279.

Strafella, A. P., Ko, J. H., & Monch, O. (2006). Therapeutic application of transcranial magnetic stimulation in Parkinson's disease: The contribution of expectation. *Neuroimage*, *31*, 1666–1672.

Wendt, L., Albring, A., Benson, S., Engler, H., Engler, A., et al. (2014). Catechol-O-methyltransferase val158met polymorphism is associated with somatosensory amplification and nocebo responses. *PLoS One*, *9*(9)e107665.

Witek, N., Stebbins, G. T., & Goetz, C. G. (2018). What influences placebo and nocebo responses in Parkinson's disease? *Movement Disorders*, *33*, 1204–1212.

Yahr, M. D., Duvoisin, R. C., Schear, M. J., Barrett, R. E., & Hoehn, M. M. (1969). Treatment of parkinsonism with levodopa. *Archives of Neurology*, *21*, 343–354.

Yavich, L., et al. (2007). Site-specific role of catechol-O-methyltransferase in dopamine overflow within prefrontal cortex and dorsal striatum. *The Journal of Neuroscience*, *27*, 10196–10209.

Yu, R., Gollub, R. L., Vangel, M., Kaptchuk, T., Smoller, J. W., & Kong, J. (2014). Placebo analgesia and reward processing: Integrating genetics, personality, and intrinsic brain activity. *Human Brain Mapping*, *35*, 4583–4593.

Zhang, J., & Tan, L. C.-S. (2016). Revisiting the medical management of Parkinson's Disease: Levodopa versus dopamine agonist. *Current Neuropharmacology*, *14*, 356–363.

CHAPTER TEN

Placebo effect in subjects with cognitive impairment

Kaori Ito[a,*], Klaus Romero[b]
[a]Applied Pharmacometrics & Research LLC, Londonderry, NH, United States
[b]Critical Path Institute (C-Path), Tucson, AZ, United States
[*]Corresponding author: e-mail address: kaori.ito.214@gmail.com

Contents

Abstract

Understanding both the placebo response and the underlying disease progression is crucial to designing clinical trials and interpreting results in Alzheimer's disease (AD) research. The disease severity at different stage of disease (e.g., mild cognitive impairment (MCI), early or late AD) is related to the rate of disease progression, which make it even difficult for AD researchers to understand the clinical trial results. A model-based meta-analysis approach using all available historical data provides quantitative understanding of placebo effect and disease progression in AD and offers a useful tool to aid in both trial design and trial interpretation. The Critical Path Institute (C-Path) is a nonprofit organization founded in 2005 as a vehicle to develop tools to accelerate drug development. The drug development tool (DDT) with AD is the first-ever quantitative DDT to be endorsed by FDA and EMA, and is publicly available to researchers through C-Path's website: https://c-path.org/ad-cts-tool-request/.

International Review of Neurobiology, Volume 153
ISSN 0074-7742
https://doi.org/10.1016/bs.irn.2020.03.032

1. Introduction

The placebo effect refers to the phenomenon that a subject receiving an inactive control experiences an improvement in their condition that is primarily attributable to the subject's personal expectations, rather than to any biological or other proposed mechanisms related to the treatment itself. The mechanism of placebo effects has been extensively studied in patients with, among others, neuropathic pain, depression, Alzheimer's disease (AD) and Parkinson's disease (PD). The placebo effect has been explained by activation of many pathways, including opioid, cannabinoid, serotoninergic and dopaminergic, which are involved in reward, expectancy, or conditioning. The placebo response may also result from the interaction of psychosocial, and cultural mechanisms, as well as human relationships. Therefore, not only the internal factors (patient's expectation from the treatment), but external factors, such as better attention from healthcare providers or care givers have great impacts on the outcomes in AD patients. In addition, there are several reasons which make it difficult to understand the placebo effect in AD. Most importantly, the disease itself progresses over time in non-linear fashion, and some of the clinical endpoints in AD are highly dependent on the patient's cognitive status across the disease continuum. In some cases, the placebo response may be limited or difficult to assess, as the disease itself may affect patients' ability to form expectations.

In this chapter, we introduce the concept of disease progression in AD, and how to quantitively understand the placebo effect in AD patients participating in clinical trials. Then, we will discuss what factors may affect the magnitude of placebo effect, as well as a tool to design clinical trials.

2. Background

Over decades, significant effort was made to investigate the placebo effect and implicated mechanisms; studies were conducted in pre-clinical animal models and in healthy volunteers (exposed to a stimulus) to understand the biological explanations of the placebo effect (Murray & Stoessl, 2013). There are now over 5000 research articles in the PubMed database that make specific reference to the placebo effect, which include demonstrations of placebo effects in, among others, pain, depression, anxiety, insomnia, immunosuppression, Attention Deficit Hyperactivity Disorder (ADHD), and Parkinson's disease (Colagiuri, Schenk, Kessler, Dorsey, & Colloca, 2015).

These studies demonstrated that the expectations generated by the novel treatments significantly alter the levels of inflammatory markers, endocannabinoids, endogenous opioids, or dopamine metabolism in both the control and active arms. However, the capacity to generate equivalent expectation may differ between specific conditions, particularly in disorders where reward signaling may be impaired. The variability of the placebo response in individuals may in part reflect variation in the activity of several neurotransmitter systems, which may, in turn, be related to a specific genetic makeup, or the specific underlying pathophysiology. While dopamine, serotonin, cholecystokinin, and opioids are key players, there are likely other neurotransmitters contributing to the placebo effect in various neurological and psychiatric conditions (Benedetti, Carlino, & Pollo, 2010). Recent advances in neuroimaging studies using positron emission tomography (PET) and functional magnetic resonance imaging (fMRI) have critically promoted the study of the placebo effect and provide some of the strongest evidence for the mechanisms of the placebo effect in various neurological disorders (Faria, Fredrikson, & Furmark, 2008).

However, these imaging studies are mostly focusing on acute neurotransmitter release, while different mechanism may contribute to longer lasting placebo effects observed in clinical trials.

In general, it is difficult to identify the disease-specific manifestations of the placebo effect, and the magnitude and duration of the placebo effect may vary considerably across different diseases, clinical measures, and therapeutic interventions, and not many studies are reported to understand the mechanism of placebo effect in AD. But, regardless of the specific mechanisms and their overlap proposed in patients with neuropathic pain, depression, and Parkinson's disease, which, may underlie the placebo effect in other neurological and psychiatric conditions including AD the empirical quantification of the placebo effect in AD remains feasible, provided that an adequate number of patient-level data sources that include relevant control arm data can be integrated and standardized.

In other words, while the mechanism of placebo effect in AD is still unknown, efforts have been made by several researchers to understand and quantifying the placebo effect in this disease.

3. Failures in recent clinical trials in AD

AD is a chronic, neurodegenerative disease that afflicts an estimated 5.8 million Americans and it is sixth leading cause of death in, 2019

(Alzheimer's disease facts and figures, 2019). The current medications approved by FDA for treatment of AD temporarily improve the syndromic presentation, but none of the pharmacologic treatments available today have been approved to slow or stop the damage and destruction of neurons that cause AD symptoms and make the disease fatal. The social and economic burden of AD and other dementias have been dramatically increased due to growing aging society.

Since 2000, Pharmaceutical Research and Manufacturers of America (PhRMA) member companies have invested more than $600 billion into research into dementia treatments. Despite the tremendous effort to combat AD, the only drugs approved for the treatment of AD over the past 30 years have been cholinesterase inhibitors and memantine, without any new approvals since 2003. Nearly 200 drugs were advanced to at least the level of phase II development, but the results from phase III trials are either negative or inconclusive. These failures have been partially attributed to inadequate clinical trial design (Schneider et al., 2014).

AD is thought to begin 20 years or more before clinical manifestation (Bateman et al., 2012; Braak, Thal, Ghebremedhin, & Tredici, 2011; Gordon et al., 2018; Jack et al., 2009; Reiman et al., 2012; Villemagne et al., 2013) without any noticeable clinical change to the person affected. Only after years of accumulated neurological damages cause individuals start to experience signs and symptoms, such as memory loss or language problems. Over time, symptoms tend to increase and start interfering with the individual's' ability to perform everyday activities, and the individual is diagnosed as having dementia due to AD only after the clinical presentation is pronounced. Theoretical models have been proposed to describe the AD pathological cascade (Jack et al., 2010, 2013), and the interest evolved in late 90s in treating AD early enough to prevent or slow down disease progression, stemming from the recognition that a risk state could be identified before the onset of clinically manifest dementia. However, it is not easy in practice to conduct clinical studies in early AD which require a longer duration of follow-up, additional clinical evaluations including biomarkers, PET or MRI for the effectiveness of a novel treatment or intervention, to determine its ability to slow disease progression.

There are several key issues researchers need to consider when evaluating the results from clinical trials in AD. Hence, it is important to quantitatively understand the rate of disease progression for individuals and separate the placebo response from the treatment effect, especially considering the large inter-individual variability in response to placebo. However, the magnitude of placebo response is altered by many factors, such as disease severity of individuals, study design, or study duration, especially given how the underlying disease progress over time is dependent on baseline disease severity.

Modeling disease progression involves the utilization of mathematical functions to quantitatively describe the time course of disease status. The first disease progression model for AD was proposed by Holford et al. (Holford & Peace, 1992a, 1992b; Peace & Holford, 1994) in the early 90s using the Alzheimer's Disease Assessment Scale-Cognitive Subscale (ADAS-Cog). The simplest model to quantitatively describe AD progression is a linear one, assuming a constant rate of change for a measure of disease status over time. Linear models have been successfully implemented to describe progression of several diseases (Cook & Bies, 2016).

4. Disease progression model in AD

The advantages of developing empirical disease progressions is to help interpret the observed data and predict the future clinical outcome using the mathematical model. Although a disease model can be developed in any diseases, it is more common to describe the slow progress of disease, such as AD. It can also lend itself to expand the description of the underlying disease process by incorporating knowledge of the disease process with more sophisticated mathematical frameworks (semi-mechanistic or physiological disease model).

Disease progression in AD is slow, and it takes long time to observe the significant cognitive or functional declines. The rate of disease progression depends on baseline severity at the start of follow-up; i.e., milder patients progress slower than severe patients. This can be easily explained by how disease progress over time; the pathological process of AD from the onset to the end-stage disease is not a linear process.

After the first proposed disease progression model for AD in early 90s, many others have expanded upon the work to describe such non-linearly (Conrado, Denney, Chen, & Ito, 2014; Rogers et al., 2012; Samtani et al., 2012). For example, Conrado et al. (2014) utilized a generalized logistic function to better fit the sigmoidal nature of disease progression in AD using pooled data from the Coalition Against Major Diseases (CAMD), now the Critical Path for Alzheimer's Disease (CPAD) database.

However, the duration of placebo controlled clinical trials in AD and mild cognitive impairment (MCI) patients are typically ranging from 6 to 18 months. This is due to the enormous cost of running clinical trials, as well as ethical reasons for patients who are going to be assigned to placebo in double-blind clinical trials. Therefore, clinical studies are relatively short duration compare to the entire disease progression process. As a result, a disease model with a linear disease progression assumption may describe the observed data reasonably well within a certain time range of clinical trials. In fact, simple linear models have been successfully implemented to describe progression of several diseases (Cook & Bies, 2016), including AD, schizophrenia, and Huntington's disease. The linear assumption is also straightforward to implement in the modeling framework, and easy to understand even for non-mathematical oriented researchers.

5. Placebo response model in AD

These mathematical modeling techniques can be also used to estimate the placebo effect in AD clinical trials. This model assumes that the placebo response in clinical trials consists of two components, (1) "*underlying disease progression*," and (2) *"placebo effect*."

The "*underlying disease progression*" describes the disease progression due to the underlying pathophysiological change/neurodegenerative process, which is progress over time, spanning 5–15 years (or longer, depending on other factors). It is noted that other conditions may influence on the apparent rate of disease progression; for example, anosognosia associated with cognitive decline, or other co-existing factors and conditions (e.g., living with caregivers or not, having other illness(es) are important to evaluate and answer to the a confounded question why dementia symptoms often fluctuate. However, this chapter will focus on the clinical trial settings in this chapter, therefore only describe the population mean response for a relatively shorter duration (up to 2–3 years).

The "*placebo effect*" in this context represents a temporal component, or temporal reaction to the placebo in the clinical trial settings, i.e., such as psychological effect, or any effect derived from the conduction of and participation in clinical trials. The temporal effect of placebo treatment is often observed in many clinical trials, commonly discussed as expectation effect by a patient or a doctor. Despite the fact that the cognitive functions decline over time with AD patients, and anosognosia associated with dementia (possibly more complicated mechanism of these interactions), the historical data shows the temporal effect of placebo treatment, and it is important to distinguish from the underlying disease progression to understand the results from clinical trials.

Therefore, the placebo response (underlying disease progression + temporal placebo effect) can be described mathematically as:

$$S(t) = S(0) + \alpha \times t + Pbo(t) + \varepsilon$$

where $S(t)$ is the disease state at time t after entry into a trial with baseline $S(0)$, α is the rate of progression of the untreated disease, $Pbo(t)$ is a function describing placebo effect, and ε is a function describing the residual error.

The onset (appearance), offset (disappearance) and overall extent of the placebo effect has been described in the past by the use of a first-order kinetics, commonly known as a Bateman-type function (Holford & Peace, 1992a; Ito et al., 2010; Lockwood, Ewy, Hermann, & Holford, 2006):

$$Pbo(t) = \beta_p \times \left(e^{-Kel_p \times t} - e^{-Keq_p \times t}\right)$$

where β_p is a factor defining the magnitude of the placebo effect, Kel_p is the first-order rate constant for the offset of the placebo effect, and Keq_p is the first-order rate constant for the onset of the placebo effect.

Fig. 1 illustrates the concept of the placebo response using ADAS-cog scores. The overall observed placebo response in a trial (C) is simply the mathematical addition of the underlying disease progression (A) and placebo effect (B).

As previously discussed, there are several key issues researchers need to be aware of when evaluating the results from clinical trials in AD. It is important to understand the rate of disease progression for individuals and separate the placebo response from the treatment effect. However, it is usually difficult to determine the disease progression and the placebo response from a single study, because results can vary from one study to another for various reasons, including confounding factors, and the different study samples used.

Therefore, the conventional disease progression model and pooled historical clinical data, which was previously used for developing the disease progression model (Ito et al., 2013) is presented in the following section as a guiding tool. The advantages of meta-analytic approaches are related to the fact that, by combining individual studies, and thus using more data, the precision and accuracy of the estimates in the individual studies can be improved. It is often statistically stronger and more generalizable than those from single studies.

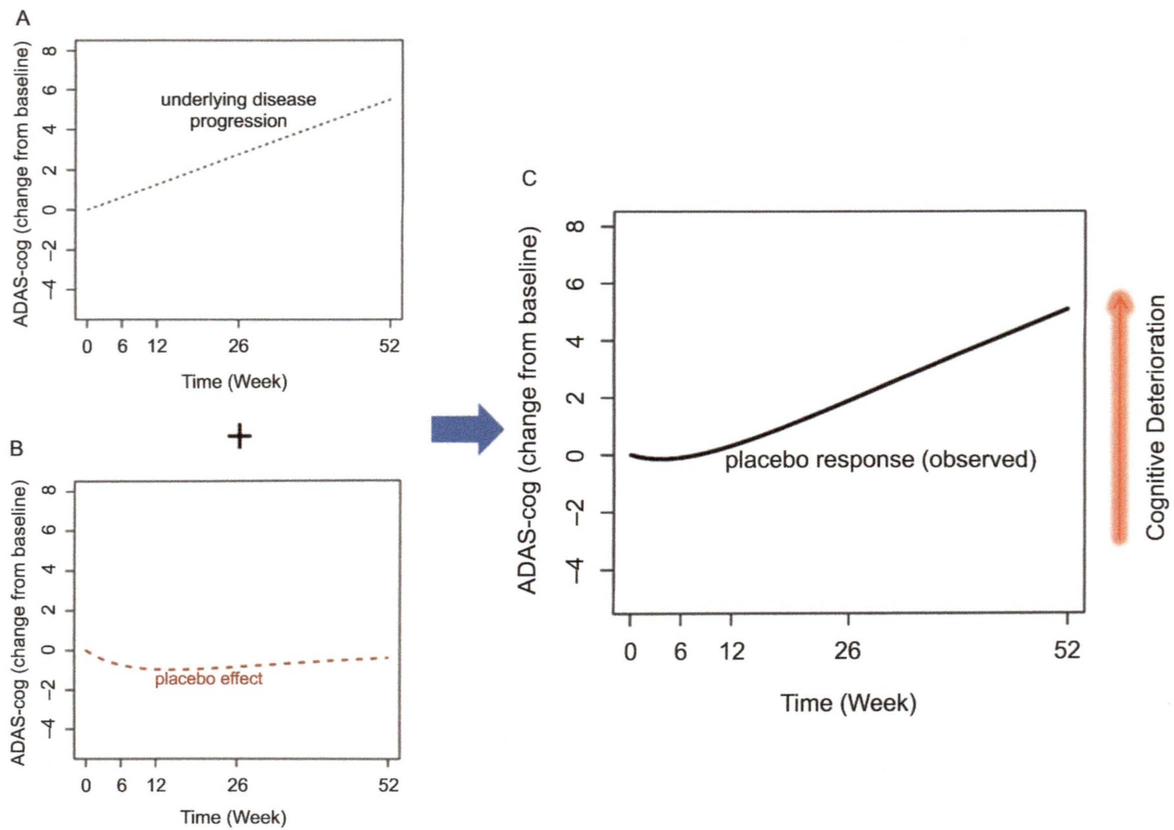

Fig. 1 Concept of placebo response in a disease progression scenario.

6. Understanding placebo effect in AD

In this section, we will look at two examples to demonstrate different placebo responses can lead to misinterpretation of the study results, and wrong decision for clinical drug development may be made. Each example is from a single clinical study and compared to the historical disease progression with added placebo effect (Ito et al., 2013). The population-level (mean) values were estimated from the conventional disease progression model previously developed for mild to moderate AD. The historical controls in this analysis were placebo arms obtained from scientific journal articles from 1990 to 2008, following a general guidance for systematic review and meta-analysis. This database contained 52 clinical trials, representing approximately 19,972 patients and more than 84,000 individual observations, who received placebo (including vitamins), acetylcholinesterase (AChE)-inhibitor treatments, or without any treatments (Ito et al., 2010). These two concepts—the disease progression model and meta-analysis database including all publicly available data—were the prototype for the C-path drug development tool (DDT).

6.1 Example 1

This example illustrates the possible issue for the dose-ascending study design, which could affect the placebo response. Fig. 2 shows the placebo data from each cohort in a phase 2, double-blind, placebo-controlled, dose-ascending cohort study (four cohorts in total; approximately 20 subjects received active drug and 10 subjects received placebo in each cohort). The observed placebo response from each cohort is overlaid against historical controls and model predictions for comparison. In this dose-ascending study, the lowest dose was administered to the first cohort, then moved to the next higher dose (second cohort) after confirming safety, followed by the next dose level (third cohort), and then the highest dose level (fourth cohort).

Interestingly, the magnitude of placebo effect looks different (and it is ordered) even though the same placebo tablets were given to patients in each cohort. In other words, the placebo response from the first cohort (the lowest dose was given to the active treatment group and placebo was given to the control group) demonstrated the most significant deviation from the expected historical placebo response (worsening than historical prediction), with each successive cohort falling closer to the median model prediction.

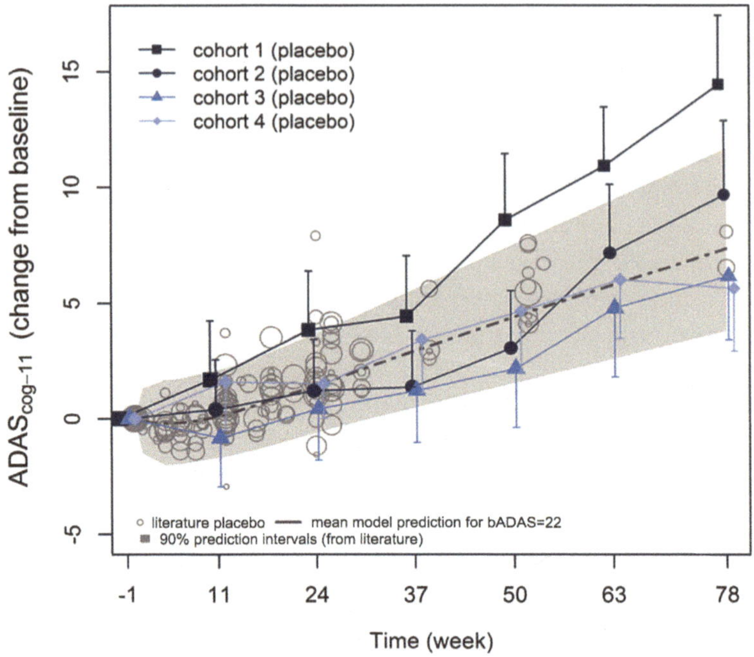

Fig. 2 Placebo response in a cohort design clinical trial (Mean ± SE). Note: The symbol size (points) for historical data is proportional to the sample size of the study, i.e., bigger symbols indicate larger sample studies. *Reprinted from Ito, K., Corrigan, B., Romero, K., Anziano, R., Neville, J., Stephenson, D., et al. (2013). Understanding placebo responses in Alzheimers disease clinical trials from the literature meta-data and CAMD database.* Journal of Alzheimers Disease, 37*(1), 173–183, with permission from IOS Press. The publication is available at IOS Press through https://doi.org/10.3233/JAD-130575.*

These results are highly suggestive that a cohort effect in the placebo response was observed. Potential causes include investigator's or rater's bias, even though the study was placebo controlled and double-blind. To elucidate, the investigators and study staffs (who administered and scored the ADAS-cog) may have been aware that cohort 1 was the lowest dose by study design, and, therefore, patients received little or no effect regardless placebo or active treatment were given. It is also possible that patients who participated in the study knew which cohort they were actually enrolled, so little expectation to the treatment (even they didn't know whether getting active or placebo tablet) resulted in worsening in their responses. While, within cohort treatment effects, treatment vs. placebo, would still be considered valid (as patients were randomized to each cohort), but caution should be made in pooling of the data for post-hoc analyses. It is also noted that the sample size for this type of study (dose ascending, early Phase 2 study) is usually small at the early stage of the clinical development (e.g., <30 patients in each cohort), therefore, more bias due to random error could occur, and interpretation of the single study result is difficult.

6.2 Example 2

Fig. 3 shows results from two Phase 2 clinical trials for Drug A and Drug B. The study designs were similar; 52 weeks, double-blind, placebo-controlled study in mild to moderate AD patients. In both studies, it appears as though a treatment effect was present as compared with the placebo group based on the change from baseline ADAS-cog scores.

However, when these two clinical trial results are compared against the historical control data overlaid along with model predictions as illustrated in Fig. 4, it appears that the placebo response in the Drug B trial was much worse than what would be predicted. Conversely, the treatment arm in Trial B appears to be where the expected response for placebo usually falls. Given this result, the clinical team concluded that the placebo response in the Drug B study was not normal and needed more data to confirm the efficacy of the Drug B before moving to Phase 3.

Case studies above demonstrate the importance of understanding the placebo response in AD clinical trials. In general, evaluation of placebo response is focused on identification of factors that influence the characteristics of

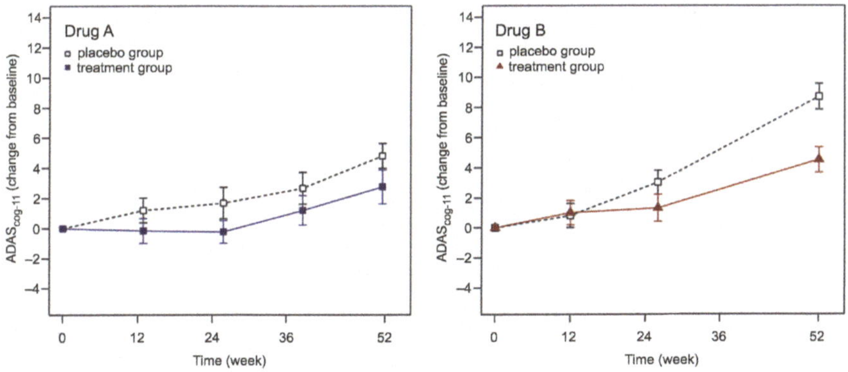

Fig. 3 Phase II clinical trial results from different studies (Mean ± SE). *Reprinted from Ito, K., Corrigan, B., Romero, K., Anziano, R., Neville, J., Stephenson, D., et al. (2013). Understanding placebo responses in Alzheimers disease clinical trials from the literature meta-data and CAMD database.* Journal of Alzheimers Disease, 37*(1), 173–183, with permission from IOS Press. The publication is available at IOS Press through https://doi.org/10.3233/JAD-130575.*

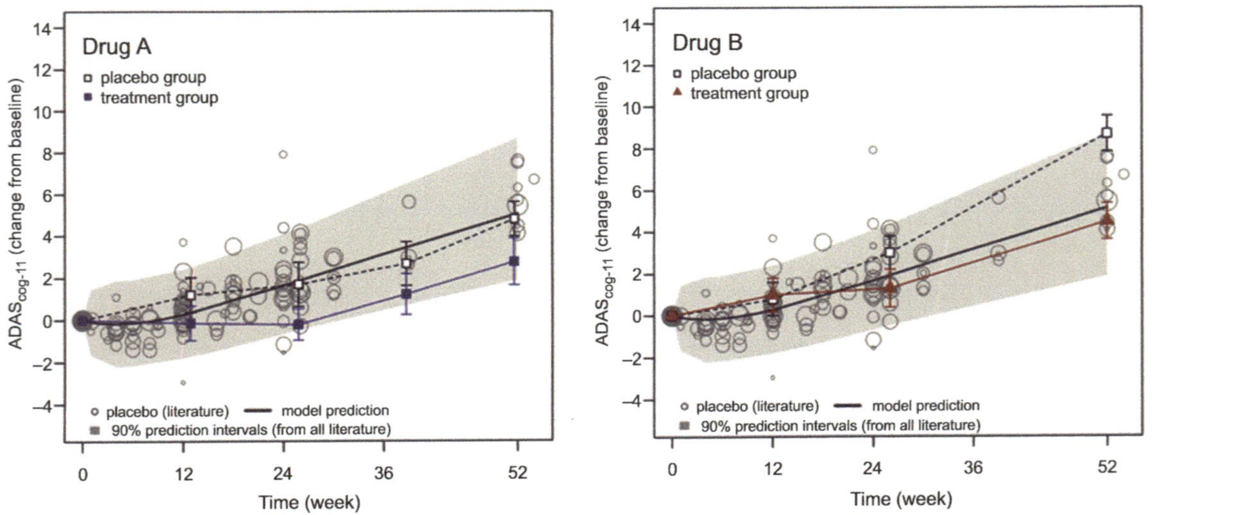

Fig. 4 Phase II clinical trial results from different studies overlaid with hisorical placebo data (Mean ± SE). Note: The same clinical data in Fig. 2 are overlaid with the historical data (literature) and its model prediction. The symbol size (points) for historical data is proportional to the sample size of the study, i.e., bigger symbols indicate larger sample studies. *Reprinted from Ito, K., Corrigan, B., Romero, K., Anziano, R., Neville, J., Stephenson, D., et al. (2013). Understanding placebo responses in Alzheimers disease clinical trials from the literature meta-data and CAMD database.* Journal of Alzheimers Disease, *37(1), 173–183, with permission from IOS Press. The publication is available at IOS Press through https://doi.org/10.3233/JAD-130575.*

magnitude and variability of response between treatment and placebo arms. Potential causes of placebo effects in Alzheimer's clinical trials are not limited to changes caused by improved standard of care due to participating in a trial, but also to investigators' or raters' expectations about scoring, as well as patients' and their caregivers' expectations around the potential treatment.

The strength of the placebo response varies since these are depending on the individual's receptiveness to the conditioning, and the greater degree of cognitive impairment in AD patients may additionally result in decreased capacity for realization of the expectation. In addition, their previous experience with therapeutic interventions may cause other mechanism of placebo or nocebo effects. These factors should be taken into consideration in the design of clinical trials as well as investigate possible causes of placebo response when analyzing the data.

The randomized, placebo controlled, double-blind study is considered a gold standard study design to compare the treatment group versus placebo group. However, the evaluation of the placebo control arm data relative to the historical placebo response provides an additional mechanism to evaluate a study's results, and to guard against Type I or false positive error. If the placebo response is different from the expected historical placebo response, and the overall disease progression is quantitatively discordant with the model predicted values, then it is important to investigate the reasons for the placebo difference prior to interpreting treatment effects.

In summary, the model prediction and placebo comparison against historical placebo data provide a way to double-check the reliability of trial results before making important drug development decisions (e.g., terminating a compound or initiating an expensive lengthy new trial in AD patients).

In the following section, we discuss the advantage to utilize the mathematical disease progression framework more details, not only quantitively assess the clinical trial results, but also to predict the future trials and improve the efficiency of the drug development.

7. History of C-Path

FDA introduced the Critical Path Initiative in 2004 to revitalize and modernize the drug approval process to fully utilize contemporary tools and approaches (Woodcock & Woosley, 2008). One key approach to achieve this necessary innovation was through the formation of public-private partnerships and working consortia to drive research (Maxfield, Buckman-Garner, & Parekh, 2017). This model brings together experts across fields in academia, industry, patient advocacy groups, and regulatory agencies to catalyze research and optimize the submission of medical devices and drug development tools for approval.

An example of one of these public-private partnerships is the Critical Path Institute (C-Path). This nonprofit organization was founded in 2005 as a vehicle to deliver on the goals of the Critical Path Initiative by facilitating collaboration between experts across fields to generate new data, measurement, and methods standards. C-Path's consortia work in the precompetitive space to advance a variety of solutions to challenges and bottlenecks in drug development, including safety biomarkers, monitoring biomarkers, novel outcome measures (including patient reported outcome measures), and mathematical models that leverage existing biomarkers and outcome measures to help understand the variability in clinical trial populations. Many of these tools are developed on a platform of shared standardized data that support the projects, and all tools are developed in the pre-competitive space by consortia of experts across multiple disciplines.

More recently, two additional pieces of legislation have sought to reinvigorate and optimize discovery and development and to focus research and development on better meeting the needs of patients. Recognizing the successes in tool development using shared data, this legislation supports use of data standards and sharing of data. The 21st Century Cures Act (December 2016) outlined provisions meant to support and expand data sharing between researchers and for the creation of an "information commons" for NIH-funded results (Majumder, Guerrini, Bollinger, Cook-Deegan, & Mcguire, 2017). This was followed by the most recent renewal of the Prescription Drug User Fee Act (PDUFA VI, reauthorized August 2017). PDUFA VI listed several priorities to support innovative drug development that include the sharing of data. PDUFA VI and the 21st Century Cures Act both support the advancement of the use of computer modeling and simulation to improve clinical trial design and inform drug development. These techniques have become increasingly important in optimizing the drug development process. That said, useful computer models require quality, available data; only by ensuring widespread data standardization and sharing can we ensure that their full value is realized. PDUFA VI also specifically pushes for the inclusion of metrics such as real-world evidence (RWE) and real-world data (RWD) to better encapsulate experiences patients have in their everyday lives (Jain, Mehrotra, Wenning, & Sinha, 2018). The utility of these types of data can be further realized when paired with other data types and innovative analyses.

The Critical Path Institute (C-Path) is a nonprofit organization founded in 2005 as a vehicle to develop tools to accelerate drug development as described in FDA's Critical Path Initiative (Maissenhaelter, Woolmore, & Schlag, 2018). C-Path has significant experience in sharing patient-level data with the drug development community and others, and using such data to develop regulatory-accepted drug development tools.

C-Path's goal is to create drug development solutions that are acceptable to the regulatory authorities and can accelerate the drug development process. Many such solutions require large amounts of patient-level data that reflect entire patient populations. C-Path's data sharing platform was built specifically for the generation of such solutions. In order to develop datasets that are optimal for this purpose, C-Path accesses data of interest from the community, curates it carefully and integrates the data into a database that can be used for analysis. Importantly, the platform allows the users to download patient-level datasets in a CDISC (clinical data interchange standards consortium) study data tabulation model (SDTM) format, which is required for FDA submissions. Data is de-identified and stored securely to comply with all international data laws.

8. AD disease progression model at C-Path (especially placebo response)

Based on CPAD's integrated patient-level databased in AD, a clinical trial simulation (CTS) tool was developed to describe disease progression based on longitudinal ADAS-Cog scores in mild-to-moderate AD (Haas et al., 2016). The model was generated following a three-stage approach: (1) definition of a proposed context of use statement (COU) and model development plan, (2) execution of model development and evaluation, (3) FDA/EMA review for endorsement (Romero et al., 2014). To capture the maximum amount of information available for development of the CTS tool, data from a variety of sources were needed, requiring a model that simultaneously fitted summary and patient-level data from observational studies and clinical trials (Rogers et al., 2012). The CPAD database consists of patient-level trial data contributed by consortium members. The CPAD database currently integrates data from over 25,000 patients across more than 30 industry trials (Neville et al., 2015). For the CTS tool, demographics, genetics, and individual items from cognitive scales (MMSE, ADAS-Cog, etc.) were included in the underlying model, which has three basic components that model drug, disease, and clinical trial feature

(Romero et al., 2014). In the model, A Bateman function describes the placebo effect as a temporary beneficial change that eventually fades, after which individuals return to the progression state and rate that would be expected under natural progression (Ito et al., 2013). The pattern associated with presumed "purely symptomatic" effects has so far been described as an overall function shift to the right (without a change in the rate of progression), such that the apparent rate of progression may change temporarily, but returns to the natural progression rate after some relatively short duration (Ito et al., 2013). By contrast, it has been generally hypothesized that the pattern associated with a "disease-modifying effect" would manifest as a change (slowing) of the progression rate, consistent with a benefit that continues to accrue over some relatively long duration (Romero et al., 2014).

This tool was the first-ever quantitative drug development tool to be endorsed by FDA in June 2013 and by EMA in September 2013, and is publicly available to researchers through C-Path's website: https://c-path.org/ad-cts-tool-request/.

References

Alzheimers Disease Facts and Figures. (2019). 2019 Alzheimer's disease facts and figures. *Alzheimers & Dementia*, *15*(3), 321–387. https://doi.org/10.1016/j.jalz.2019.01.010.

Bateman, R. J., Xiong, C., Benzinger, T. L., Fagan, A. M., Goate, A., Fox, N. C., et al. (2012). Clinical and biomarker changes in dominantly inherited Alzheimer's disease. *New England Journal of Medicine*, *367*(9), 795–804. https://doi.org/10.1056/NEJMoa1202753.

Benedetti, F., Carlino, E., & Pollo, A. (2010). How placebos change the patients brain. *Neuropsychopharmacology*, *36*(1), 339–354. https://doi.org/10.1038/npp.2010.81.

Braak, H., Thal, D. R., Ghebremedhin, E., & Tredici, K. D. (2011). Stages of the pathologic process in Alzheimer disease: Age categories from 1 to 100 years. *Journal of Neuropathology & Experimental Neurology*, *70*(11), 960–969. https://doi.org/10.1097/nen.0b013e318232a379.

Colagiuri, B., Schenk, L., Kessler, M., Dorsey, S., & Colloca, L. (2015). The placebo effect: From concepts to genes. *Neuroscience*, *307*, 171–190. https://doi.org/10.1016/j.neuroscience.2015.08.017.

Conrado, D. J., Denney, W. S., Chen, D., & Ito, K. (2014). An updated Alzheimer's disease progression model: Incorporating non-linearity, beta regression, and a third-level random effect in NONMEM. *Journal of Pharmacokinetics and Pharmacodynamics*, *41*(6), 581–598. https://doi.org/10.1007/s10928-014-9375-z.

Cook, S. F., & Bies, R. R. (2016). Disease progression Modeling: Key concepts and recent developments. *Current Pharmacology Reports*, *2*(5), 221–230. https://doi.org/10.1007/s40495-016-0066-x.

Faria, V., Fredrikson, M., & Furmark, T. (2008). Imaging the placebo response: A neurofunctional review. *European Neuropsychopharmacology*, *18*(7), 473–485. https://doi.org/10.1016/j.euroneuro.2008.03.002.

Gordon, B. A., Blazey, T. M., Su, Y., Hari-Raj, A., Dincer, A., Flores, S., et al. (2018). Spatial patterns of neuroimaging biomarker change in individuals from families with autosomal dominant Alzheimers disease: A longitudinal study. *The Lancet Neurology*, *17*(3), 241–250. https://doi.org/10.1016/s1474-4422(18)30028-0.

Haas, M., Stephenson, D., Romero, K., Gordon, M. F., Zach, N., & Geerts, H. (2016). Big data to smart data in Alzheimers disease: Real-world examples of advanced modeling and simulation. *Alzheimers & Dementia*, *12*(9), 1022–1030. https://doi.org/10.1016/j.jalz.2016.05.005.

Holford, N. H., & Peace, K. E. (1992a). Methodologic aspects of a population pharmacodynamic model for cognitive effects in Alzheimer patients treated with tacrine. *Proceedings of the National Academy of Sciences*, *89*(23), 11466–11470. https://doi.org/10.1073/pnas.89.23.11466.

Holford, N. H., & Peace, K. E. (1992b). Results and validation of a population pharmacodynamic model for cognitive effects in Alzheimer patients treated with tacrine. *Proceedings of the National Academy of Sciences*, *89*(23), 11471–11475. https://doi.org/10.1073/pnas.89.23.11471.

Ito, K., Ahadieh, S., Corrigan, B., French, J., Fullerton, T., & Tensfeldt, T. (2010). Disease progression meta-analysis model in Alzheimers disease. *Alzheimers & Dementia*, *6*(1), 39–53. https://doi.org/10.1016/j.jalz.2009.05.665.

Ito, K., Corrigan, B., Romero, K., Anziano, R., Neville, J., Stephenson, D., et al. (2013). Understanding placebo responses in Alzheimers disease clinical trials from the literature meta-data and CAMD database. *Journal of Alzheimers Disease*, *37*(1), 173–183. https://doi.org/10.3233/jad-130575.

Jack, C. R., Knopman, D. S., Jagust, W. J., Petersen, R. C., Weiner, M. W., Aisen, P. S., et al. (2013). Tracking pathophysiological processes in Alzheimers disease: An updated hypothetical model of dynamic biomarkers. *The Lancet Neurology*, *12*(2), 207–216. https://doi.org/10.1016/s1474-4422(12)70291-0.

Jack, C. R., Knopman, D. S., Jagust, W. J., Shaw, L. M., Aisen, P. S., Weiner, M. W., et al. (2010). Hypothetical model of dynamic biomarkers of the Alzheimers pathological cascade. *The Lancet Neurology*, *9*(1), 119–128. https://doi.org/10.1016/s1474-4422(09)70299-6.

Jack, C. R., Lowe, V. J., Weigand, S. D., Wiste, H. J., Senjem, M. L., Knopman, D. S., et al. (2009). Serial PIB and MRI in normal, mild cognitive impairment and Alzheimers disease: Implications for sequence of pathological events in Alzheimers disease. *Brain*, *132*(5), 1355–1365. https://doi.org/10.1093/brain/awp062.

Jain, L., Mehrotra, N., Wenning, L., & Sinha, V. (2018). PDUFA VI: It is time to unleash the full potential of model-informed drug development. *CPT: Pharmacometrics & Systems Pharmacology*, *8*(1), 5–8. https://doi.org/10.1002/psp4.12365.

Lockwood, P., Ewy, W., Hermann, D., & Holford, N. (2006). Application of clinical trial simulation to compare proof-of-concept study designs for drugs with a slow onset of effect; an example in Alzheimers disease. *Pharmaceutical Research*, *23*(9), 2050–2059. https://doi.org/10.1007/s11095-006-9048-8.

Maissenhaelter, B. E., Woolmore, A. L., & Schlag, P. M. (2018). Real-world evidence research based on big data. *Der Onkologe*, *24*(S2), 91–98. https://doi.org/10.1007/s00761-018-0358-3.

Majumder, M. A., Guerrini, C. J., Bollinger, J. M., Cook-Deegan, R., & Mcguire, A. L. (2017). Sharing data under the 21st century cures act. *Genetics in Medicine*, *19*(12), 1289–1294. https://doi.org/10.1038/gim.2017.59.

Maxfield, K. E., Buckman-Garner, S., & Parekh, A. (2017). The role of public–private partnerships in catalyzing the critical path. *Clinical and Translational Science*, *10*(6), 431–442. https://doi.org/10.1111/cts.12488.

Murray, D., & Stoessl, A. J. (2013). Mechanisms and therapeutic implications of the placebo effect in neurological and psychiatric conditions. *Pharmacology & Therapeutics*, *140*(3), 306–318. https://doi.org/10.1016/j.pharmthera.2013.07.009.

Neville, J., Kopko, S., Broadbent, S., Avilés, E., Stafford, R., Solinsky, C. M., et al. (2015). Development of a unified clinical trial database for Alzheimers disease. *Alzheimers & Dementia*, *11*(10), 1212–1221. https://doi.org/10.1016/j.jalz.2014.11.005.

Peace, K., & Holford, N. (1994). The effect of tacrine and lecithin in Alzheimers disease. *European Journal of Clinical Pharmacology*, *47*(1). https://doi.org/10.1007/bf00193473.

Reiman, E. M., Quiroz, Y. T., Fleisher, A. S., Chen, K., Velez-Pardo, C., Jimenez-Del-Rio, M., et al. (2012). Brain imaging and fluid biomarker analysis in young adults at genetic risk for autosomal dominant Alzheimers disease in the presenilin 1 E280A kindred: A case-control study. *The Lancet Neurology*, *11*(12), 1048–1056. https://doi.org/10.1016/s1474-4422(12)70228-4.

Rogers, J. A., Polhamus, D., Gillespie, W. R., Ito, K., Romero, K., Qiu, R., et al. (2012). Combining patient-level and summary-level data for Alzheimer's disease modeling and simulation: A beta regression meta-analysis. *Journal of Pharmacokinetics and Pharmacodynamics*, *39*(5), 479–498. https://doi.org/10.1007/s10928-012-9263-3.

Romero, K., Ito, K., Rogers, J., Polhamus, D., Qiu, R., Stephenson, D., et al. (2014). The future is now: Model-based clinical trial design for Alzheimers disease. *Clinical Pharmacology & Therapeutics*, *97*(3), 210–214. https://doi.org/10.1002/cpt.16.

Samtani, M. N., Farnum, M., Lobanov, V., Yang, E., Raghavan, N., Dibernardo, A., et al. (2012). An improved model for disease progression in patients from the Alzheimers disease neuroimaging initiative. *The Journal of Clinical Pharmacology*, *52*(5), 629–644. https://doi.org/10.1177/0091270011405497.

Schneider, L. S., Mangialasche, F., Andreasen, N., Feldman, H., Giacobini, E., Jones, R., et al. (2014). Clinical trials and late-stage drug development for Alzheimers disease: An appraisal from 1984 to 2014. *Journal of Internal Medicine*, *275*(3), 251–283. https://doi.org/10.1111/joim.12191.

Villemagne, V. L., Burnham, S., Bourgeat, P., Brown, B., Ellis, K. A., Salvado, O., et al. (2013). Amyloid β deposition, neurodegeneration, and cognitive decline in sporadic Alzheimers disease: A prospective cohort study. *The Lancet Neurology*, *12*(4), 357–367. https://doi.org/10.1016/s1474-4422(13)70044-9.

Woodcock, J., & Woosley, R. (2008). The FDA critical path initiative and its influence on new drug development. *Annual Review of Medicine*, *59*(1), 1–12. https://doi.org/10.1146/annurev.med.59.090506.155819.

CHAPTER ELEVEN

Placebo in epilepsy

Daniel M. Goldenholz*, Shira R. Goldenholz
Beth Israel Deaconess Medical Center, Boston, MA, United States
*Corresponding author: e-mail address: daniel.goldenholz@bidmc.harvard.edu

Contents

International Review of Neurobiology, Volume 153
ISSN 0074-7742
https://doi.org/10.1016/bs.irn.2020.03.033

Abstract

Placebos impact epilepsy in a number of ways. Through randomized clinical trials, explicit clinical use, and also through implicit clinical use, placebos play a role in epilepsy. This chapter will discuss the reasons placebo is used, the determinants of placebo response in epilepsy, observations about placebo specific to epilepsy, and ways in which clinical trial design is impacted by placebo.

1. Introductory remarks

Placebos have played a role in epilepsy long before they were formally studied. Indeed, epilepsy is one of the oldest recorded human diseases (Panteliadis, Vassilyadi, Fehlert, & Hagel, 2017), yet highly effective treatments did not become available until around the 20th century (Perucca, 2019). Prior to that time, it must be presumed that the treatments provided were of lower efficacy, or entirely related to placebo. Even today, the impact of placebo on epilepsy is widespread, poorly understood, and underappreciated.

2. Basic definitions of terms

In the history of science (medicine?) many researchers have put forth various ideas for the concept of a "placebo effect" or "placebo effects." However, a fairly straightforward definition of placebo within a clinical trial should encompass the following (Kienle & Kiene, 1997):

1. A placebo of some kind must be given.
2. The event or effect observed must be a result of the placebo treatment; that is, the event would not have happened without placebo administration.
3. The event or effect noted must be relevant for the disease or symptom, i.e., it had to be a therapeutic event.

Additionally, we define the "placebo response" as any response observed during the trial in the placebo arm, regardless of cause. It is noteworthy that with these definitions, one must be careful not to mis-attribute "placebo effect" to what is, in actuality, a "placebo response". For example, suppose that a randomized trial of drug X were conducted to see if patients would have less seizures compared to a placebo. If 20% of patients in the placebo arm of the trial were responders that would be called the "placebo response." Careful analysis might reveal that due to improper study design, the majority of the responders actually have large excursions (highs and lows) in their typical seizure frequency, and so they appeared as responders by chance alone, and not due to any kind of "placebo effect" at all. Conversely, if one discovered that the percentage of patients responding in the placebo arm varied between 0 and 20% depending only on the presence or absence of a scientist wearing a white coat that would strongly argue for a "placebo effect" that accounts for the "placebo response."

In the context of this chapter, the terms "natural fluctuations" and "natural variability" will refer to the increases and decreases in monthly seizure counts that occur due to the disease itself, and not some external "effect". For instance, some patients typically have 3 seizures per month on average. With further questioning (or a detailed diary), one finds that sometimes those patients have 1 per month and others months have 5. It is these excursions above and below the typical seizure rate that are referred to be these terms "fluctuations" and "variability".

3. Placebo use in epilepsy

Epilepsy affects at least 2.2 million Americans (Institute of Medicine (U.S.), Committee on the Public Health Dimensions of the Epilepsies, & England, 2012). Approximately 1 out of 3 patients are uncontrolled on anti-seizure medications (Kwan & Brodie, 2000). After surgery for some of these uncontrolled patients, 40%–50% achieve long-term seizure freedom (Englot, Wang, Rolston, Shih, & Chang, 2012; Malmgren & Edelvik, 2017; Schmidt & Stavem, 2009). To combat these challenges, modern epilepsy trials (similar to clinical trials in other fields of medicine) frequently compare therapy to placebo (Perucca, 2012). Sometimes treatments are given as monotherapy and compared with placebo, although this study design is not the norm. Rather, most studies use a treatment versus placebo paradigm as adjuvant to a baseline treatment regimen because it is generally agreed upon that untreated, epilepsy (placebo assignment) is too dangerous (Perucca, 2012). Remarkably, the response to placebo often is quite positive (Rheims, Cucherat, Arzimanoglou, & Ryvlin, 2008; Rheims, Perucca, Cucherat, & Ryvlin, 2011), making an efficacious therapy more difficult to validate (Goldenholz, Moss, Scott, Auh, & Theodore, 2015).

Placebos have also been used as a treatment for seizures in clinical practice, outside the realm of trials. A number of patients have episodes called "non-epileptic events" or NEE, which are sometimes mistaken by physicians and/or patients to be seizures. NEE have previously been called "pseudoseizures" or "psychogenic non-epileptic seizures"; however, the newer term is both more descriptive and less pejorative (Morgan, Dvorchik, Williams, Jarrar, & Buchhalter, 2013). When faced with the diagnostic dilemma of a patient with either epilepsy or NEE, some physicians have found the use of placebo, such as intravenous saline or color patch placement, to be helpful. The rationale, presumably, is that a patient with a psychogenic cause for seizure-like behaviors will be more responsive to placebo than if the underlying cause was epilepsy.

Other physicians, however, have argued that this approach may be unethical (Chen, Dave, Gadelmola, Jeroudi, & Fadipe, 2018).

In summary, placebos have found their way into epilepsy trials as a comparator for novel treatments, and also into the clinical practice of epilepsy when diagnostic uncertainty exists about the underlying disease.

For the purposes of the remainder of this chapter, we will emphasize the use of placebo as a tool to prevent or reduce occurrence of seizures. Other conceptualizations of placebo are of course plausible; however, this particular use has practical value in clinical practice and clinical trials.

4. Determinants of placebo response

Three main categories are assumed to play a key role in placebo response for epilepsy trials (Goldenholz & Goldenholz, 2016): inherent, methodological and psychological causes (Fig. 1).

4.1 Inherent factors

There are a number of reasons why the natural course of seizure rates may fluctuate over time. These changes in turn could be mis-interpreted as a "therapeutic response" if a treatment were given when the rate was by chance high, and then the response was measured when the rate was by chance low. It has been shown that such natural fluctuations, in the right circumstances, can reproduce placebo responses seen in historical trials (Goldenholz et al., 2015; Romero, Larimer, Chang, Goldenholz, & Goldenholz, 2020).

Temporary remission from seizures is one form of fluctuation that has been studied in several contexts. In one prospective study, a cohort of 225 drug-resistant patients was followed for 4 years without surgical intervention to determine prognosis. Each year, the percentages of patients who ended the year with at least 6 months of remission (terminal remission) were: 3%, 9%, 13% and 14% at 1, 2, 3, and 4 years respectively (Callaghan,

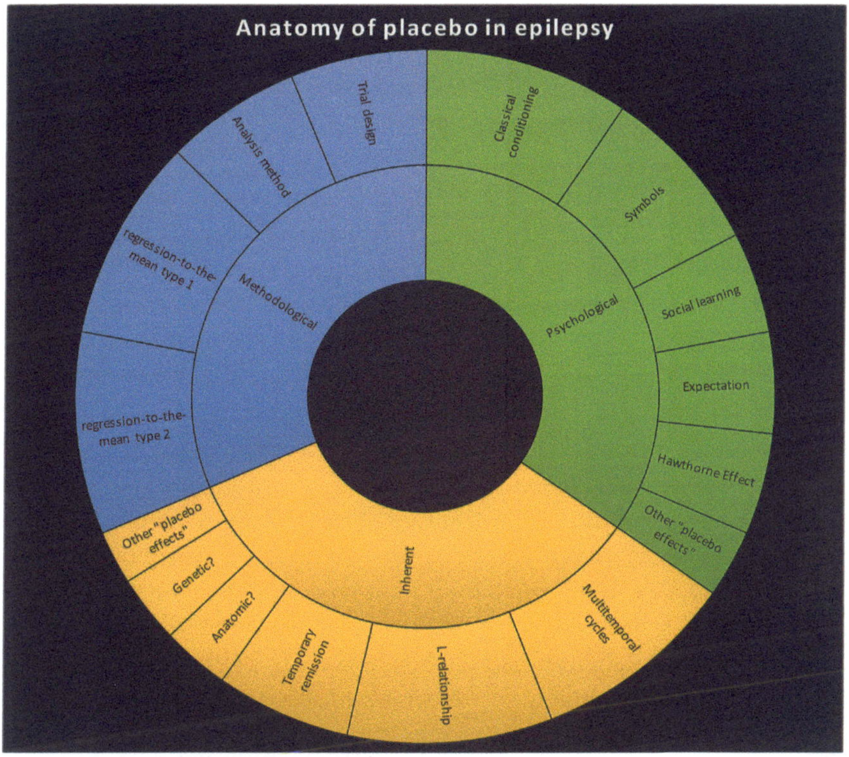

Fig. 1 Anatomy of the placebo response in epilepsy.

Anand, Hesdorffer, Hauser, & French, 2007). The same patients were followed for an additional 3 years, and the cumulative probability of 12-months seizure remission was found to be 33%. The risk of a relapse after a 12-month remission was 71% at 5 years (Callaghan et al., 2011). Similar rates have been reported by others (Choi et al., 2011; Luciano & Shorvon, 2007).

In children, the expected fluctuations seen with total remission are perhaps even larger. A cohort of 413 patients was followed for an average of 14.8 years Geerts et al., 2010). In 25%, seizures went into remission in <1 year, while in another 25% the patients suffered from seizures for over 12 years. Over 70% achieved 5-year terminal remission (Geerts et al., 2010).

The subtype of epilepsy may play a role in possible relative likelihood of various clinical courses, with idiopathic epilepsy tending toward more favorable outcomes. A group of 144 children aged 15 or under were followed prospectively for an average of 37 years. That study suggested several patterns: early complete remission (16%), early remission with relapse (16%), delayed remission (mean delay of 9 years, 32%), worsening (16%), and relapsing-remitting (19%) (Sillanpää & Schmidt, 2006). A similar set of patterns has been observed in a group of 1098 newly diagnosed patients ages 9–93 (Brodie, Barry, Bamagous, Norrie, & Kwan, 2012).

The natural fluctuations seen in epilepsy occur after surgery as well. For example, a cohort of 915 post-surgical patients was followed for a minimum of 1 year (mean 6.6 years) (Jehi, Sarkis, Bingaman, Kotagal, & Najm, 2010). The study authors found a relationship between the number of post-operative seizures the patient suffered and the risk of recurrence. A single post-operative seizure was associated with a 66% risk of recurrence; 2 or 3 post-op seizures were associated with a 75% and 80% risk of recurrence, respectively. However, 20%–34% of patients will have 1–3 seizures after surgery and then remain seizure-free thereafter. These results indicate that even post-operatively, the natural fluctuations of the disease remain relevant.

It is possible that natural variability may account for some degree of placebo response. This has been shown in several realistic simulation studies. A patient-centric database of 684,825 seizures reported by 8228 patients was used to simulate clinical trials (Goldenholz et al., 2015). The study lacked all the traditional psychological influences (discussed in Section 4.3), and accounted for regression to the mean by extending the duration of the study at multiple time points up to 2 years. Nevertheless, the simulation found "placebo responders" at levels comparable to actual clinical trials (rates between 20% and 30%). Looking at the results in more detail, the study found that patients who could be categorized as "50% responders," "worseners," etc., would, over the course of months, change their categorization repeatedly. As a result, studies that identify "responders" in a 3-month follow-up may well discover that the same patients become "non-responders" later on, and vice versa. A simulation across multiple databases similarly found that typical 50%-responder rates could be generated in both a forward and reverse time calculation, suggesting that natural variability in seizure rates is sufficient in realistic settings to produce large placebo responses (Goldenholz, Strashny, Cook, Moss, & Theodore, 2017). A recent simulation study using longitudinal intracranial monitoring data from 14 patients found that a "50% responder" categorization would be expected to very frequently misclassify patients unless extremely long trial durations (longer than 1 year) were used (Karoly, Romero, Cook, Freestone, & Goldenholz, 2019).

One surprising aspect of the natural fluctuations in seizure rates is that there may be predictable variability. A study examined this using three independent datasets, one from chronically implanted electrodes from NeuroVista (Cook et al., 2013), one from patient reported diaries in https://www.SeizureTracker.com/ (Ferastraoaru et al., 2018), and one from a physician curated diary system from newly diagnosed people with epilepsy (The Epilepsy Study Consortium, n.d.). The study found that there was a very reliable relationship between the average seizure rate of an individual, and the standard deviation of the seizure rate (Goldenholz et al., 2018). Specifically, the logarithms of both quantities are highly linearly related, shown in Fig. 2, called the "L-relationship." Simply put, as the average seizure rate increases, there is a predictable increase in seizure rate variability as well.

There is compelling evidence that there may be multiple cycles within the seemingly random seizure timing. One study of 37 patients with longitudinally implanted electrodes for a responsive neurostimulator found cycles in the interictal epileptiform activity (IEA) that respected a circadian cycle and simultaneously multiday cycles of different lengths (Baud et al., 2018). That study further noted that seizures were more likely to occur at certain phases of those IEA cycles. Another study looking at chronically implanted electrodes and independently at patient-reported diaries similarly found simultaneous circadian and multiday cycles in seizure occurrence as well (Karoly, Goldenholz, Freestone, et al., 2018). Taken together, these multitemporal cycles may underlie the natural fluctuations in seizure rates seen in other studies.

4.2 Methodological factors

Clinical trials in epilepsy are typically organized as follows. All modern RCTs in epilepsy follow a parallel design in order to avoid therapy carryover effects. There is a "baseline period" (typically 6–8 weeks) during which seizure counts are carefully documented. Patients are kept on a stable treatment regimen. Some trials require that the number of seizures captured during the baseline must exceed some minimum, others require a similar minimum to have occurred prior to the study start. Then there is a "titration period" which can last between several days to 2 months, during which a patient gradually builds up a steady state of the therapy (typically when the blood levels reach equilibrium at the target range). After this, they enter the "treatment period" during which steady doses of treatment are given to the treatment group and steady amounts of placebo are given to the placebo group. In both groups, some trials allow a stable baseline

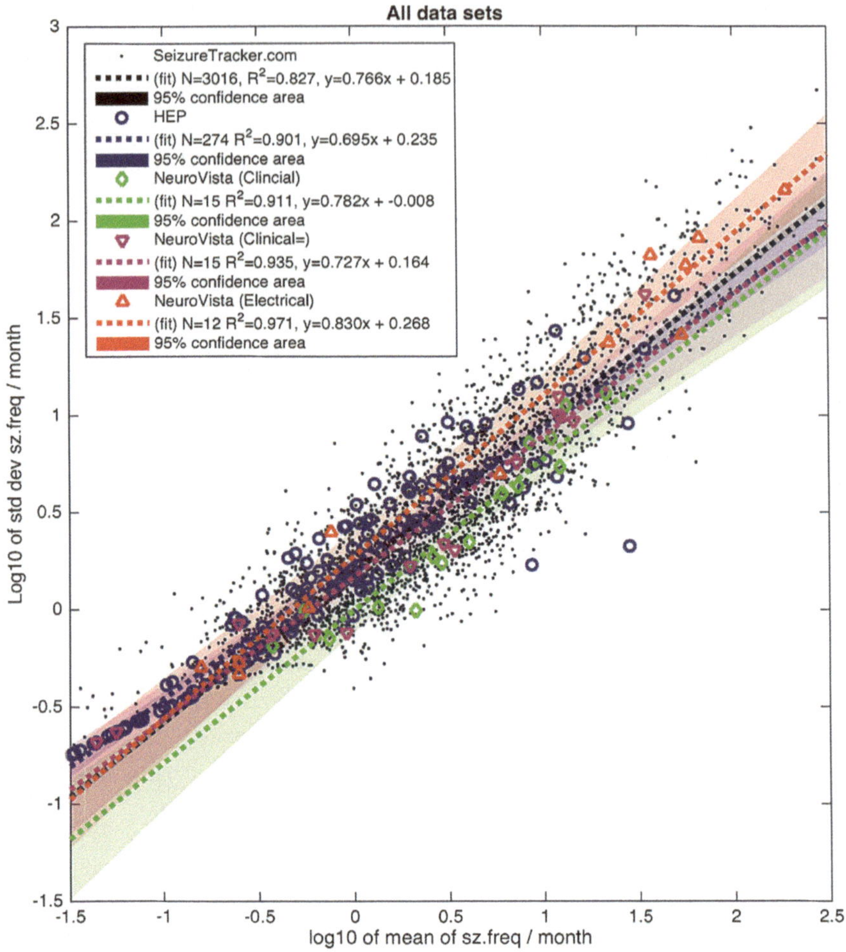

Fig. 2 The L-relationship. Log-log plot of mean and standard deviation of seizure frequency for each patient, described previously (Goldenholz et al., 2018). Each patient is represented by a single point on this plot. Linear fit lines (with confidence regions) are drawn for each of the representative datasets: the three subtypes of NeuroVista data (Cook et al., 2013), the Human Epilepsy Project (HEP) data (The Epilepsy Study Consortium, n.d.), and the SeizureTracker data (Casassa, Rathbun Levit, & Goldenholz, 2018; Ferastraoaru et al., 2018). Of note, all the different fit lines appear to overlap, suggesting a common trend for all the datasets.

treatment regimen, thus making the trial an "add-on" therapy trial. Sometimes a monotherapy trial is conducted, though this is much less common now that there are many anti-seizure medications approved on the market. Typically, the percentage change of monthly seizure counts between the "baseline" and "treatment" periods represents the outcome for the individual.

The trial design can also impact the magnitude of the placebo response, as has been shown in several studies investigating shorter and longer durations of baseline and treatment periods. In a Monte Carlo simulation involving 6.7 million trials, it was shown that longer baselines and shorter treatment periods would each independently increase the placebo response (Goldenholz, Tharayil, Moss, Myers, & Theodore, 2017). A study pooling data from 4 clinical trials comprising 899 patients similarly reported that trial durations impact statistical efficacy as a result of variability in seizure frequency (French, Cabrera, Emir, Whalen, & Lu, 2013).

Choice of analysis methods can dramatically impact placebo response. Whereas selecting a 50%-responder rate (RR50) method typically results in large placebo responses, lower placebo responses are expected from a median percentage change method (MPC) (Goldenholz, Tharayil, Moss, et al., 2017).

An often-cited cause of placebo response is a concept known as "regression to the mean" (Morton & Torgerson, 2005). The term generically refers to the idea that numbers typically go back to their average value. Because of ambiguity in the terminology, it can be helpful to further subdivide the term. The phenomenon of patients enrolling in a study at a period where seizure frequency is higher than typical for the patient (such as during acute illness) and then experiencing a return to baseline rate is referred to in this chapter as "regression to the mean type 1," or RTM1. Conversely, those who have a "normal" fluctuation (i.e. without any external cause) first higher in rate then back to a lower rate is referred to in this chapter as "regression to the mean type 2," or RTM2.

RMT1 refers to the idea that patients who perceive themselves to be "sicker" than usual are more likely to volunteer to participate in research studies, and as they gradually return to their usual state of being "less sick," that change is sometimes mis-attributed to the placebo effect. For example, patients whose treatment and overall health are relatively stable may notice an increase in monthly rate seizures compared to their "typical" monthly rate during temporal periods of illness (such as having a common cold). In this situation, they might volunteer for a study when they notice this increase. Such patients would be expected to "naturally" resume their mean seizure frequency after some time and would thus potentially appear to improve with whatever treatment they were given. As typical clinical trials are not designed to detect the difference between placebo effects and regression-to-the-mean, the two effects are often grouped together. Of note, although regression-to-the-mean is considered an important effect in epilepsy, it has not been clinically studied directly.

In simulations based on patient-centered seizure diaries, RTM1 effects appeared to decrease within 3–6 months (Goldenholz et al., 2015).

A broader interpretation of the term regression-to-the-mean can be used to encompass all natural fluctuations as well (Morton & Torgerson, 2005). In this view, selecting a strict eligibility criterion will artificially include patients that only temporarily are qualified for a trial. It is convenient to refer to this as RTM2. As an example of this, consider a trial that requires a minimum of four seizures per month for a patient to be included in the trial. Further, consider a patient who typically has two seizures per month with some associated variability. Perhaps 2 months after the imagined trial begins, the patient, by chance, has four seizures in a month. This patient may be included in the study, and then would quickly appear to "get better" regardless of treatment assignment. The influence of this effect was noted in a simulation study across multiple datasets (Goldenholz, Strashny, et al., 2017).

To summarize the difference between RTM1 and RTM2, RTM1 effects would be expected when a patient experiences an unusual event that temporarily changes what their brain does, and RTM2 effects would be expected when eligibility criteria are temporarily satisfied when a "usual" natural fluctuation is at a higher than average value. Both effects are possible in patients simultaneously, and each should be considered separately when considering methods for mitigation.

It is currently unknown to what extent regression-to-the-mean contribute to overall placebo response in epilepsy.

4.3 Psychological factors

In other diseases, the psychological underpinnings of placebo effects include classical conditioning, the Hawthorne effect, symbols and expectations, and social learning. In epilepsy, it remains unknown to what extent each play a role, given that none of them have been formally studied in epilepsy.

Ivan Pavlov's "classical conditioning" connects a stimulus to a response. Suppose stimulus A is known to cause a response. If stimulus A is repeatedly paired with stimulus B, after a while the subject may produce the response to B even in the absence of A. The well-cited example connects the sound of a bell to a dog salivating in the absence of food (Jarius & Wildemann, 2015). Some patients may connect the stimulus of swallowing a pill with the response of feeling better. Alternatively, some may connect swallowing experimental pills with side effects (Zaccara et al., 2014).

When improvement is seen in groups of people being observed, regardless of whether there is actually any intervention, this has been called the "Hawthorne effect." There was a telephone manufacturing company in the town of Hawthorne called Western Electric. The company noted that close supervision without any specific intervention, resulted in higher employee productivity (McCambridge, Witton, & Elbourne, 2014).

Symbols and expectation have been described to impact placebo response (Benedetti, 2014). Generic pain pills were deemed less effective than brand-name pills (Branthwaite & Cooper, 1981) in general, however, in epilepsy so far this distinction has not been seen (Holtkamp & Theodore, 2018). In a migraine study, falsely labeling rizatriptan as "placebo" versus as "rizatriptan" decreased the efficacy of the drug—while falsely labeling placebo as "rizatriptan" (versus "placebo") increased the placebo effect (Kam-Hansen et al., 2014). A study of people with Parkinson's disease (PD) found that higher drug cost increased objective efficacy measures (Espay et al., 2015). Studies in people with PD receiving treatment for pain, anxiety, and PD symptoms demonstrated that masked administration of therapy decreased efficacy compared with open administration. Also, open withdrawal of therapy worsened symptoms more than hidden withdrawal (Benedetti et al., 2003). Expectation may explain the "social placebo effect": watching others benefit from a placebo can "infect" subjects with placebo potency (Frisaldi, Carlino, Lanotte, Lopiano, & Benedetti, 2014). A recent review suggested that patient awareness of which trial arm they were assigned to (i.e., breaking of blinding) impacted the results of randomized trials, creating the illusion of drug efficacy but actually revealing the power of placebo (Kirsch, 2019).

Thus, it seems evident that knowledge, expectation, or belief about the treatment modified the treatment result (Benedetti, 2014). It is important to be aware that some patients might be able to distinguish which treatment they were assigned to in a double-blind study or in clinical settings that use placebos. If blinding may have been incomplete, the results should be interpreted with caution (Kirsch, 2019; Margraf & Ehlers, 1991). This problem may directly relate to epilepsy trials, because many anti-seizure drugs are associated with side effects and drug interactions. If a patient in a double-blinded study becomes aware of typical side effects, such as dizziness or excessive fatigue, blinding itself or the perception of blinding may be compromised, thereby impact the result of the RCT evaluation of an antiepileptic treatment.

One important puzzle to consider is how psychological influences can cause or quell seizures at all. Under the assumption that seizures are an organic event that results either from pathological firing of small groups of neurons, or due to a network behavior (Mahoney et al., 2019), it is difficult to conceive of how psychological effects should impact seizures. Conversely, some evidence that stress can modify seizure risk has been reported (McKee & Privitera, 2017), therefore perhaps future studies may untangle this question. Moreover, many patients have difficulty distinguishing between seizures and non-epileptic events (Syed et al., 2011), therefore perhaps a portion of the psychologically mediated placebo response reflects events that are not seizures at all.

5. Modeling placebo response

At present, no comprehensive model of placebo response in epilepsy exists. However, based on multiple studies, it may be possible to construct a crude outline of what such a model would include. Fig. 1 shows the three main components as described above (inherent, methodological, and psychological). Each of these components represents multiple sub-processes, however when considered as only three main components, certain patterns can be observed.

It can be very instructive to imagine a world in which time moved backwards instead of forwards. Although impossible, the thought experiment allows one to imagine odd things like scrambled eggs going back into a shell, or patients clinically worsening after they take pills out of their mouth. In this imaginary reversal of time, seizure counts can be considered, and interesting effects emerge.

For example, placebo response due to inherent causes would be expected regardless of the flow of time (Fig. 3) (Goldenholz, Strashny, et al., 2017). If time were to suddenly flow backwards, the natural variability in seizure frequency would still cause placebo response, particularly in the 50%-responder method (RR50). Indeed, two of the three datasets showed larger placebo response when time was reversed. The reversibility of time is not true of methodological or psychological causes for placebo response.

Similarly, changing variables in how the trial is constructed or analyzed would primarily affect the methodological causes. Changing the symbols presented to patients would be expected to modify the psychological aspects alone. In this way, manipulations within specific placebo components could be made to increase or decrease specific contributions.

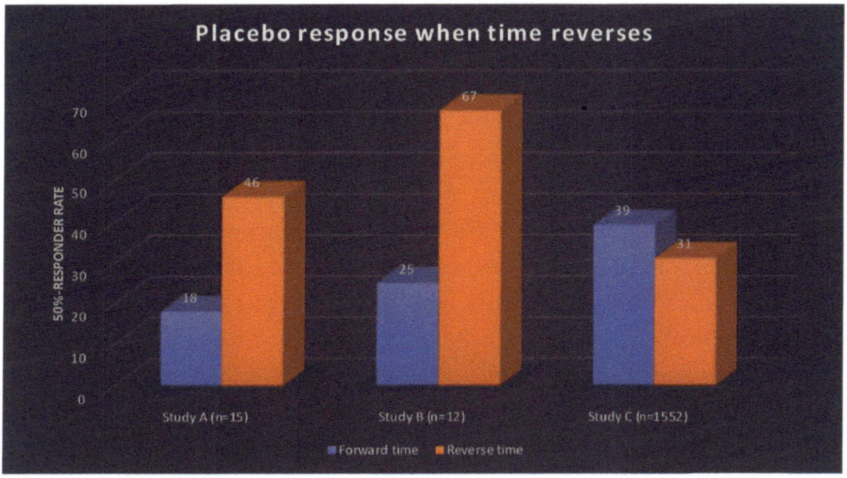

Fig. 3 Reversing time. A study evaluating three different datasets in epilepsy showed that if the flow of time were to reverse, a still large (and sometimes larger) placebo response would still be observed (Goldenholz, Strashny, et al., 2017).

It may be possible to produce a model of placebo response that makes reasonable predictions about the outcome of clinical trials, even in the absence of a comprehensive degree of detail about all aspects described in Fig. 1. Rather, a model that has a reasonable degree of detail about natural fluctuations and trial design with only a crude psychological component may be sufficient.

Early developments in this area have produced some interesting results (Romero et al., 2020). Using a statistical model that: (1) matches patient diaries (Tharayil et al., 2017), (2) has a heterogenous set of seizure frequencies that match previously study populations (Ferastraoaru et al., 2018), and (3) includes the L-relationship between mean and standard deviation in seizure frequency (Goldenholz et al., 2018), we developed a model we call "NV model 1" that can produce synthetic seizure diaries of any length from artificial patients, and then construct clinical trials around them. NV model 1 can reproduce the placebo and drug response rates seen in 23 clinical trials (Romero et al., 2020). NV model 1 was developed based on a patient population (Ferastraoaru et al., 2018) that generally was less "sick" than the typical patients in those 23 RCTs (median seizure rate was about 3/month in the patient population, but about 9/month in the RCTs). Simply increasing the median seizure frequency of the NV model population and re-optimizing parameters resulted in a surprising finding: the sicker population, which we called "NV model 2", produced a much smaller placebo response than NV model 1 due to natural variability alone. To adjust for this, we introduced a normally distributed correction factor which corresponds to the psychological component of placebo response. Selecting parameters for this correction factor to match observed values in 23 clinical trials, we found that the psychological component needed to be approximately 21% with a standard deviation of 10%. The psychological component encompasses multiple psychological aspects of placebo as outlined earlier in this chapter, as these currently are not separable. These numbers appear large because NV model 1 did not require any correction factor to produce reasonable placebo responses of similar magnitude. The difference between NV model 1 and 2 lies primarily in the seizure frequency—whereas NV model 1 patients have lower seizure frequencies and are more similar to patient reported diaries that have been previously studied, NV model 2 patients have higher rates and may be more similar to the RCTs that we tried to match. Higher seizure frequencies are expected to be associated with higher variability as described above due to the "L-relationship" (Goldenholz et al., 2018), so the need for a correction factor was unexpected. On the other hand, because of the L-relationship, there may be a predictable balance between the helpful impact of having higher seizure rates and the harmful impact of higher variability such that one can optimally choose a patient population a priori. Further work remains to determine the significance and/or clinical relevance of these simulation studies.

6. Clinical observations about placebo in epilepsy

6.1 Animal studies

There is little in the medical literature on placebo effects in animals with epilepsy. One study summarizing three different trials with dogs found evidence of placebo-treated 50% responder rates. The rates ranged between 0% and 60%, depending on the trial. Seizure-free rates were much lower, ranging 0%–7% (Muñana, Zhang, & Patterson, 2010). The same group was followed up with a randomized drug trial that failed to show a significant difference between levetiracetam and placebo in dogs (Thomas et al., 2012). This is a somewhat surprising result given the success of levetiracetam in humans (Mbizvo, Dixon, Hutton, & Marson, 2012), though it has been less impressive in dogs (Fredsø, Sabers, Toft, Møller, & Berendt, 2016; Hardy, Patterson, Cloyd, Hardy, & Leppik, 2012). The latter study (Hardy et al., 2012) compared 19 dogs in status epilepticus. They were given a double-blinded intravenous levetiracetam or intravenous placebo, finding no statistical difference between complete response rates at stopping seizures (drug = 56%, placebo = 10%, $p = 0.06$).

Another study compared medium chain triglyceride diet (MCTD) to placebo diet in a double-blinded crossover design (Law et al., 2015). The study reported improvement of seizure rate with the MCTD over placebo ($p = 0.02$), however, they did not report a baseline seizure rate, so it is difficult to ascertain the magnitude of the placebo response.

6.2 Effect size in human trials

Several meta-analyses of randomized controlled trials (RCTs) report on the responses to placebo in epilepsy. Estimates of the placebo response magnitude (for 50% responders) in drug trials range from 4% to 19% (Burneo, Montori, & Faught, 2002; Cramer, Fisher, Ben-Menachem, French, & Mattson, 1999; Guekht, Korczyn, Bondareva, & Gusev, 2010; Rheims et al., 2008; Zaccara, Giovannelli, Cincotta, Loiacono, & Verrotti, 2015). These rates have been increasing in recent years (Rheims et al., 2011).

In device studies, a similar range exists for 50%-responders. In a review of trials in transcranial magnetic stimulation (TMS), placebo responder rates were 16%–20% (Bae et al., 2011). In trigeminal nerve stimulation (TNS), a phase II RCT found the 25 placebo-arm patients to respond at a rate of 21.1% (DeGiorgio et al., 2013). A trial of responsive neuro-stimulation (RNS) found 27% response rate in the 93 patients assigned to sham stimulation (Morrell, 2011). The large vagal nerve stimulation (VNS) trial found that the 60 patients assigned to placebo (in this case "low stimulation") had a 13% responder rate (Vagus Nerve Stimulation Study Group, 1995). In a deep brain stimulation (DBS) RCT of 109 patients (54 stimulated, 55 control), the authors found no "statistically significant treatment group difference" for 50%-responders during the blinded phase, but did not actually report the number (Fisher et al., 2010).

Seizure freedom rates in RCTs are much lower than the 50% responder rates: 8.2% for drug and 2.1% for placebo in one meta-analysis (Beyenburg, Stavem, & Schmidt, 2010). In another study, drug-treated seizure-free rates were 4.5% and 2.8% in children and adults, respectively; placebo-treated children and adults achieved rates of 0.6% and 0.4%, respectively (Rheims et al., 2008). Several studies raised the point that use of the last observation carried forward (LOCF) method (rather than using full completers) artificially increase seizure freedom rates and responder rates in both placebo and active drug-treated patients (Gazzola, Balcer, & French, 2007; Rheims et al., 2011).

Device trials reported 0% "placebo" seizure-free responders for TMS (Bae et al., 2011), TNS (DeGiorgio et al., 2013), RNS (Morrell, 2011), and VNS (Vagus Nerve Stimulation Study Group, 1995). A randomized trial of DBS reported 1.8% (1 out of 55) "placebo" patients and 0% "active treatment" patients became seizure-free during the blinded phase (Fisher et al., 2010).

Taken altogether across drugs and devices, seizure freedom rates on placebo are quite low (0%–2.8%), while 50%-responder rates on placebo are quite a bit larger (4%–27%), with further increases seen in more recent years.

In addition to the increasing rates noted above, a statistically significant publication bias in epilepsy clinical trials was reported in a meta-analysis of 55 trials (Beyenburg et al., 2010). Therefore, the magnitude of the placebo response may be even higher than the levels reported here.

Interestingly, a meta-analysis found a difference in placebo response sizes between RCTs of five different anti-seizure medications ($p=0.01$, chi squared) suggesting heterogeneity of placebo responses between trials (Cramer et al., 1999). Also, the rates of response between placebo arm and drug arm were correlated in a large meta-analysis ($r=0.498$, $p<0.001$) (Zaccara, Giovannelli, Cincotta, et al., 2015). The implication of these findings may be that differing RCT designs and differing populations may influence the magnitude of placebo response.

6.3 Rising placebo response

The size of the placebo response has been "drifting" upwards over the past three decades in epilepsy trials (Fig. 4) (Khan, Fahl Mar, Schilling, & Brown, 2018b; Rheims et al., 2011; Zaccara, Giovannelli, & Schmidt, 2015) as well as in other illnesses. The year of publication appears to correlate positively with responder rate ($r=0.65$, $p=0.021$) in placebo, whereas the arm with anti-seizure drugs had a non-significant association (Rheims et al., 2011). A similar trend has been observed in drugs trials for many conditions, including mania (Sysko & Walsh, 2007), depression (Walsh, Seidman, Sysko, & Gould, 2002), psychosis (Agid et al., 2013), dementia (Hyde, May, Xue, & Zhang, 2017), hypertension (Khan, Fahl Mar, Schilling, & Brown, 2018a), and diabetes (Khan, Mar, Schilling, & Brown, 2018).

The cause for this increase over decades remains unknown in epilepsy. One proposed possibility is the increase in the number of international sites typically found in more modern studies is higher, while the number of patients per site is lower. Both of those relate to the increasing difficulty

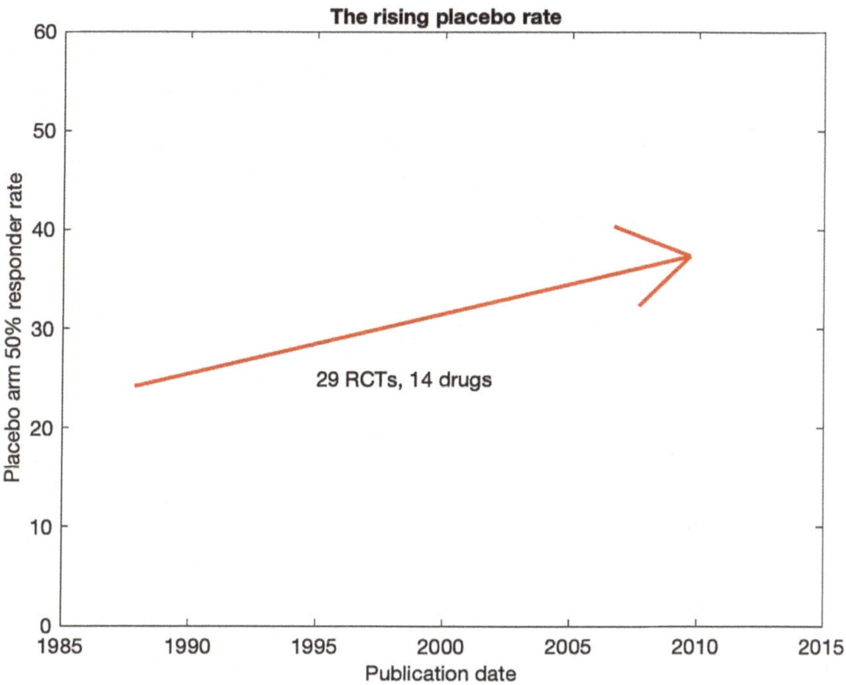

Fig. 4 The rising placebo response. This figure summarizes findings from a meta-analysis of 29 randomized controlled trials (RCTs), comprising 14 anti-seizure medications (Rheims et al., 2011). Similar conclusions have been reached by others (Khan, Fahl Mar, et al., 2018b; Zaccara, Giovannelli, Cincotta, et al., 2015).

in enrolling eligible patients in clinical epilepsy trials given the much larger set of available drugs on the market. Related to this idea, eligibility criteria have become stricter over time, again making recruitment more challenging. Rising eligibility criteria also increase the chance of regression-to-the-mean type 2.

Regardless of the cause, the rise in placebo likely has practical implications for costs. Overall costs of clinical trials have been rising over decades (Pharmaceutical Research and Manufacturers of America, 2015): costs of moving one drug to the market are currently in the multi-billion dollar range. Some trials have found enrollment time to increase dramatically over the years (for instance, 4 months in the 1988 to 42 months in 2014 (Auvin, French, et al., 2019)). In light of the rising costs and rising placebo response, it is instructive to consider the impact of higher placebo rates on a trial. Consider a typical 5-month trial that gives half of the patients a drug and half a placebo. Assume that the drug is 20% more effective than placebo

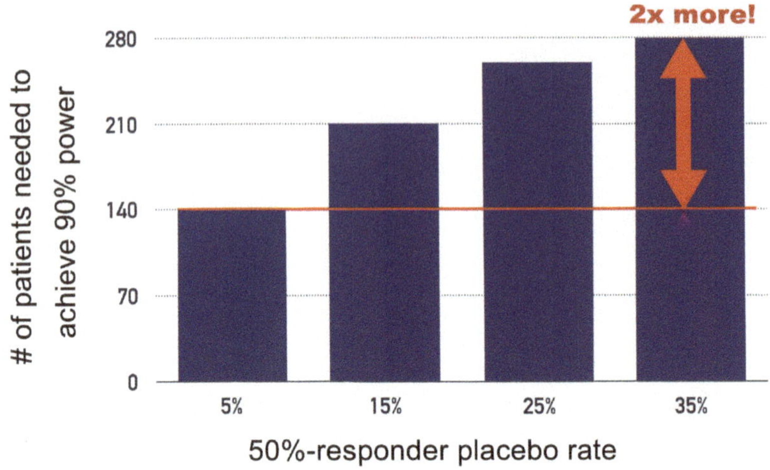

Fig. 5 Number of patients needed as placebo rate goes up. Shown here, between 5% and 35% placebo rate, the number of patients required to have a 90% powered study doubled. Obviously, higher placebo rates would have very detrimental effects on cost.

(Rheims et al., 2011). Now, suppose further that the standard 50%-responder rate method is used, and the study must be powered at 90%. One can calculate the number of patients needed to power such a study for different given values of placebo response, and thereby see what the impact of higher placebo rates would be on cost. Shown in Fig. 5, increasing placebo response from 5% to 35% doubles the required number of patients in the study. Based on numbers from an economic analysis of trial parameters (Goldenholz, Tharayil, Moss, et al., 2017), the study in Fig. 5 with 35% placebo will cost $906,500 more than the one with 5% placebo. The financial implications are clear—higher placebo rates can dramatically increase the costs of clinical trials.

6.4 Patient demographics and geography

Placebo response differentially impacts children and therefore trials for pediatric indications of seizure therapy. When ages were separated in a meta-analysis of 32 trials, 50%-responders to placebo were seen in 19% ± 2.3% of children (age < 18 years) and 9.9% ± 4.6% of adults (Rheims et al., 2008). This difference affected statistical comparison of drug effectiveness—adults had seemingly higher effectiveness for the same drugs as children. The difference was driven by equal rates of drug responders but higher placebo responder rates in children (Goldenholz & Goldenholz, 2016; Rheims et al., 2008). Similarly, the natural variability in seizure rates differ between children and adults, and in simulation the placebo rates were higher in children even when the only possible cause was natural fluctuations in seizure rates (Goldenholz et al., 2015).

There appear to be some additional baseline characteristics of patients that correlate with placebo response. For example, median seizure frequency seems to be inversely proportional to placebo response ($r = -0.312$, $p = 0.018$) (Zaccara, Giovannelli, Cincotta, et al., 2015). A review of 3 studies in lacosamide found that >6 anti-seizure drug exposures ($p < 0.001$) and >10 seizures/28 days ($p < 0.001$), and history of respective surgery ($p = 0.003$) all predicted low placebo response rates (Schmidt, Beyenburg, D'Souza, & Stavem, 2013).

Placebo response also may vary with geographic region, though this is more controversial. For example, in a study of perampanel, placebo responses were found to be higher in Central and South America (33% placebo responders) when compared to North America (21.9% placebo responders) (French et al., 2012). A review of lacosamide drug trials found a similar, though not statistically significant trend (Schmidt et al., 2013). Conversely, a meta-analysis did not find a systematic regional difference in placebo response rates (Zaccara, Giovannelli, Cincotta, et al., 2015).

6.5 Placebos in surgery

Before modern imaging, Penfield and Steelman (1947) compared seizure outcomes after craniotomies without resections and craniotomies with full resections. They found that 5 out of 16 (31%) patients achieved a 50% reduction in seizures after craniotomy even without resection. Also known as "trepanation," this technique had been in use throughout the world, dating at least as far back as Hippocrates or perhaps even to the Neolithic period (Asadi-Pooya & Rostami, 2017). Interestingly, animal data support the idea that seizures are reduced in the short term after trepanation (Forcelli, Kalikhman, & Gale, 2013), and there is evidence that temporary placement of intracranial electrodes alone results in seizure freedom in some patients (Schulze-Bonhage et al., 2010). The placebo effect is difficult to study in surgery in modern times because of ethical considerations (Samuel Wiebe, 2003). Epilepsy surgery was demonstrated in two RCTs to be superior to medical management (Engel et al., 2013; Wiebe, Blume, Girvin, Eliasziw, & Effectiveness and Efficiency of Surgery for Temporal Lobe Epilepsy Study Group, 2001). Interestingly, both RCTs did not use a true placebo as a comparator to surgery. Medical therapy and surgery therapy are known to be effective, and epilepsy carries high morbidity and mortality. Therefore, comparison surgery to medical therapy was previously possible when the relative effectiveness was unknown, but both were considered to be effective. Therefore, given the lack of equipoise with surgery, it would be challenging to design a trial that included sham surgery.

6.6 Trial parameters

Duration of test period influences the proportion of placebo responders in epilepsy trials. A meta-analysis took the ramp up (titration) period and included or excluded it from the placebo test period, finding 14.5% responders when the titration phase was included and 18.3% responders when it was excluded ($p=0.005$) (Rheims et al., 2011). On the other hand, a different analysis found that longer study duration correlated with larger placebo response rates ($r=0.264$, $p=0.030$) (Zaccara, Giovannelli, Cincotta, et al., 2015). It is possible that small numbers of patients and trials make these types of analyses underpowered. Use of realistic statistical models may be beneficial to clarify what to expect in the general case.

A simulation study found that various levels of placebo response would be expected depending on the duration of the baseline and test periods (Goldenholz, Tharayil, Moss, et al., 2017). By using realistic diaries, it was hoped that natural variability associated placebo response could be characterized (Fig. 6). Specifically, longer baselines correlated with larger placebo response in RR50 and median percent change (MPC). Similarly, shorter test periods correlated with larger placebo response in RR50 and MPC. The results of these simulations must be viewed with some skepticism however, because the diary data was patient-reported diaries (Casassa et al., 2018; Victor Ferastraoaru et al., 2016). Although these diaries do not likely differ qualitatively from diaries used in RCTs, the patients generally had lower seizure rates than typical RCT patients (Romero et al., 2020). Thus, trial parameters would be expected to have the impact noted in the simulation study if a trial used patients that had seizure rates that were similar to https://www.SeizureTracker.com/ users. Otherwise, the full extent that trial parameters have on placebo response is under-explored (French et al., 2013).

6.7 Sudden unexpected death in placebo-controlled epilepsy trials (SUDEP)

A meta-analysis of placebo-arm patients in epilepsy clinical trials found a statistically significant increase in SUDEP (Ryvlin, Cucherat, & Rheims, 2011).

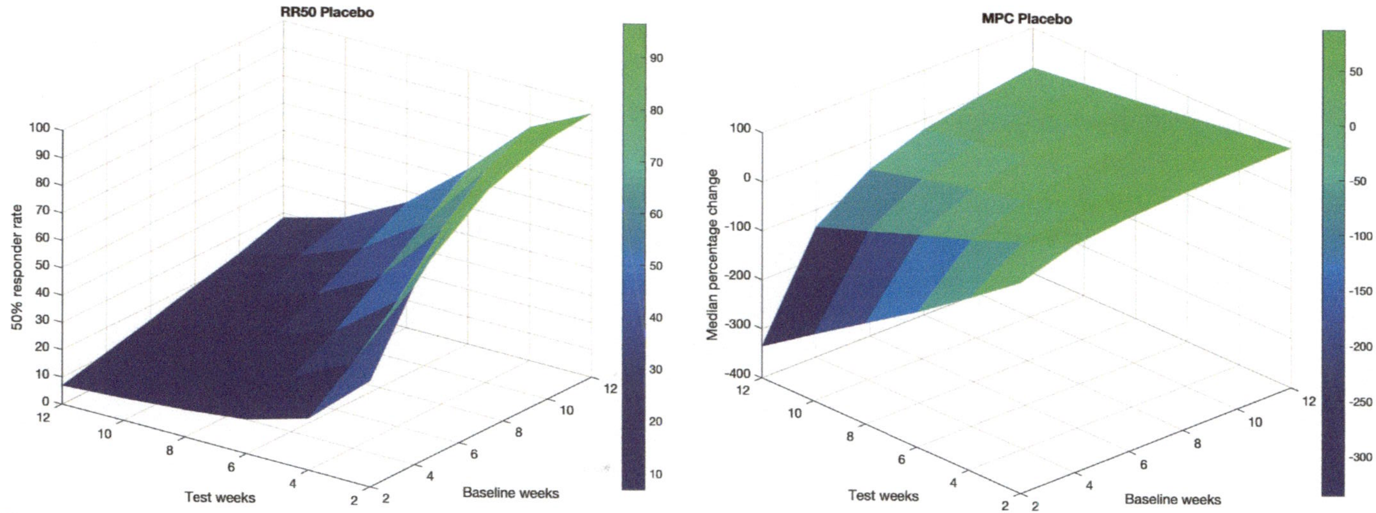

Fig. 6 Expected placebo rates with different trial parameters. Based on Monte Carlo simulations of patient-reported data, trials with 900 patients (50% having placebo) showed a reliable relationship between trial duration and expected placebo response when the only contribution was natural variability (Goldenholz, Tharayil, Moss, et al., 2017).

The rate per 1000 person-years of definite or probable SUDEP was 6.9 in the placebo arm and 0.9 in the efficacious drug arm; calculations were based on 112 trials (21,224 patients over 5589 patient-years). In 14 of 112 trials that reported a total of 20 SUDEP cases (11 definite, 7 probable, and 2 possible), there were 14 placebo-arm deaths (7 definite SUDEP, 7 probable SUDEP), 3 efficacious drug-treatment arm deaths (3 definite SUDEP), and 3 non-efficacious drug-treatment arm deaths (1 definite SUDEP, 2 possible SUDEP). Of note, "non-efficacious drug treatment" refers to a dose of a medication that does not appear significantly different from placebo. From this report, inclusion of a placebo arm in a randomized controlled study for epilepsy has been questioned. However, there are several concerns about this meta-analysis. First, the authors found that death under-reporting bias was unclear or present in 15 studies (13%). Additionally, up to 40 studies that may have been relevant were excluded from analysis. Also, in the absence of complete information, the probable and possible SUDEP cases may have been mis-classified. Therefore, a more measured approach is to consider the small absolute number of events, the under-reporting bias, the uncertainty in diagnosis, and the extremely large number of patients who were safe. If drug trials no longer included a placebo arm (and therefore were unable to distinguish effective therapies as clearly), millions of patients would be affected. Even modifying trials to minimize placebo exposure risk may be an extreme reaction in light of the multiple uncertainties. With all these mitigating factors, it may be prudent to interpret the results of the meta-analysis cautiously (Romero et al., 2020).

6.8 Nocebo effects

Placebos can also result in adverse consequences, a concept termed "nocebo." Interestingly, this phenomenon is common: A recent study found that nearly two-thirds (60%) of placebo-treated patients reported an adverse event. The most common adverse events experienced were headache (12.4%), somnolence (8.6%), dizziness (8.2%), and fatigue (7.9%) (Zaccara, Giovannelli, Cincotta, et al., 2015). The nocebo effect is not limited to adult patients. In a meta-analysis of 32 RCTs, 0.7% of adults and 5.6% children had worsening of seizures while in the placebo arm of studies (Rheims et al., 2008). The same study found that 7% of adults and 13.3% of children withdrew from trials while taking placebo, making it clear that nocebo effects can result in early dropout from a study. This early dropout can, in turn, result in inflated effect size when using the LOCF method as mentioned above (Gazzola et al., 2007). Therefore, a deeper understanding of how to minimize nocebo effects could have sizeable effects on clinical trial power and effect size.

7. Trial design and analysis considerations

7.1 Standard epilepsy RCT designs

Clinical trials have significantly evolved over the past century (Perucca, 2012). Many years ago, trials were not blinded, and randomized controls were certainly not the norm. For example, in 1912, Alfred Hauptmann monitored the effects of phenobarbital on patients for many months, noting that the drug was likely effective in bromide-resistant patients (Perucca, 2012). Thanks in part to regulatory changes from the Kefauver–Harris Drug Amendment of 1962, trials began to improve over time. Between 1970 and 1990, crossover designs that included placebo were commonly used, but these fell out of favor due to concerns over carry-over effects. From 1990 onward, clinical trials in epilepsy began to favor a parallel design randomized controlled trials, which consequently meant larger sample sizes. Because of this larger sample size requirement, multi-site studies have become increasingly common, raising problems of comparing groups from different settings. Fundamental barriers to developing new efficacious treatments for epilepsy—including large sample sizes, high costs, differences in regulatory requirements in the United States versus Europe (and elsewhere), and the increasing magnitude of the placebo response—all of these challenges appear to conspire against drug and device companies. Indeed, the individual efficacy rates of each of the whole shelf full of over 20 approved anti-seizure medications are, sadly, not dramatically different than the phenobarbital tested by Dr. Hauptmann more than 100 years ago.

The standard modern parallel design is shown in Fig. 7. The trial is typically divided into half of the patients randomized to drug (or other therapy), and half to placebo. Both groups are monitored on stable baseline therapy (adjuvant trial) or no therapy (monotherapy trial) during the baseline period to determine the baseline seizure rate, typically expressed in number of seizures per 28 days. Then, both groups are given a ramp up period, during which they receive increasing doses of a blinded treatment. During the test period, patients get either drug or placebo in a blinded fashion. It is assumed that after the ramp period, the drug achieves steady state. Typical trial analysis then summarizes each patient in terms of percentage change from baseline or time to prerandomization baseline rate (French et al., 2015;

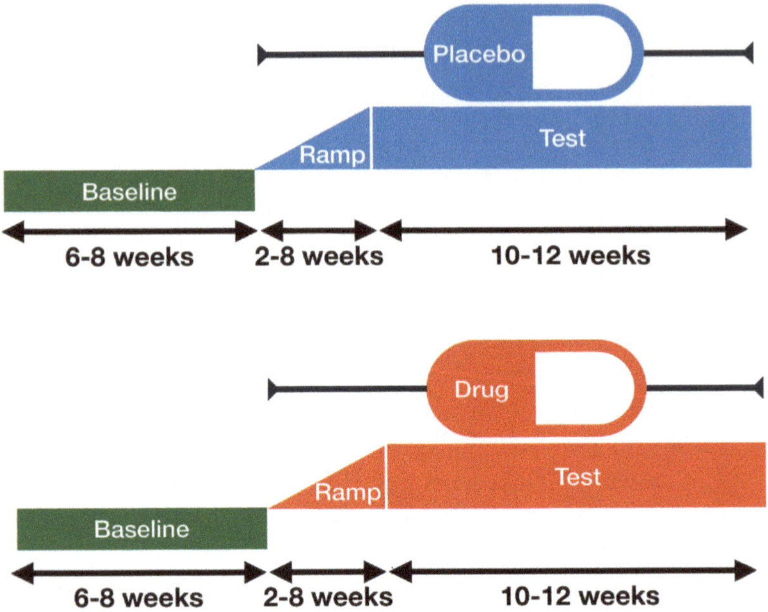

Fig. 7 Typical trial design. Patients are randomized to either placebo or drug arm typically in a 50%/50% ratio. Both groups are monitored during a baseline period. There is a ramp period when a blinded treatment is given in potentially increasing dose strengths. Then during the testing period, the blinded treatment is given at a steady dose.

Siddiqui & Hershkowitz, 2010). In the RR50, the percentage change values are binarized to 1 if ≥50%, and 0 if <50%. Then the proportions of responders in both arms are tallied and compared using a Fisher Exact test (Siddiqui & Hershkowitz, 2010). In the median percent change (MPC) method, the percentage change values from both groups are compared using a non-parametric test, the Wilcoxon rank sum test (Siddiqui & Hershkowitz, 2010).

7.2 Proposed alternative trial design and analysis methods

There are a number of interesting trial designs which may mitigate placebo response and have been proposed or used outside epilepsy. These include placebo run-in, sequential parallel comparison design, two way enriched design, and others (Fava, Evins, Dorer, & Schoenfeld, 2003; Goldenholz & Goldenholz, 2016; Ivanova, Qaqish, & Schoenfeld, 2011). It is unknown if these designs would be helpful in epilepsy.

7.3 Responder enriched design

A "responder enriched" design (also called randomized withdrawal) exposes patients to a drug in an open label fashion. Responders to the drug are then randomized to drug or placebo in a blinded fashion. Such a technique was used in several pediatric studies. The first required children age 3–15 to become seizure-free (i.e., 100%-responders) to enter randomization. That study found seizure freedom in placebo (21%, $n=14$) and drug (62%, $n=14$), which represents a very dramatic difference compared to traditional trials (Frank et al., 1999). A second trial in patients age 1–24 months found a nonsignificant effect size (overall $n=38$), but also used enrichment was based on a less rigorous 40%-responders rate, perhaps diminishing the power of the study (Piña-Garza et al., 2008).

7.4 Delayed start and cohort reduction

Several proposed simple placebo reduction methods have been simulated. The "delayed start" method (Goldenholz et al., 2015) was designed to mitigate RTM1 by enrolling patients but not starting a trial for 6 months. Although financially and logistically challenging, simulation over patient-reported diaries suggest this method would dramatically decrease placebo response. The "cohort reduction" method (Goldenholz et al., 2015) removed the "worst offenders" of seizure rate fluctuations from further analysis, thus decreasing the natural fluctuation impact on placebo response. This method has not been rigorously evaluated, and is presumed impractical.

7.5 Variability based method: Z_V

A statistically based analysis (Z_V) approach to natural variability was validated across 3 diverse datasets (Goldenholz, Goldenholz, Moss, et al., 2017). Unlike delayed start or cohort reduction, the method can be readily applied to existing trial data and uses information from all patients. This method was designed explicitly to reduce the impact of natural variability of seizure frequency on trial outcomes, and currently is the only validated method available in epilepsy to do so. When seizure count variability is significantly high, Z_V is expected to outperform RR50, but not MPC in terms of statistical power. Further studies are needed to confirm the relative merits of this method in populations with high seizure rates typically seen in RCTs (Fig. 8).

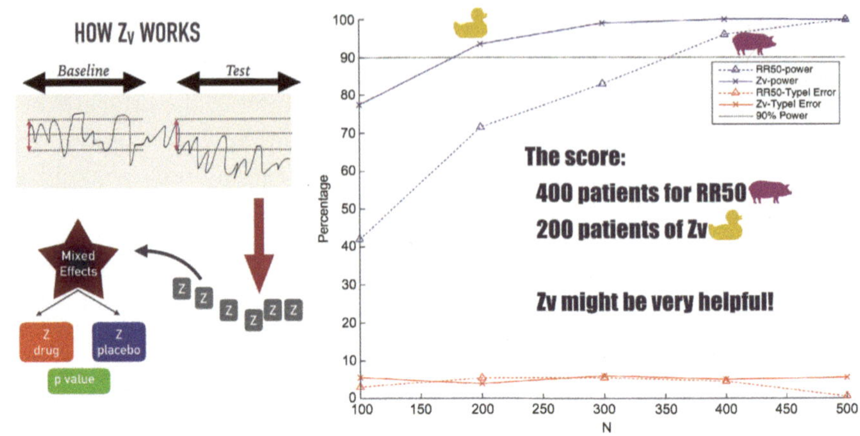

Fig. 8 Z_V, a proposed method for improving statistical power by accounting for variability in seizure counts (Goldenholz, Goldenholz, Moss, et al., 2017). On the left is a cartoon representation of the method, which converts seizure counts during the test period into Z scores normalized to the baseline period. The Z scores are used as input to a mixed effects model that accounts for dropout and patient random effects. The p value reported measures the chance that the two arms of the trial differ by chance alone. On the right is a comparison between Z_V and RR50. Highlighted with a duck, the Z_V achieves >90% power with 200 patients while the same trial parameters require 400 patients with RR50 to achieve >90% power (highlighted with a pig).

7.6 Time to prerandomization

A recently accepted alternative is the "time to prerandomization baseline" method (TTP) (French et al., 2015). The endpoint is the time it takes for a patient to reach their baseline number of seizures. The technique was validated over three clinical trials with perampanel, and may be used in future trials as well. The advantages of this method include shorter study duration, decreased exposure to risk—both from placebo and from drug—the possibility of including patients with lower seizure frequency, and endpoints that are clinically meaningful. The disadvantages include less observation time for adverse events, and unequal exposure times between patients with frequent seizures compared to those with infrequent seizures. Simulations from our lab suggests that TTP has higher statistical efficiency than RR50 but less than MPC, and that an optimal MPC trial would still expose less patients to placebo than TTP. Further study of the potential value of this trial method is clearly warranted.

7.7 Time to first seizure

Similar to TTP's use of an event model, "time to first seizure" (TFS) was proposed in 2007 (Cowling, Shaw, Hutton, & Marson, 2007). The authors note that in a meta-analysis, their method uncovered a dependence on age and epilepsy type in terms of efficacy of carbamazepine and valproate that were not noted with traditional techniques. It is unclear if the time to first seizure method would have sufficient reproducibility and statistical power to be more widely adopted. Moreover, there are regulatory concerns because this method would result in a very short testing phase, preventing assessment of side effects, tolerability and maintenance of efficacy (Baulac & Cascino, 2015).

7.8 Sustained 50%-responder status

Klein et al. proposed a "time to onset of sustained 50%-responder status" (SRS) method (Klein, Johnson, Schiemann, & Whitesides, 2017), which shares features with TTP and TFS. As with the other event-based approaches, a Kaplan Meier analysis is used to estimate a time. In this case, low numbers are preferable, with 1 day being the highest achievable value. Unlike TTP or TFS, this approach does not decrease the number of trial days. SRS is presumed to be more insensitive to the random fluctuations in seizure count compared with RR50, however that claim has yet to be more extensively studied. Unfortunately, the statistical approach to multiplicity is expected to lead to excessive type 1 errors, which could present additional obstacles. At present, this approach has not been widely adopted.

7.9 Variable baseline time to event

Another time-to-event model was proposed for pediatric studies in an effort to increase the opportunities for enrollment and decreasing costs and exposure time (Auvin, Williams, et al., 2019). The method calls for variable baseline periods (defined based on several baseline seizure rate categories). The trial proceeds to treatment phase until a patient experiences their *N*th seizure, where N is the number of seizures in the baseline. Like TTP, these trials would be analyzed using survival analysis. Also, like TTP, such trials may benefit from the decreased exposure time and overall trial duration. Additionally, the variable baseline and stop criteria allows for less strict inclusion criteria, which may facilitate enrollment. This method has not yet been tested in simulation nor in formal RCTs.

7.10 Human photosensitivity screening protocol

An alternative strategy for evaluating drugs in a proof of concept study has been developed, called "Human Photosensitivity Screening Protocol" (French, Krauss, Kasteleijn, DiVentura, & Bagiella, 2014). This technique uses a crossover design to compare the frequency range of photosensitive seizures patients are susceptible to with administration of multiple drugs and placebo. Mixed effects modeling was used to address the complex nature of the study design. According to their model results, placebo did not appear statistically different from baseline. Early evidence exists that this trial design may be broadly applicable to novel treatments for focal and generalized epilepsy (Kasteleijn-Nolst Trenite, Genton, Brandt, & Reed, 2017).

7.11 Impact of intracranial EEG

All drugs that have US FDA approval for adult use have based their applications on clinical trials with patient reported diaries. Indeed, such seizure counting is the dominant method of determining the degree of "sickness" clinically (Karoly, Goldenholz, & Cook, 2018). However, it is well-known that patient-reported diaries suffer from both under-reporting and over-reporting, sometimes as much as by a factor 10 times different than the true rate (Cook et al., 2013). The use of chronic implanted intracranial electrodes from several devices (Cook et al., 2013; Fisher et al., 2010; Sun & Morrell, 2014) has opened the possibility for a higher level of sensitivity and specificity to seizures.

Given that intracranial recordings typically detect more seizures (when one includes electrographic only seizures that have no clinical manifestations), and that higher seizure frequencies are expected to have higher variability (due to the L-relationship) (Goldenholz et al., 2018), it is uncertain if using intracranial seizures as a more reliable proxy for drug efficacy would improve the statistical efficacy of drug RCTs. A simulation study using Monte Carlo techniques and data from chronically implanted electrodes (Goldenholz, Tharayil, Kuzniecky, et al., 2017) found that use of intracranial seizure data would dramatically improve statistical efficiency. For example, when simulating 200 patients, half getting placebo (with no correction factor, i.e. due to natural variability alone) and half getting a drug of 20% efficacy, using only clinical seizure reporting resulted in statistical power of approximately 60% compared to using all intracranial seizures in the same group resulted in nearly 90% power. From this it may be suspected that as more implanted devices as well as non-invasive but accurate wearables (Onorati et al., 2017) become more widely used, further improvements in trial accuracy is possible that may overcome some of the limitations of natural variability.

8. Conclusions

The importance of placebo in epilepsy can be seen through the lens of historical treatments of epilepsy, modern trials, and clinical treatment as well. With increasing sophistication, it may be possible to anticipate portions or eventually all the aspects of placebo response. Early work is now delineating the aspects of placebo response attributable to mathematical features of natural history and regression to the mean. Future studies will need to quantify the psychological components of placebo response in epilepsy. As our understanding of placebo response grows, we can leverage that knowledge to helping improve the lives of people with epilepsy through improved clinical utilization, and optimized clinical trials.

References

Agid, O., Siu, C. O., Potkin, S. G., Kapur, S., Watsky, E., Vanderburg, D., et al. (2013). Meta-regression analysis of placebo response in antipsychotic trials, 1970-2010. *American Journal of Psychiatry, 170*(11), 1335–1344. https://doi.org/10.1176/appi.ajp.2013.12030315.

Asadi-Pooya, A. A., & Rostami, C. (2017). History of surgery for temporal lobe epilepsy. *Epilepsy & Behavior : E&B, 70*(Pt A), 57–60. https://doi.org/10.1016/j.yebeh.2017.02.020.

Auvin, S., French, J., Dlugos, D., Knupp, K. G., Perucca, E., Arzimanoglou, A., et al. (2019). Novel study design to assess the efficacy and tolerability of anti-seizure medications for focal-onset seizures in infants and young children: A consensus document from the Regulatory Task Force and the Pediatric Commission of the International League against Epilepsy (ILAE), in collaboration with the Pediatric Epilepsy Research Consortium (PERC). *Epilepsia Open*, epi4.12356. https://doi.org/10.1002/epi4.12356.

Auvin, S., Williams, B., McMurray, R., Kumar, D., Perdomo, C., & Malhotra, M. (2019). Novel seizure outcomes in patients with Lennox-Gastaut syndrome: Post hoc analysis of seizure-free days in rufinamide study 303. *Epilepsia Open, 4*(2), 275–280. https://doi.org/10.1002/epi4.12314.

Bae, E. H., Theodore, W. H., Fregni, F., Cantello, R., Pascual-Leone, A., & Rotenberg, A. (2011). An estimate of placebo effect of repetitive transcranial magnetic stimulation in epilepsy. *Epilepsy and Behavior, 20*(2), 355–359. https://doi.org/10.1016/j.yebeh.2010.12.005.

Baud, M. O., Kleen, J. K., Mirro, E. A., Andrechak, J. C., King-Stephens, D., Chang, E. F., et al. (2018). Multi-day rhythms modulate seizure risk in epilepsy. *Nature Communications, 9*(1), 1–10. https://doi.org/10.1038/s41467-017-02577-y.

Baulac, M., & Cascino, G. D. (2015). Antiepileptic drug trials: Less placebo exposure for safer studies? *Neurology, 84*(20), 2010–2011. https://doi.org/10.1212/WNL.0000000000001597.

Benedetti, F. (2014). Placebo effects: From the neurobiological paradigm to translational implications. *Neuron, 84*(3), 623–637. https://doi.org/10.1016/j.neuron.2014.10.023.

Benedetti, F., Maggi, G., Lopiano, L., Lanotte, M., Rainero, I., Vighetti, S., et al. (2003). Open versus hidden medical treatments: The patient's knowledge about a therapy affects

the therapy outcome. *Prevention & Treatment*, *6*(1), 1–19. https://doi.org/10.1037/1522-3736.6.0001a.
Beyenburg, S., Stavem, K., & Schmidt, D. (2010). Placebo-corrected efficacy of modern antiepileptic drugs for refractory epilepsy: Systematic review and meta-analysis. *Epilepsia*, *51*(1), 7–26. https://doi.org/10.1111/j.1528-1167.2009.02299.x.
Branthwaite, A., & Cooper, P. (1981). Analgesic effects of branding in treatment of headaches. *British Medical Journal (Clinical Research Ed.)*, *282*(6276), 1576–1578. https://doi.org/10.1136/bmj.282.6276.1576.
Brodie, M. J., Barry, S. J. E., Bamagous, G. A., Norrie, J. D., & Kwan, P. (2012). Patterns of treatment response in newly diagnosed epilepsy. *Neurology*, *78*(20), 1548–1554. https://doi.org/10.1212/WNL.0b013e3182563b19.
Burneo, J. G., Montori, V. M., & Faught, E. (2002). Magnitude of the placebo effect in randomized trials of antiepileptic agents. *Epilepsy and Behavior*, *3*(6), 532–534. https://doi.org/10.1016/S1525-5050(02)00531-0.
Callaghan, B. C., Anand, K., Hesdorffer, D., Hauser, W. A., & French, J. A. (2007). Likelihood of seizure remission in an adult population with refractory epilepsy. *Annals of Neurology*, *62*(4), 382–389. https://doi.org/10.1002/ana.21166.
Callaghan, B., Schlesinger, M., Rodemer, W., Pollard, J., Hesdorffer, D., Hauser, W. A., et al. (2011). Remission and relapse in a drug-resistant epilepsy population followed prospectively. *Epilepsia*, *52*(3), 619–626. https://doi.org/10.1111/j.1528-1167.2010.02929.x.
Casassa, C., Rathbun Levit, E., & Goldenholz, D. M. (2018). Opinion and special articles: Self-management in epilepsy: Web-based seizure tracking applications. *Neurology*, *91*(21). https://doi.org/10.1212/WNL.0000000000006547.
Chen, D. K., Dave, H., Gadelmola, K., Jeroudi, M., & Fadipe, M. (2018). Provocative induction of psychogenic nonepileptic seizures: Noninferiority of an induction technique without versus with placebo. *Epilepsia*, *59*(11), e161–e165. https://doi.org/10.1111/epi.14570.
Choi, H., Heiman, G. A., Munger Clary, H., Etienne, M., Resor, S. R., & Hauser, W. A. (2011). Seizure remission in adults with long-standing intractable epilepsy: An extended follow-up. *Epilepsy Research*, *93*(2–3), 115–119. https://doi.org/10.1016/j.eplepsyres.2010.11.005.
Cook, M. J., O'Brien, T. J., Berkovic, S. F., Murphy, M., Morokoff, A., Fabinyi, G., et al. (2013). Prediction of seizure likelihood with a long-term, implanted seizure advisory system in patients with drug-resistant epilepsy: A first-in-man study. *The Lancet Neurology*, *12*(6), 563–571. https://doi.org/10.1016/S1474-4422(13)70075-9.
Cowling, B. J., Shaw, J. E. H., Hutton, J. L., & Marson, A. G. (2007). New statistical method for analyzing time to first seizure: Example using data comparing carbamazepine and valproate monotherapy. *Epilepsia*, *48*(6), 1173–1178. https://doi.org/10.1111/j.1528-1167.2007.01036.x.
Cramer, J. A., Fisher, R., Ben-Menachem, E., French, J., & Mattson, R. H. (1999). New antiepileptic drugs: Comparison of key clinical trials. *Epilepsia*, *40*(5), 590–600. https://doi.org/10.1111/j.1528-1157.1999.tb05561.x.
DeGiorgio, C. M., Soss, J., Cook, I. A., Markovic, D., Gornbein, J., Murray, D., et al. (2013). Randomized controlled trial of trigeminal nerve stimulation for drug-resistant epilepsy. *Neurology*, *80*(9), 786–791. https://doi.org/10.1212/WNL.0b013e318285c11a.
Engel, J., Mcdermott, M. P., Wiebe, S., Langfitt, J. T., Stern, J. M., Dewar, S., et al. (2013). Early surgical therapy for drug-resistant temporal lobe epilepsy. *Survey of Anesthesiology*, *57*(3), 121–122. https://doi.org/10.1097/01.UC.0000428885.17795.75.
Englot, D. J., Wang, D. D., Rolston, J. D., Shih, T. T., & Chang, E. F. (2012). Rates and predictors of long-term seizure freedom after frontal lobe epilepsy surgery: A systematic

review and meta-analysis. *Journal of Neurosurgery, 116*(5), 1042–1048. https://doi.org/10.3171/2012.1.JNS111620.

Espay, A. J., Norris, M. M., Eliassen, J. C., Dwivedi, A., Smith, M. S., Banks, C., et al. (2015). Placebo effect of medication cost in Parkinson disease: A randomized double-blind study. *Neurology, 84*(8), 794–802. https://doi.org/10.1212/WNL.0000000000001282.

Fava, M., Evins, A. E., Dorer, D. J., & Schoenfeld, D. A. (2003). The problem of the placebo response in clinical trials for psychiatric disorders: Culprits, possible remedies, and a novel study design approach. *Psychotherapy and Psychosomatics, 72*(3), 115–127. https://doi.org/10.1159/000069738.

Ferastraoaru, V., Goldenholz, D. M., Chiang, S., Moss, R., Theodore, W. H., & Haut, S. R. (2018). Characteristics of large patient-reported outcomes: Where can one million seizures get us? *Epilepsia Open, 3*(3). https://doi.org/10.1002/epi4.12237.

Ferastraoaru, V., Schulze-Bonhage, A., Lipton, R. B., Dümpelmann, M., Legatt, A. D., Blumberg, J., et al. (2016). Termination of seizure clusters is related to the duration of focal seizures. *Epilepsia, 57*(6), 889–895. https://doi.org/10.1111/epi.13375.

Fisher, R., Salanova, V., Witt, T., Worth, R., Henry, T., Gross, R., et al. (2010). Electrical stimulation of the anterior nucleus of thalamus for treatment of refractory epilepsy. *Epilepsia, 51*(5), 899–908. https://doi.org/10.1111/j.1528-1167.2010.02536.x.

Forcelli, P. A., Kalikhman, D., & Gale, K. (2013). Delayed effect of craniotomy on experimental seizures in rats. *PLoS One, 8*(12), 8–13. https://doi.org/10.1371/journal.pone.0081401.

Frank, L. M., Enlow, T., Holmes, G. L., Manasco, P., Concannon, S., Chen, C., et al. (1999). Lamictal (lamotrigine) monotherapy for typical absence seizures in children. *Epilepsia, 40*(7), 973–979. https://doi.org/10.1111/j.1528-1157.1999.tb00805.x.

Fredsø, N., Sabers, A., Toft, N., Møller, A., & Berendt, M. (2016). A single-blinded phenobarbital-controlled trial of levetiracetam as mono-therapy in dogs with newly diagnosed epilepsy. *Veterinary Journal (London, England: 1997), 208*, 44–49. https://doi.org/10.1016/j.tvjl.2015.10.018.

French, J. A., Cabrera, J., Emir, B., Whalen, E., & Lu, F. (2013). Designing a new proof-of-principle trial for treatment of partial seizures to demonstrate efficacy with minimal sample size and duration-a case study. *Epilepsy Research, 106*(1–2), 230–236. https://doi.org/10.1016/j.eplepsyres.2013.04.008.

French, J. A., Gil-Nagel, A., Malerba, S., Kramer, L., Kumar, D., & Bagiella, E. (2015). Time to prerandomization monthly seizure count in perampanel trials: A novel epilepsy endpoint. *Neurology, 84*(20), 2014–2020. https://doi.org/10.1212/wnl.0000000000001585.

French, J. A., Krauss, G. L., Biton, V., Squillacote, D., Yang, H., Laurenza, A., et al. (2012). Adjunctive perampanel for refractory partial-onset seizures: Randomized phase III study 304. *Neurology, 79*(6), 589–596. https://doi.org/10.1212/WNL.0b013e3182635735.

French, J. A., Krauss, G. L., Kasteleijn, D., DiVentura, B. D., & Bagiella, E. (2014). Effects of marketed antiepileptic drugs and placebo in the human photosensitivity screening protocol. *Neurotherapeutics, 11*(2), 412–418. https://doi.org/10.1007/s13311-013-0243-0.

Frisaldi, E., Carlino, E., Lanotte, M., Lopiano, L., & Benedetti, F. (2014). Characterization of the thalamic-subthalamic circuit involved in the placebo response through single-neuron recording in Parkinson patients. *Cortex, 60*, 3–9. https://doi.org/10.1016/j.cortex.2013.12.003.

Gazzola, D. M., Balcer, L. J., & French, J. A. (2007). Seizure-free outcome in randomized add-on trials of the new antiepileptic drugs. *Epilepsia, 48*(7), 1303–1307. https://doi.org/10.1111/j.1528-1167.2007.01136.x.

Geerts, A. T., Arts, W. F., Stroink, H., Peeters, E., Brouwer, O., Peters, B., et al. (2010). Course and outcome of childhood epilepsy: A 15-year follow-up of the Dutch study of epilepsy in childhood. *Epilepsia, 51*(7), 1189–1197. https://doi.org/10.1111/j.1528-1167.2010.02546.x.

Goldenholz, D. M., & Goldenholz, S. R. (2016). Response to placebo in clinical epilepsy trials-old ideas and new insights. *Epilepsy Research, 122*, 15–25. https://doi.org/10.1016/j.eplepsyres.2016.02.002.

Goldenholz, D. M., Goldenholz, S. R., Moss, R., French, J., Lowenstein, D., Kuzniecky, R., et al. (2017). Does accounting for seizure frequency variability increase clinical trial power? *Epilepsy Research, 137*, 145–151. https://doi.org/10.1016/j.eplepsyres.2017.07.013.

Goldenholz, D. M., Goldenholz, S. R., Moss, R., French, J., Lowenstein, D., Kuzniecky, R., et al. (2018). Is seizure frequency variance a predictable quantity? *Annals of Clinical and Translational Neurology, 5*(2), 201–207. https://doi.org/10.1002/acn3.519.

Goldenholz, D. M., Moss, R., Scott, J., Auh, S., & Theodore, W. H. (2015). Confusing placebo effect with natural history in epilepsy: A big data approach. *Annals of Neurology, 78*(3), 329–336. https://doi.org/10.1002/ana.24470.

Goldenholz, D. M., Strashny, A., Cook, M., Moss, R., & Theodore, W. H. (2017). A multi-dataset time-reversal approach to clinical trial placebo response and the relationship to natural variability in epilepsy. *Seizure, 53*, 31–36. https://doi.org/10.1016/j.seizure.2017.10.016.

Goldenholz, D. M., Tharayil, J. J., Kuzniecky, R., Karoly, P., Theodore, W. H., & Cook, M. J. (2017). Simulating clinical trials with and without intracranial EEG data. *Epilepsia Open, 2*(2), 156–161. https://doi.org/10.1002/epi4.12038.

Goldenholz, D. M., Tharayil, J., Moss, R., Myers, E., & Theodore, W. H. (2017). Monte Carlo simulations of randomized clinical trials in epilepsy. *Annals of Clinical and Translational Neurology, 4*(8), 544–552. https://doi.org/10.1002/acn3.426.

Guekht, A. B., Korczyn, A. D., Bondareva, I. B., & Gusev, E. I. (2010). Placebo responses in randomized trials of antiepileptic drugs. *Epilepsy and Behavior, 17*(1), 64–69. https://doi.org/10.1016/j.yebeh.2009.10.007.

Hardy, B. T., Patterson, E. E., Cloyd, J. M., Hardy, R. M., & Leppik, I. E. (2012). Double-masked, placebo-controlled study of intravenous Levetiracetam for the treatment of status epilepticus and acute repetitive seizures in dogs. *Journal of Veterinary Internal Medicine, 26*(2), 334–340. https://doi.org/10.1111/j.1939-1676.2011.00868.x.

Holtkamp, M., & Theodore, W. H. (2018). Generic antiepileptic drugs-safe or harmful in patients with epilepsy? *Epilepsia, 59*(7), 1273–1281. https://doi.org/10.1111/epi.14439.

Hyde, A. J., May, B. H., Xue, C. C., & Zhang, A. L. (2017). Variation in placebo effect sizes in clinical trials of oral interventions for management of the behavioral and psychological symptoms of dementia (BPSD): A systematic review and Meta-analysis. *The American Journal of Geriatric Psychiatry : Official Journal of the American Association for Geriatric Psychiatry, 25*(9), 994–1008. https://doi.org/10.1016/j.jagp.2017.02.022.

Institute of Medicine (U.S.), Committee on the Public Health Dimensions of the Epilepsies, & England, M. J. (2012). *Epilepsy across the spectrum promoting health and understanding*. Washington, DC: National Academies Press.

Ivanova, A., Qaqish, B., & Schoenfeld, D. A. (2011). Optimality, sample size, and power calculations for the sequential parallel comparison design. *Statistics in Medicine, 30*(23), 2793–2803. https://doi.org/10.1002/sim.4292.

Jarius, S., & Wildemann, B. (2015). And Pavlov still rings a bell: Summarising the evidence for the use of a bell in Pavlov's iconic experiments on classical conditioning. *Journal of Neurology, 262*(9), 2177–2178. https://doi.org/10.1007/s00415-015-7858-5.

Jehi, L., Sarkis, R., Bingaman, W., Kotagal, P., & Najm, I. (2010). When is a postoperative seizure equivalent to "epilepsy recurrence" after epilepsy surgery? *Epilepsia, 51*(6), 994–1003. https://doi.org/10.1111/j.1528-1167.2010.02556.x.

Kam-Hansen, S., Jakubowski, M., Kelley, J. M., Kirsch, I., Hoaglin, D. C., Kaptchuk, T. J., et al. (2014). Altered placebo and drug labeling changes the outcome of episodic migraine attacks. *Science Translational Medicine, 6*(218). https://doi.org/10.1126/scitranslmed.3006175.

Karoly, P., Goldenholz, D. M., & Cook, M. (2018). Are the days of counting seizures numbered? *Current Opinion in Neurology, 31*(2), 162–168. https://doi.org/10.1097/WCO.0000000000000533.

Karoly, P. J., Goldenholz, D. M., Freestone, D. R., Moss, R. E., Grayden, D. B., Theodore, W. H., et al. (2018). Circadian and circaseptan rhythms in human epilepsy: A retrospective cohort study. *The Lancet Neurology, 17*(11), 977–985. https://doi.org/10.1016/S1474-4422(18)30274-6.

Karoly, P. J., Romero, J., Cook, M. J., Freestone, D. R., & Goldenholz, D. M. (2019). When can we trust responders? Serious concerns when using 50% response rate to assess clinical trials. *Epilepsia, 60*(9), e99–e103. https://doi.org/10.1111/epi.16321.

Kasteleijn-Nolst Trenite, D., Genton, P., Brandt, C., & Reed, R. C. (2017). The "photosensitivity model" is (also) a model for focal (partial) seizures. *Epilepsy Research, 133*, 113–120. https://doi.org/10.1016/j.eplepsyres.2016.11.012.

Khan, A., Fahl Mar, K., Schilling, J., & Brown, W. A. (2018a). Does the rising placebo response impact antihypertensive clinical trial outcomes? An analysis of data from the Food and Drug Administration 1990-2016. *PLoS One, 13*(2). e0193043https://doi.org/10.1371/journal.pone.0193043.

Khan, A., Fahl Mar, K., Schilling, J., & Brown, W. A. (2018b). Magnitude and pattern of placebo response in clinical trials of antiepileptic medications: Data from the Food and Drug Administration 1996–2016. *Contemporary Clinical Trials, 64*, 95–100. https://doi.org/10.1016/j.cct.2017.10.017.

Khan, A., Mar, K. F., Schilling, J., & Brown, W. A. (2018). Magnitude and pattern of placebo response in clinical trials of oral antihyperglycemic agents: Data from the U.S. food and drug administration, 1999-2015. *Diabetes Care, 41*(5), 994–1000. https://doi.org/10.2337/dc17-1316/-/DC1.

Kienle, G. S., & Kiene, H. (1997). The powerful placebo effect: Fact or fiction? *Journal of Clinical Epidemiology, 50*(12), 1311–1318. https://doi.org/10.1016/S0895-4356(97)00203-5.

Kirsch, I. (2019). Placebo effect in the treatment of depression and anxiety. *Frontiers in Psychiatry, 10*, 407. https://doi.org/10.3389/fpsyt.2019.00407.

Klein, P., Johnson, M. E., Schiemann, J., & Whitesides, J. (2017). Time to onset of sustained ≥50% responder status in patients with focal (partial-onset) seizures in three phase III studies of adjunctive brivaracetam treatment. *Epilepsia, 58*(2), e21–e25. https://doi.org/10.1111/epi.13631.

Kwan, P., & Brodie, M. J. (2000). Early identification of refractory epilepsy. *New England Journal of Medicine, 342*(5), 314–319.

Law, T. H., Davies, E. S. S., Pan, Y., Zanghi, B., Want, E., & Volk, H. A. (2015). A randomised trial of a medium-chain TAG diet as treatment for dogs with idiopathic epilepsy. *British Journal of Nutrition, 114*(9), 1438–1447. https://doi.org/10.1017/S000711451500313X.

Luciano, A. L., & Shorvon, S. D. (2007). Results of treatment changes in patients with apparently drug-resistant chronic epilepsy. *Annals of Neurology, 62*(4), 375–381. https://doi.org/10.1002/ana.21064.

Mahoney, J. M., Mills, J. D., Muhlebner, A., Noebels, J., Potschka, H., Simonato, M., et al. (2019). 2017 WONOEP appraisal: Studying epilepsy as a network disease using systems biology approaches. *Epilepsia, 60*(6), 1045–1053. https://doi.org/10.1111/epi.15216.

Malmgren, K., & Edelvik, A. (2017). Long-term outcomes of surgical treatment for epilepsy in adults with regard to seizures, antiepileptic drug treatment and employment. *Seizure, 44*, 217–224. https://doi.org/10.1016/j.seizure.2016.10.015.

Margraf, J., & Ehlers, A. (1991). How "Blind" are double-blind studies? *Journal of Consulting andClinical Psychology*, *59*(I), 184–187.

Mbizvo, G. K., Dixon, P., Hutton, J. L., & Marson, A. G. (2012). Levetiracetam add-on for drug-resistant focal epilepsy: An updated Cochrane review. *Cochrane Database of Systematic Reviews*, (9), CD001901. https://doi.org/10.1002/14651858.cd001901.pub2.

McCambridge, J., Witton, J., & Elbourne, D. R. (2014). Systematic review of the Hawthorne effect: New concepts are needed to study research participation effects. *Journal of Clinical Epidemiology*, *67*(3), 267–277. https://doi.org/10.1016/j.jclinepi.2013.08.015.

McKee, H. R., & Privitera, M. D. (2017). Stress as a seizure precipitant: Identification, associated factors, and treatment options. *Seizure*, *44*, 21–26. https://doi.org/10.1016/j.seizure.2016.12.009.

Morgan, L. A., Dvorchik, I., Williams, K. L., Jarrar, R. G., & Buchhalter, J. R. (2013). Parental ranking of terms describing nonepileptic events. *Pediatric Neurology*, *48*(5), 378–382. https://doi.org/10.1016/j.pediatrneurol.2012.12.029.

Morrell, M. J. (2011). Responsive cortical stimulation for the treatment of medically intractable partial epilepsy. *Neurology*, *77*(13), 1295–1304. https://doi.org/10.1212/WNL.0b013e3182302056.

Morton, V., & Torgerson, D. J. (2005). Regression to the mean: Treatment effect without the intervention. *Journal of Evaluation in Clinical Practice*, *11*(1), 59–65. https://doi.org/10.1111/j.1365-2753.2004.00505.x.

Muñana, K. R., Zhang, D., & Patterson, E. E. (2010). Placebo effect in canine epilepsy trials. *Journal of Veterinary Internal Medicine*, *24*(1), 166–170. https://doi.org/10.1111/j.1939-1676.2009.0407.x.

Onorati, F., Regalia, G., Caborni, C., Migliorini, M., Bender, D., Poh, M. Z., et al. (2017). Multicenter clinical assessment of improved wearable multimodal convulsive seizure detectors. *Epilepsia*, *58*(11), 1870–1879. https://doi.org/10.1111/epi.13899.

Panteliadis, C. P., Vassilyadi, P., Fehlert, J., & Hagel, C. (2017). Historical documents on epilepsy: From antiquity through the 20th century. *Brain and Development*, *39*(6), 457–463. https://doi.org/10.1016/j.braindev.2017.02.002.

Penfield, W., & Steelman, H. (1947). The treatment of focal epilepsy by cortical excision. *Annals of Surgery*, *126*(5), 740–762. https://doi.org/10.1097/00000658-194711000-00008.

Perucca, E. (2012). What clinical trial designs have been used to test antiepileptic drugs and do we need to change them? *Epileptic Disorders*, *14*(2), 124–131. https://doi.org/10.1684/epd.2012.0511.

Perucca, E. (2019). Antiepileptic drugs: Evolution of our knowledge and changes in drug trials. *Epileptic Disorders : International Epilepsy Journal with Videotape*, *21*(4), 319–329. https://doi.org/10.1684/epd.2019.1083.

Pharmaceutical Research and Manufacturers of America. (2015). *2015 Biopharmaceutical Research Industry Profile*.

Piña-Garza, J. E., Levisohn, P., Gucuyener, K., Mikati, M. A., Warnock, C. R., Conklin, H. S., et al. (2008). Adjunctive lamotrigine for partial seizures in patients aged 1 to 24 months. *Neurology*, *70*(22 Pt 2), 2099–2108.

Rheims, S., Cucherat, M., Arzimanoglou, A., & Ryvlin, P. (2008). Greater response to placebo in children than in adults: A systematic review and meta-analysis in drug-resistant partial epilepsy. *PLoS Medicine*, *5*(8), 1223–1237. https://doi.org/10.1371/journal.pmed.0050166.

Rheims, S., Perucca, E., Cucherat, M., & Ryvlin, P. (2011). Factors determining response to antiepileptic drugs in randomized controlled trials. A systematic review and meta-analysis. *Epilepsia*, *52*(2), 219–233. https://doi.org/10.1111/j.1528-1167.2010.02915.x.

Romero, J., Larimer, P., Chang, B., Goldenholz, S. R., & Goldenholz, D. M. (2020). Natural variability in seizure frequency: Implications for trials and placebo. *Epilepsy Research*, *162*, 106306. https://doi.org/10.1016/j.eplepsyres.2020.106306.

Ryvlin, P., Cucherat, M., & Rheims, S. (2011). Risk of sudden unexpected death in epilepsy in patients given adjunctive antiepileptic treatment for refractory seizures: A meta-analysis of placebo-controlled randomised trials. *The Lancet Neurology*, *10*(11), 961–968. https://doi.org/10.1016/S1474-4422(11)70193-4.

Schmidt, D., Beyenburg, S., D'Souza, J., & Stavem, K. (2013). Clinical features associated with placebo response in refractory focal epilepsy. *Epilepsy and Behavior*, *27*(2), 393–398. https://doi.org/10.1016/j.yebeh.2013.01.015.

Schmidt, D., & Stavem, K. (2009). Long-term seizure outcome of surgery versus no surgery for drug-resistant partial epilepsy: A review of controlled studies. *Epilepsia*, *50*(6), 1301–1309. https://doi.org/10.1111/j.1528-1167.2008.01997.x.

Schulze-Bonhage, A., Dennig, D., Wagner, K., Cordeiro, J., Carius, A., Fauser, S., et al. (2010). Seizure control resulting from intrahippocampal depth electrode insertion. *Journal of Neurology, Neurosurgery and Psychiatry*, *81*(3), 352–353. https://doi.org/10.1136/jnnp.2009.180075.

Siddiqui, O., & Hershkowitz, N. (2010). Primary efficacy endpoint in clinical trials of anti-epileptic drugs: Change or percentage change. *Therapeutic Innovation & Regulatory Science*, *44*(3), 343–350. https://doi.org/10.1177/009286151004400316.

Sillanpää, M., & Schmidt, D. (2006). Natural history of treated childhood-onset epilepsy: Prospective, long-term population-based study. *Brain*, *129*(3), 617–624. https://doi.org/10.1093/brain/awh726.

Sun, F. T., & Morrell, M. J. (2014). The RNS system: Responsive cortical stimulation for the treatment of refractory partial epilepsy. *Expert Review of Medical Devices*, *11*(6), 563–572. https://doi.org/10.1586/17434440.2014.947274.

Syed, T. U., Lafrance, W. C., Kahriman, E. S., Hasan, S. N., Rajasekaran, V., Gulati, D., et al. (2011). Can semiology predict psychogenic nonepileptic seizures? A prospective study. *Annals of Neurology*, *69*(6), 997–1004. https://doi.org/10.1002/ana.22345.

Sysko, R., & Walsh, B. T. (2007). A systematic review of placebo response in studies of bipolar mania. *Journal of Clinical Psychiatry*, *68*(8), 1213–1217. https://doi.org/10.4088/JCP.v68n0807.

Tharayil, J. J., Chiang, S., Moss, R., Stern, J. M., Theodore, W. H., & Goldenholz, D. M. (2017). A big data approach to the development of mixed-effects models for seizure count data. *Epilepsia*, *58*(5), 835–844. https://doi.org/10.1111/epi.13727.

The Epilepsy Study Consortium, The Human Epilepsy Project. (n.d.). Retrieved from http://www.humanepilepsyproject.org/ August 28, 2019

Thomas, W. B., Inzana, K. D., Mclucas, K. J., Olby, N. J., Mariani, C. J., & Early, P. J. (2012). Evaluation of Levetiracetam as adjunctive treatment for refractory canine epilepsy: A randomized, placebo-controlled. Crossover Trial. *Journal of Veterinary Internal Medicine*, *26*(2), 341–348.

Vagus Nerve Stimulation Study Group. (1995). A randomized controlled trial of chronic vagus nerve stimulation for treatment of medically intractable seizures. *Neurology*, *45*, 224–230 Retrieved from http://scholar.google.com/scholar?hl=en&btnG=Search&q=intitle:A+randomized+controlled+trial+of+chronic+vagus+nerve+stimulation+for+treatment+of+medically+intractable+seizures#0%5Cnhttp://www.scopus.com/inward/record.url?eid=2-s2.0-0028819092&partnerID=40&.

Walsh, B. T., Seidman, S. N., Sysko, R., & Gould, M. (2002). Placebo response in studies of major depression variable, substantial, and growing. *JAMA*, *287*(14), 1840–1847. https://doi.org/10.1001/jama.287.14.1840.

Wiebe, S. (2003). Randomized controlled trials of epilepsy surgery. *Epilepsia*, *44*(SUPPL. 7), 38–43. https://doi.org/10.1046/j.1528-1157.44.s7.9.x.

Wiebe, S., Blume, W. T., Girvin, J. P., Eliasziw, M., & Effectiveness and Efficiency of Surgery for Temporal Lobe Epilepsy Study Group. (2001). A randomized, controlled trial of surgery for temporal-lobe epilepsy. *The New England Journal of Medicine*, *345*(5), 311–318. https://doi.org/10.1056/NEJM200108023450501.

Zaccara, G., Giovannelli, F., Cincotta, M., Loiacono, G., & Verrotti, A. (2015). Adverse events of placebo-treated, drug-resistant, focal epileptic patients in randomized controlled trials: A systematic review. *Journal of Neurology*, *262*(3), 501–515. https://doi.org/10.1007/s00415-014-7391-y.

Zaccara, G., Giovannelli, F., Franco, V., Cincotta, M., Tramacere, L., & Verrotti, A. (2014). Adverse events, placebo and nocebo effects in placebo-treated paediatric patients with refractory focal epilepsies. Analysis of double-blind studies. *Epilepsy Research*, *108*(10), 1685–1693. https://doi.org/10.1016/j.eplepsyres.2014.09.015.

Zaccara, G., Giovannelli, F., & Schmidt, D. (2015). Placebo and nocebo responses in drug trials of epilepsy. *Epilepsy and Behavior*, *43*, 128–134. https://doi.org/10.1016/j.yebeh.2014.12.004.

FSC
www.fsc.org
MIX
Papier | Fördert gute Waldnutzung
FSC® C083411

Kolibri 360 GmbH
Ferdinand-Jühlke-Straße 7
99095 Erfurt, Deutschland
produktsicherheit@kolibri360.de